The Anomeric Effect and Associated Stereoelectronic Effects

ACS SYMPOSIUM SERIES **539**

The Anomeric Effect and Associated Stereoelectronic Effects

Gregory R. J. Thatcher, EDITOR

Queen's University

Developed from a symposium sponsored
by the Division of Carbohydrate Chemistry
at the 204th National Meeting
of the American Chemical Society,
Washington, DC,
August 23–28, 1992

American Chemical Society, Washington, DC 1993

Library of Congress Cataloging-in-Publication Data

The Anomeric effect and associated stereoelectronic effects / [edited by]
Gregory R. J. Thatcher.

p.   cm.—(ACS symposium series, ISSN 0097–6156; 539)

"Developed from a symposium sponsored by the Division of
Carbohydrate Chemistry at the 204th meeting of the American Chemical
Society, Washington, DC, August 23–28, 1992."

Includes bibliographical references and index.

ISBN 0–8412–2729–2

1. Isomerism. 2. Stereochemistry.

I. Thatcher, Gregory Robert James, 1959–   . II. American Chemical
Society. Division of Carbohydrate Chemistry. III. American Chemical
Society. Meeting (204th: 1992: Washington, D.C.) IV. Series.

QD471.A56   1993
541.2'252—dc20   93–34423
                                                          CIP

The paper used in this publication meets the minimum requirements of American National
Standard for Information Sciences—Permanence of Paper for Printed Library Materials, ANSI
Z39.48–1984.   ∞

Foreword

THE ACS SYMPOSIUM SERIES was first published in 1974 to provide a mechanism for publishing symposia quickly in book form. The purpose of this series is to publish comprehensive books developed from symposia, which are usually "snapshots in time" of the current research being done on a topic, plus some review material on the topic. For this reason, it is necessary that the papers be published as quickly as possible.

Before a symposium-based book is put under contract, the proposed table of contents is reviewed for appropriateness to the topic and for comprehensiveness of the collection. Some papers are excluded at this point, and others are added to round out the scope of the volume. In addition, a draft of each paper is peer-reviewed prior to final acceptance or rejection. This anonymous review process is supervised by the organizer(s) of the symposium, who become the editor(s) of the book. The authors then revise their papers according to the recommendations of both the reviewers and the editors, prepare camera-ready copy, and submit the final papers to the editors, who check that all necessary revisions have been made.

As a rule, only original research papers and original review papers are included in the volumes. Verbatim reproductions of previously published papers are not accepted.

M. Joan Comstock
Series Editor

Contents

Preface ix

Introduction xi

1. Anomeric Effect: How It Came To Be Postulated 1
   John T. Edward

2. Anomeric and Associated Stereoelectronic Effects: Scope
   and Controversy 6
   Gregory R. J. Thatcher

3. Intramolecular Strategies and Stereoelectronic Effects:
   Glycosides and Orthoesters Hydrolysis Revisited 26
   P. Deslongchamps

4. Anomeric and Gauche Effects: Some Basic Stereoelectronics 55
   Anthony J. Kirby and Nicholas H. Williams

5. Anomeric Effects: An Iconoclastic View 70
   Charles L. Perrin

6. No Kinetic Anomeric Effect in Reactions of Acetal Derivatives ... 97
   Michael L. Sinnott

7. Involvement of nσ* Interactions in Glycoside Cleavage 114
   C. Webster Andrews, Bert Fraser-Reid, and J. Phillip Bowen

8. X–C–Z Anomeric Effect and Y–C–C–Z Gauche Effect
   (X,Y = O,S; Z = O,N): Evaluation of the Steric, Electrostatic,
   and Orbital Interaction Components 126
   B. Mario Pinto and Ronald Y. N. Leung

9. Origin and Quantitative Modeling of Anomeric Effect 156
   Peter A. Petillo and Laura E. Lerner

10. Application of Quantum Theory of Atoms in Molecules
    to Study of Anomeric Effect in Dimethoxymethane 176
    N. H. Werstiuk, K. E. Laidig, and J. Ma

11. **Anomeric and Reverse Anomeric Effect in Acetals and Related Functions** 205
  F. Grein

12. **Glycosylmanganese Complexes and Anomeric Anomalies: The Next Generation?** 227
  Philip DeShong, Thomas A. Lessen, Thuy X. Le,
  Gary Anderson, D. Rick Sidler, Greg A. Slough,
  Wolfgang von Philipsborn, Markus Vöhler, and Oliver Zerbe

13. **Do Stereoelectronic Effects Control the Structure and Reactivity of Trigonal-Bipyramidal Phosphoesters?** 240
  Philip Tole and Carmay Lim

14. **Stereoelectronic Effects in Pentaoxysulfuranes: Putative Intermediates in Sulfuryl-Group Transfer** 256
  Dale R. Cameron and Gregory R. J. Thatcher

15. **O–C–N Anomeric Effect in Nucleosides: A Major Factor Underlying the Experimentally Observed Eastern Barrier to Pseudorotation** 277
  Ravi K. Jalluri, Young H. Yuh, and E. Will Taylor

INDEXES

Author Index 297

Affiliation Index 297

Subject Index 297

Preface

Two WIDELY READ TEXTS, *The Anomeric Effect and Related Stereoelectronic Effects at Oxygen* by Anthony J. Kirby, and *Stereoelectronic Effects in Organic Chemistry* by P. Deslongchamps, represent landmark treatises of a decade ago. The frequent use of these texts for upper-year undergraduate and graduate courses attests to the vitality and breadth of the science of stereoelectronic effects.

In this area, the domains of synthetic, physical, inorganic, and theoretical chemistry and enzymology meet and overlap. No theory is allowed to persist to the point of dogma unchallenged and without rigorous examination. Thus, this area is controversial and at times adversarial. In this field, often for the first time, students are confronted by a number of competing hypotheses with no established dogma to be safely memorized. Students are required to question theories, criticize conclusions, and understand the scope and limitations of a large number of modern experimental and computational methods. Furthermore, the "true picture" presented by the pervasive and seductive images of computer-generated molecular modeling must be queried when confronted with stereoelectronic effects.

The object of this book is to provide a forum for competing viewpoints rather than to resolve the controversies surrounding the anomeric effect. Some of these viewpoints remain diametrically opposed. Although each chapter contains original research, the style is usually broader than that found in a traditional journal research paper. The range of techniques presented includes synthesis, kinetic analysis, crystallography, NMR analysis, enzymology, and ab initio and molecular mechanics calculations.

The book begins by describing the origins of the field with Lemieux's work in carbohydrate chemistry and the Edward–Lemieux effect. The second chapter sets the ground for the controversies to come. This chapter importantly emphasizes the significant scope and implications of the proposed stereoelectronic effects discussed in this book and provides references to those researchers who are not contributors. The following four chapters provide a forum for the major protagonists (or antagonists!) in this field who rely largely on the results of experimental approaches and have been intimately associated with the anomeric effect over the years. Two chapters follow from groups that integrate experimental and theoretical methods. In subsequent chapters, three very different ab initio based theoretical methods are described, including

Weinhold's natural bond orbital analysis and Bader's theory of atoms in molecules. Although the anomeric effect specifically refers to carbohydrate chemistry, the scope of the associated stereoelectronic effects extends far beyond carbohydrate chemistry. Examples of applications in organometallic chemistry and for derivatives of sulfur and phosphorus oxyacids are presented in Chapters 12–14. The final chapter deals with the breakdown of molecular modeling methods in compounds influenced by stereoelectronic effects and subsequent efforts in reparameterizing molecular mechanics force fields. It is hoped that the intensity and vitality of the presentations and the ensuing heated discussion at the symposium are faithfully mirrored in this book.

Acknowledgments

Of the many people involved with this book, I thank, in addition to the authors: Vernon Box, Russ Boyd, Bruce Branchaud, Chris Cramer, Richard Franck, Alfred French, Moses Kaloustian, Victor Marquez, Morten Meldal, Martin Tessler, Kenneth Wiberg, Steve Withers, and, in particular, Walter Szarek. I also thank the Division of Carbohydrate Chemistry of the American Chemical Society for sponsorship and the donors of the Petroleum Research Fund of the American Chemical Society for partial funding of the symposium.

GREGORY R. J. THATCHER
Department of Chemistry
Queen's University
Kingston, Ontario K7L 3N6
Canada

July 6, 1993

Introduction

In 1977 a symposium entitled "Origin and Consequences of the Anomeric Effect" was organized by Derek Horton and me and cosponsored by the Division of Carbohydrate Chemistry of the American Chemical Society (ACS) and the Organic Chemistry Division of the Chemical Institute of Canada (CIC). The symposium was part of the Second Joint Conference of the ACS and CIC in Montreal, Quebec, Canada. This symposium was the first devoted exclusively to a discussion of the anomeric effect, and it provided a mechanism for interactions among a variety of chemists, namely, theoreticians, structural chemists, physical organic chemists, and synthetic chemists. The material presented at that symposium, together with some new interpretations by the speakers that resulted from these interactions, formed the basis of ACS Symposium Series 87, *Anomeric Effect: Origin and Consequences*.

Fifteen years after the Montreal symposium, the ACS Division of Carbohydrate Chemistry sponsored a second symposium on the anomeric effect. Greg Thatcher organized this symposium as part of the 204th ACS National Meeting in Washington, DC. The Washington symposium again involved diverse types of chemists, and the presentations again evoked vigorous discussions among the participants.

The 15-year period between the two symposia has witnessed many developments in the area of the anomeric effect involving not only new theoretical interpretations and computational approaches, but also structure and reactivity relationships that have far-reaching significance, for example, in enzymology and synthetic chemistry. However, the dominant feature that emerged from the Washington symposium was that our current degree of theoretical understanding of the anomeric effect still does not provide a totally integrated and quantitatively predictive theory. This book provides a stimulating overview of recent developments and captures the spirit of the continuing investigation of an important phenomenon.

Walter A. Szarek
Department of Chemistry
Queen's University
Kingston, Ontario K7L 3N6
Canada

Received July 19, 1993

Chapter 1

Anomeric Effect

How It Came To Be Postulated

John T. Edward

Department of Chemistry, McGill University, Montreal, Quebec H3A 2K6, Canada

The dissemination within the carbohydrate community in the 50's of Hassel and Barton's ideas on conformational analysis is reviewed briefly. The superior stability of axial over equatorial electronegative groups at the anomeric position remained an anomaly, and required recognition of the operation of an additional factor, the anomeric effect.

"It pays to speculate as widely and wildly as possible; people remember only when you are right"[1].

After receiving a D. Phil for research in alkaloid chemistry in 1949, I dabbled in several other fields, hoping to find an important but tractable problem. At the University of Birmingham, at the suggestion of Professor Maurice Stacey, I worked briefly on glycoproteins, and learned something of carbohydrate and protein chemistry. Then in 1952 I joined Professor Wesley Cocker's department at Trinity College, Dublin, and became involved in his researches on sesquiterpene.

Terpene chemists at that time were becoming aware that the stereochemistry of six-membered rings was being revolutionized by Barton. I had heard him lecture on the stereochemistry of triterpenes at both Birmingham and Dublin in the early 50's and had read A.J. Birch's judgment that "conformational analysis for the study of .... cyclic systems promises to have the same degree of importance as the use of resonance in aromatic systems"[2]. But I don't remember reading Barton's epoch-making 1950 paper in , *Experientia*[3], and fully comprehended the new theories only after studying his Tilden Lecture[4] in the more accessible *Journal of the Chemical Society* in 1953. It is difficult nowadays to understand the excitement and satisfaction which came with the new ideas, which made so much of the chemistry of steroids and terpenes intelligible for the first time[5].

0097–6156/93/0539–0001$06.00/0

Barton's ideas found immediate application in Professor Cocker's researches[6, 7]. Furthermore, with J.C.P. (Peter) Schwarz, the other lecturer in Organic Chemistry at Trinity, I tried to apply them to carbohydrate chemistry. However, when we looked up the literature we were astonished to find that the "new ideas" had been around for a long time. Already in 1929 W.N. Haworth[8] had pointed out that pyranose rings could exist in two chair and four "classical" boat conformations (the flexible boat had not yet been thought of), and in the 30's and 40's other workers[9, 10] (particularly Hassel and Ottar[11]) had given reasons to believe most pyranose rings existed in one of the two possible chair conformations.

This idea received convincing support from the work of Reeves[12] on the shifts in optical rotation which occurred when the cuprammonium complexes of pyranosides were formed. His 1950 paper completing and summarizing the work from four preceding papers, classifies the possible chair and boat conformations, and gives a lucid discussion on why one chair conformation is favoured.

The one point on which Peter and I found Reeve's paper a bit unclear concerned the equilibrium between α and ß-anomers of pyranose sugars in water. Most D-hexoses have the C1 conformation (in Reeves' designation), so that the ß-anomer has an equatorial anomeric hydroxyl and should be more stable than the α-anomer. But the reverse is usually true.[13] Some new factor must be operating to overcome the steric effect favouring the equatorial configuration. Just at that time Corey[14] published a paper showing that bromine atoms of α-bromocyclohexanones also preferred an axial orientation, and explained the result by the electrostatic repulsion between the C=O dipole and the equatorial C-Br dipole; this repulsion should be lessened when the bromine was axial. Eureka! A similar explanation could account for the fact that electronegative substituents such as Cl, OH, and OR at C-1 of pyranosides preferred an axial orientation.

This idea lay fallow for some years until I wrote a paper[15] in 1955 entitled: "Stability of glycosides to acid hydrolysis. A conformational analysis." (Note the title: in 1955 the last phrase made the paper chic!). The paper was provoked by a note by G.N. Richards, offering an explanation for the lability of glycosides of 2-deoxy sugars to acid hydrolysis, which I thought must be wrong. The mechanism for acid hydrolysis which I developed depended on the formation of a half-chair oxocarbonium ion from the pyranoside chair in the rate-determining step, and accounted for the different rates of hydrolysis of glycosides of glucose, galactose etc., by the varying resistance to the formation of the half-chair conformer caused by the variously-disposed hydroxyl groups. The mechanistic reasoning made use of very recent publications: on glycoside hydrolysis[17], on the conformations of the half-chair ring[18], and on the effect of differing hydroxyl configurations of quercitols and inositols on the ease of forming ketals.[19]

On re-reading my paper after a lapse of almost forty years, I find its arguments still plausible, although I have no idea how they are regarded in the carbohydrate community. However, my explanation of the great acid-lability of glycosides of 2-deoxy sugars (the original reason for writing the paper) is clearly wrong, because it does not take account of the importance of inductive effects revealed in a paper of Kreevoy and Taft[20] which came out about the same time. These effects were correctly recognized as important by Shafizadeh and Thompson[21] in a paper appearing early in 1956. Curiously, this paper covered the same experimental data as mine, but offered completely different explanations, and yet failed to cite my paper and point out their differences. Probably the writing of it was completed before my paper came into their hands.

As a separate and minor issue in my paper I had to explain why α-D-glycosides generally hydrolysed more rapidly than ß-D-glycosides, and made use of fallow idea mentioned above.

Some time after returning to Canada in 1956 I met Raymond Lemieux and found out that he had come to the same conclusions regarding the stability of α-and ß-anomers, as a result of the experimental work of N.J. Chu[22], and (most important) had given the effect a name: the anomeric effect. Since that time he has continued studying the effect, and discovered two further manifestations of it: the reversed and the exo-anomeric effect. And, of course, others have extended it into the other fields of chemistry. Lemieux has been meticulous in citing ref.15 in his various publications, and others have followed him (often, I suspect, without actually reading the paper). It is ironic that the paper is almost always cited for the brief section on the anomeric effect,and not for the much longer discussion of mechanism.

So I can really thank Raymond Lemieux for whatever recognition the paper has received. How did the other chemists whose papers are mentioned above fare?

Hassel and Barton shared the Nobel Prize in 1969 for "developing and applying the principles of conformation in chemistry", and Barton's 1950 paper[3] was considered his key contribution[23]. This paper was written after he had arrived at Harvard in 1949 for a one-year visiting lectureship. He heard a seminar by Louis Fieser pointing out the difficulties in devising an explanation for the relative case of oxidation and esterification of hydroxyl groups at various positions on the steroid skeleton. Barton, because of Hassel's papers and his own researches,[14,25] had already been thinking and talking about the conformational analysis of polycyclic systems, and saw immediately the explanation for Fieser's data. He[3] and Fieser[26] each published a paper in *Experientia*, a Swiss journal. Barton's paper was immediately recognized for its importance by perceptive chemists, so that by the time he published his Tilden Lecture[4] in 1953 a very considerable amount of new material buttressing his theories had appeared, and still more in 1956 when he wrote an important review[27] with Cookson. By 1965 the field of conformational analysis required a whole book by four authors.[28]

However, Barton had not been alone. As we saw above, in 1950 Reeves[10,29] independently introduced Hassel's ideas to carbohydrate chemists, and in 1952 Angyal[19] did the same. (The latter's paper introduced the terms "polar" and "equatorial"; "polar" was soon replaced by "axial"[30]). But these chemists never received the recognition that they deserved. Both had small research groups, and perhaps lacked the panache of Barton in propagating their ideas (though I never heard Reeves). And furthermore, the reception of the new ideas by the carbohydrate community was very tepid. In Birmingham from 1950 to 1952 I cannot remember anyone mentioning Reeves. On the other hand, I do remember Ted Bourne giving a lecture outlining the empirical rules for the favoured ring structures of the acetals and ketals of tetritols, pentitols, and hexitols[31]. Here could have been an occasion for some organic chemist, like Barton at Harvard, to apply conformational analysis. But none of us did. The first person to provide an explanation was David Whiffen,[32] an infrared spectroscopist very ingenious in solving organic problems (see his later work on optical rotation[33]). Whiffen explained the results on the basis of the distance between hydroxyl oxygens when the carbon chains of the polyols were in their fully-extended conformations. An explanation based on conformational analysis, and more to the taste of organic chemists, came later from Mills.[38]

The slowness of the carbohydrate chemists to take up the new ideas is shown by the fact that Reeves' 1950 paper[12] was cited only 25 times in the five years up to 1954, while Barton's was cited 108 times; for the ten years 1955-1964 the citation frequencies were 108 and 133 respectively. Even more striking evidence comes from Barton and Cookson's 39 page review[27] in 1956: only one page (discussing chiefly Reeves' work[12,29]) was devoted to carbohydrate chemistry, and many pages to steroids, terpenes, alkaloids, etc. The proselytizers for the new ideas were young men on the fringes such as Angyal, Mills,[35] and Lemieux,[36] and not those in the established schools of carbohydrate chemistry.

After 1956 I had the good luck to supervise three gifted and hard-working graduate students who over 10 years studied the anomeric effect in cyclic hemiacetals derived from steroids. Their papers[37-39] seem to me to be interesting and important, but up to 1989 had been cited 36,16, and 7 times(respectively), respectable but not outstanding numbers[40]. By contrast, my 1955 paper,[15] which required only a few weeks library work, had been cited 304 times, which puts it in the top 0.3% of all scientific papers[40]. The moral is given by Barton in the quotation which precedes this essay: luck counts for more that virtue in this unfair world.

<u>Literature Cited</u>

1.   Barton, D.H.R., cited by R.B. Woodword, "Further Perspectives in Organic Chemistry", Ciba Foundation Symposium 53, Elsevier, Amsterdam, 1978, p.50.
2.   Birch, A.J. *Ann. Reps. Prog. Chem.,* **1951**, 45, 192
3.   Barton, D.H.R. *Experientia,* **1950**, 6, 316-320.
4.   Barton, D.H.R., *J. Chem. Soc.,* **1953**, 1027-1040.
5.   Eliel, E.L. *Science,* **1969**, 166, 718-720.
6.   Chopra, N.M., Cocker W., Edward, J.T. *Chem. Ind.* (London), **1954**, 1535.
7.   Cocker, W., Edward, J.T., Holley, T.F. *Chem. Ind.* (London), **1954**, 1561-1562.
8.   Harworth, W.N., "The Constitution of Sugars", E. Arnold   and Co., London, 1929, p.91.
9.   Scattergood, A., Pacsu, E. *J. Am. Chem. Soc.,* **1940**, 62, 903-910.

10. Gorin, E., Kauzmann, W., Walter, J. *J. Chem. Phys.,* **1939**, 7, 327-338.
11. Hassel, O., Ottar, B. *Acta Chem. Scand.,* **1947**, 1,   929-942.
12. Reeves, R.E., *J. Am Chem. Soc.*, **1950**, 72, 1499-1506.
13. Montgomery, R., Smith, F. *Ann. Rev. Biochem.*, **1952**, 21, 79.
14. Corey, E.J., *J. Am. Chem. Soc.*, **1953**, 75, 2301-2304.
15. Edward, J.T. *Chem. Ind. (London)*, **1955**, 1102-1104.
16. Richards, G.N. *Chem. Ind.,(London)*, **1955**, 228.
17. Bunton, C.A., Lewis, T.A., Llewellyn, D.R., Tristram, H., Vernon, C.A. *Nature (London)*, **1954**, 174, 500.
18. Barton, D.H.R., Cookson, R.C, Klyne, W., Shoppee, C.W. *Chem. Ind. (London)*, **1954**, 21.
19. Angyal, S.J., Macdonald, C.G., *J. Chem. Soc.*, **1952**, 686-695.
20. Kreevoy, M.M., Tft, R.W., *J. Am. Chem. Soc.*, **1955**, 77, 3146-3148.
21. Shafizadeh, F., Thompson, A. *J. Org. Chem.*, **1956**, 21, 1059-1062.
22. Lemieux, R.U., Chu, N.J., American Chemical Society 1309 Meeting, San Francisco, California, April 13-18, 1958, abst. papers; Chu, N.J.,   Ph.D. Thesis University of Ottawa, Ottawa, Canada, 1959, p. 97; Lemieux,R.U. "Molecular Rearrangements", Ed. P. de Mayo, Interscience Publishers, Inc., New York, 1964, Part II, p.735.
23. Ramsay, O.B., "Stereochemistry", Heyden, London, 1981, pp. 1-5.
24. Barton, D.H.R., *J. Chem. Soc.* **1948**, 340-342.
25. Barton, D.H.R., Schmiedler, J.A., *J. Chem. Soc.*, **1948**, 1197-1203.
26. Fieser, L.F., *Experientia.*, **1950**, 6, 312-315.
27. Barton, D.H.R., Cookson, R.C. *Quart. Rev.*, **1956**, 10, 44-82.
28. Eliel, E.L., Allinger, N.L., Angyal, S.J., and Morrison, G.A., "Conformational Analysis. Interscience, New York, 1965, Xiii + 524pp.
29. Reeves, R.E., *Adv. Carb. Chem.*, **1951**, 6, 107-134.
30. Barton, D.H.R., Hassel, O., Pitzer, K.S., Prelog, V. *Science,* **1954**, 119, 49.
31. Barker, S.A., and Bourne, E.J., *Adv. Carb. Chem.*, **1952**, 7, 137-207.
32. Barker, S.A., Bourne, E.J., Whiffen, D.H., *J.Chem.Soc.*, **1952**, 3865-3870.
33. Whiffen, D.H., *Chem. Ind. (London)*, **1956**, 964-968.
34. Mills, J.A., *Chem. Ind. (London)*, **1954**, 633-634.
35. Mills, J.A., *Adv. Carb. Chem.*, **1955**, 10, 1-53.
36. Lemieux, R.U., Huber, G., *J. Am. Chem. Soc.*, **1956**, 78, 4117-4119.
37. Edward, J.T., Morand, P.F. Puskas, I., *Can. J. Chem.*, **1961**, 39, 2069-2085.
38. Edward, J.T., Puskas, I. *Can. J. Chem.*, **1968**, 40, 711-717.
39. Edward, J.T. Ferland, J.-M., *Can. J. Chem.*, **1966**, 44, 1299-1309.
40. Edward, J.T. *Chemtech*, **1992**, 22, 534-539.

RECEIVED May 13, 1993

Chapter 2

Anomeric and Associated Stereoelectronic Effects

Scope and Controversy

Gregory R. J. Thatcher

Department of Chemistry, Queen's University, Kingston, Ontario K7L 3N6, Canada

The anomeric and associated stereoelectronic effects applied to structure and conformation comprise the Edward-Lemieux effect, generalized anomeric effect (GAE), exo-anomeric effect, gauche effect and reverse anomeric effect. Extrapolation of these conformational effects to reactivity has yielded related stereoelectronic effects that have been labeled Deslongchamp's theory of stereoelectronic control, the kinetic anomeric effect and anti-periplanar lone pair hypothesis (ALPH). In much of the literature these effects are accepted as having profound effects on structure and reactivity. In many important enzyme systems catalysis is proposed to require stereoelectronic assistance. However, strong objections have been published as to the veracity of experimental and theoretical data and its rationale in support of these stereoelectronic effects. The basis for this criticism and the various areas of controversy, past and present, are discussed. An attempt is made to clarify the definitions of the various related stereoelectronic effects and the various hypotheses proposed in this area. The application of ALPH to the mechanism of ribonuclease is discussed.5

It has been stated that a wealth of evidence indicates that stereoelectronic stabilizing interactions are important in transition state structures and influence the course of chemical reactions involving acetals, esters, amides, phosphate esters and triple bonds[1]. The enzymes ribonuclease[2], chymotrypsin[3], lysozyme[4], carboxypeptidase A[5], alcohol dehydrogenase[6] and ribozymes [2] have been proposed to rely on stereolectronic effects to catalyze reaction. Furthermore, owing to the importance of stereoelectronic effects in enzymology, it has been suggested that some well-accepted catalytic principles will require reconceptualization[7]. Proposed stereoelectronic effects have been equally well applied in synthetic organic chemistry to explain and predict product distribution in a large range of reactions[8]. These stereoelectronic effects are, as proposed, intimately related to the anomeric effect. The anomeric effect describes the empirical observation of a preference for exocyclic substituents (O or halogen) to occupy the axial over the equatorial position at the

0097–6156/93/0539–0006$06.00/0

anomeric carbon of a pyranose ring system[9]. It is the rapid extrapolation of the simple, phenomenological anomeric effect to yield stereoelectronic effects of enormous scope and impact on the reactivity of organic and inorganic molecules that has provoked deserved questioning and controversy. An effect of such importance requires rigorous examination before being established as dogma.

Purpose

The purpose of this overview chapter is not to provide a comprehensive review of the literature, nor a comprehensive list of references, but to set the stage for the papers to follow. The need then is to introduce many of the controversies that continue to embroil the study of the anomeric and related effects. As a fortuitous coincidence, an excellent comprehensive review on the structural and conformational aspects of the anomeric effect appeared some months before the ACS Symposium[10].

1. STRUCTURE AND CONFORMATION

1.1. Definition and Clarification

Much confusion and some of the controversy in this area has resulted from the use of the phrases stereoelectronic effect and anomeric effect to imply many different meanings. *Stereoelectronic effects* are variously defined: "...those factors concerned with the conformational requirements of the groups involved in the reaction with respect to the electron orientations in the transition state"[11]; "Pertaining to the dependence of the properties (especially the energy) of a molecular entity in a particular electronic state (or of a transition state) on relative nuclear geometry"[12]; "The term stereoelectronic refers to the effect of orbital overlap requirements on the steric course of reaction"[13]; "Stereoelectronic effects arise from the different alignment of electronic orbitals in different arrangements of nuclear geometry"[11]. These definitions are sufficiently broad as to include effects attributed to interactions involving bonding and non-bonding electrons without defining the cause nor rationale of the effect. We may usefully decompose a phenomenological stereolectronic effect into its causal components which may be electrostatic (in particular dipole-dipole interactions), steric or electronic (interactions between bonding, non-bonding or anti-bonding electronic orbitals). In solution, solvation must also be considered.

The term anomeric effect, originally defined by Lemieux (see above), would be

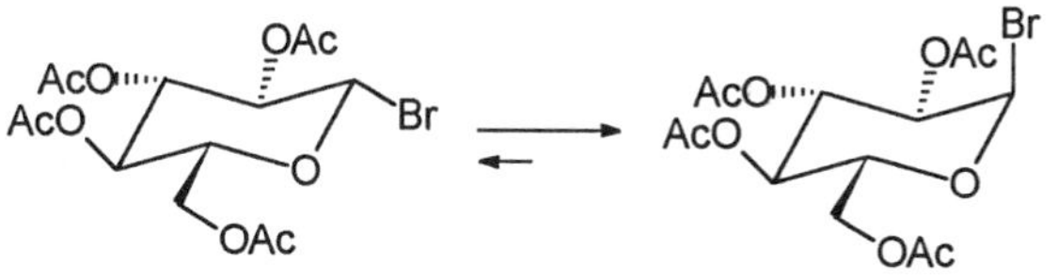

Scheme 1. The Edward-Lemieux Effect

better substituted by the *Edward-Lemieux effect[14]*, since more recent usage of the term anomeric effect goes far beyond pyranose conformation[15]. The effect describes the relative stabilization of α-glycosides (Scheme 1). The *generalized anomeric effect* (GAE) describes the preference for synclinal (sc, gauche) over antiperiplanar (ap, trans) conformations in the molecular fragment W-X-Y-Z, where X possesses one or more pairs of non-bonding electrons, Z is electron-withdrawing and W and Y are of intermediate electronegativity (Y is usually C, P, Si or S). Confusion occurs since recent definitions have: (1) specifically defined the GAO in terms of p-type electron lone pairs on X (although many researchers invoke sp³ hybrid orbitals, *vide infra*); (2) inferred a specific cause for the observed preference; or (3) dropped the adjective "general" entirely[*1,10*].

Unlike the anomeric effect, the *exo-anomeric effect* has remained closely associated with carbohydrate chemistry. As originally defined by Lemieux, the result of this stereoelctronic effect on glycoside conformation is specifically to restrict the orientation of the acetal aglycon[*17*]. The introduction of this stabilising effect requires a complementary *endo-anomeric effec*t in glycosides. The Edward-Lemieux effect can thus be energetically quantified as the difference between the sum of the endo- and exo-anomeric effects for one of the anomers and the sum for the other glycoside anomer[*17*]. A further useful quantitative expression is *anomeric stabilization* defined as non-steric stabilization of the gauche conformer. This stabilization term unmasks electronic and electrostatic effects which might otherwise be overwhelmed by steric effects. It is not restricted to glycosides, but is of major application in glycosides and related cyclic systems, and may be expressed:

$$\text{anomeric stabilization} = \Delta G°(\text{heterocycle}) - \Delta G°(\text{steric}) \qquad \{1\}$$

$\Delta G°$(steric) may use A-values from non-anomeric systems (e.g. cyclohexanes) and should be corrected for structural modifications in the heterocyclic system[*14,19*]. *Thus the GAE requires that anomeric stabilization be the dominant influence on conformation.*

The *gauche effect* applies to a molecular fragment R-X-Y-R or X-C-C-Y, where the central atoms are two heteroatoms or two carbons with electronegative substituents. It has been defined as "...a tendency to adopt that structure which has the maximum number of gauche interactions between the adjacent electron pairs and/or polar bonds"[*20*]. Kirby has pointed out the limitations of this definition with respect to experimentally determined conformational preferences, and supported by the work of Epiotis[*21*], has substituted an alternative definition: "There is a stereoelectronic preference for conformations in which the best donor lone pair or bond is antiperiplanar to the best acceptor bond"[*22*].

Finally we have the *reverse anomeric effect*. This term has been used by Wolfe *et al.* to describe a specific hyperconjugative interaction[*23*] and by Gorenstein and co-workers to account for observations on reactivity in direct contradiction to a proposed kinetic anomeric effect (*vide infra*)[*24*]. However, the accepted and more general definition simply describes a conformational preference opposite to the GAE (fragment W-X-Y-Z, where Z is usually N⁺), and in the specific case of glycosides the preference of the aglycon for the equatorial position. However, the use of this term specifically implies an *anomeric destabilization* contribution in addition to simple steric effects. An early example is the equatorial-substituted boat conformer observed for an imidazole-appended pyranoside implied by NMR studies (Scheme 2)[*25*].

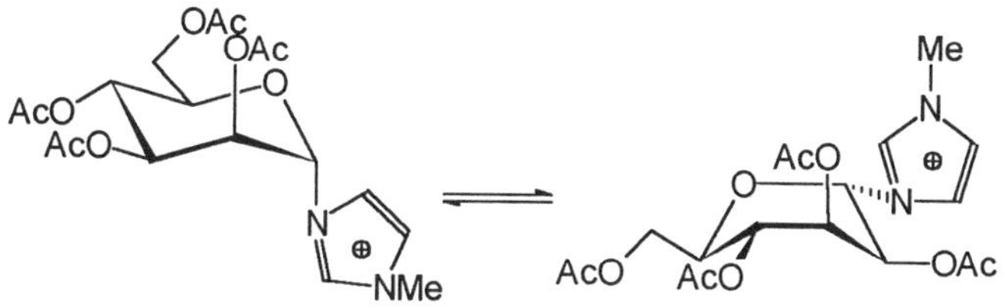

Scheme 2. Reverse anomeric effect

Thus the Edward-Lemieux effect, GAE, exo-anomeric effect, gauche effect and reverse anomeric effect are phenomenological effects devoid of implicit theoretical rationale. Two questions must then be answered:

   1. Is there sufficient rigorous experimental evidence to support these effects as phenomena that require explanation?

   2. If so, then what is the underlying theoretical explanation?

1.2. Experimental Evidence

X-ray crystallography and variable temperature NMR provide the abundance of relevant structural information.   Early ¹H NMR studies provided quantification of conformational and anomerization equilibria, indicating the requirement for a stereoelectronic contribution and providing the stimulus for the various stereoelectronic effects. However, crystallography provided an additional feature associated with the GAE, namely variation in bond length and bond angle correlated with torsion angle. Indeed, the bond length and in particular, bond angle changes associated with the GAE are now widely seen as more reliably symptomatic of the presence of a stereoelectronic effect than conformational preferences: the electronic effects influencing conformational energy may be masked by steric and electrostatic effects. Many crystal structure determinations suffer from large standard errors, resulting in differences in bond lengths which when examined closely are insignificant with respect to the experimental error[26].   However, the wealth of crystal structure data on carbohydrates, to which Jeffrey's group has made a major contribution[27], allows correlation and analysis of large data samples. These correlations have negated many of the criticisms founded on the use of selected crystal structures[28].   In addition, Kirby and co-workers have demonstrated striking trends in bond lengths associated with varying the exocyclic substituent in pyranoside derivatives[22,29].   Nevertheless, criticism of the relationship between crystal structure data and stereoelectronic influences persists[30,31].   It is argued: (1) that crystal packing forces and dipolar interactions are dominant in the solid phase; (2) intermolecular interactions are ignored when analysing data for stereoelectronic effects; (3) molecular structure and conformation differs not only between solution and solid phase, but between different crystal forms.   Furthermore, the extensive data correlations and individual structure analysis have demonstrated some anomolous relationships with the GAE[10,28c,30].

NMR studies including recent work by Booth *et al.* have demonstrated conformational preferences in solution[32] in accord with the Edward-Lemieux effect. Alkoxy substituents in tetrahydropyrans strongly favor axial placement, whereas amino substituents may strongly favor the equatorial.   Rotameric preferences also support thepresence of an exo-anomeric effect with both oxy and amino substituents[32].

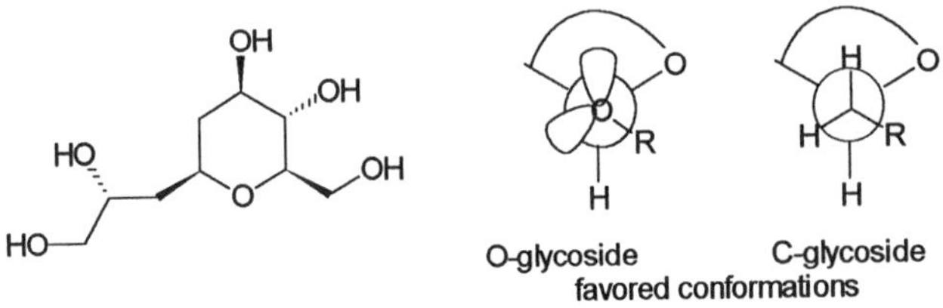

O-glycoside   C-glycoside
favored conformations
Scheme 3

However, anomolous structural information has been revealed in NMR studies. Kishi and co-workers examined the conformational equilibria of *C*-glycosides, which were revealed to be very similar to the parent glycosides, despite the presence of the exo-anomeric effect in the glycoside and its absence in the carbon homologue (Scheme 3)[*33*]. This work casts doubt on the need for the term "exo-anomeric effect", if indeed the conformational preference is not peculiar to acetals. Deslongchamps has countered that steric effects are not comparable in the two pyranoside systems studied by Kishi and has provided the observed conformational preference of spirocyclic systems as support for the exo-anomeric effect[*34*].

Experimental data directed at the reverse anomeric effect has in comparison been very scant, based on N-heterocyclic ammonium glycosides. The elegant, "textbook" experiment relies on protonation to drive the axial⇔equatorial equilibrium (Scheme 4) in the direction of the equatorial imidazolium[*35*]. Indeed, it is the beauty of this experiment that has largely sustained faith in the reverse anomeric effect, despite the failure to obtain experimental support in glycoside systems bearing carbon substituents with partial positive charge.

Scheme 4

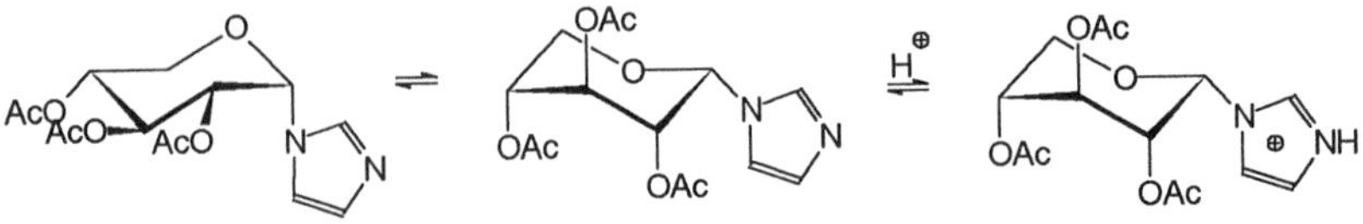

Scheme 4

Although the experimental assessment of conformation largely focusses on conformational energies, the barrier to conformational change has also been proposed to be accentuated owing to the GAE[*19*]. Conversely, NMR studies have indicated a reduced barrier to ring inversion in systems subject to the GAE[*36*].

1.3 Theoretical Rationale
Original explanations for the Edward-Lemieux effect focussed on unfavorable dipole interactions in the β-anomer[*37*]. The classic no-bond⇔double bond resonance argument was introduced by Altona and Havinga[*38*]. These early postulates form the two camps to which almost all rationale for the GAE may be assigned, to this day (Scheme 5). The dipole repulsion, electrostatic interaction camp has been joined by lone pair-lone pair repulsion (often described as Eliel's rabbit ear effect[*39*]) and 4e⁻ orbital mixing destabilizing effects based on interactions between non-bonding

electrons. The electron delocalization camp of no-bond↔double bond resonance may be seen to include hyperconjugation and other 2e⁻ stabilizing orbital mixing interactions.

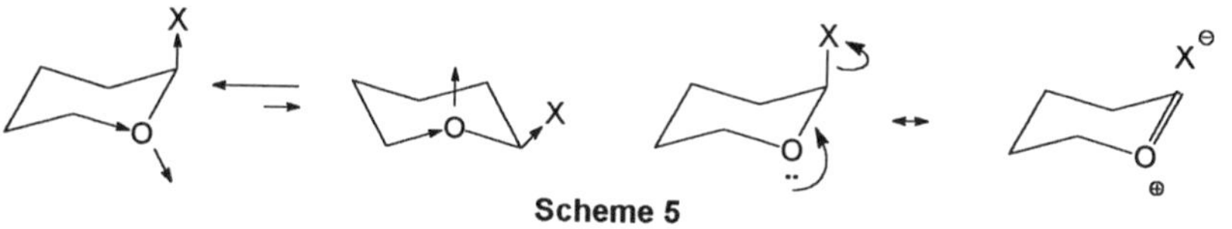

Scheme 5

Within these two camps, many of these interactions are viewed as functionally equivalent. Early suggestions that the GAE may be based on contributions from 4e⁻/electrostatic and 2e⁻/delocalization interactions[40] are muted amidst the clamour for dominance between the two camps. The perceived inability of the 4e⁻ model to account for and predict the changes in bond length and angle associated with the GAE has formed the strongest argument against this model[42]. The large observed dependency of the GAE on solvent, compatible with the 4e⁻ model, initially questioned the delocalization model. More importantly, the glaring anomaly of the reverse anomeric effect is difficult to rationalize in the context of a delocalization model. However, it must be stated that delocalization is now firmly in the ascendancy. It is widely held that stabilizing n→σ* orbital mixing (hyperconjugation) is the root of the GAE. This is supported by recent theoretical work employing natural bond orbital analysis[41]. Such belief has gained sufficient ground that the terms anomeric effect and n→σ* hyperconjugation are erroneously treated as interchangeable and synonomous. Indeed, the importance of negative hyperconjugation has been questioned by Wiberg [43a]: "...observation of structural or energetic consequences of rotation about a bond or atom having a lone pair does not, by itself, provide evidence for negative hyperconjugation". Comparison of calculational with experimental data provides strong evidence for intramolecular electrostatic stabilization as a dominant influence on structure and energy. Delocalization through negative hyperconjugation is ascribed a modest role in FCH_2NH_2. It should also be noted that the cause of anomeric stabilization (n→σ* or otherwise) as derived from molecular orbital calculations, depends upon the choice of energy decomposition algorithm[43b].

1.4 Lone Pairs of Electrons

Box has suggested that a 4e⁻ destabilization between electron lone pairs on oxygen correctly predicts the rotameric and conformational preferences of the (exo and endo) GAE in glycosides (Scheme 6)[30]. By definition the GAE requires the participation of electron lone pairs. In the knowledge of experimental evidence for the inequivalence of non-bonding electronic orbitals on oxygen (recently discussed by Juaristi and Cuevas[10]), Kirby has justified the standard usage of sp³-hybrid lone pairs as a valid approximation for examination of the influence of the GAE (Scheme 7a,b)[21]. This justification is based on the assumption that the gauche conformation is preferred owing to a combination of n_σ→σ* and n_π→σ* hyperconjugation (Scheme 7a,b). Thus the use of sp³-hybrids correctly predicts optimum conformational stabilization. There are problems inherent in the use of sp³-hybrids. For example, in a recent high level,

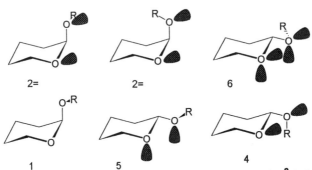

Scheme 6. Destabilising 4e⁻ interactions in α and β-glycosides (sp³ hybrid orbitals shown). Conformational preferences are qualitatively ranked according to number of interactions and overlap (1=most favorable).

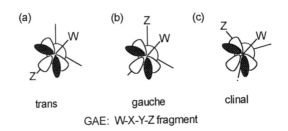

GAE: W-X-Y-Z fragment

Scheme 7. W-X-Y-Z fragment, sp3 hybrids shaded and superimposed on n_σ and n_π atomic orbitals

theoretical examination of anomeric stabilization in the conformational energy profile of $HOCH_2OH$ and $HOCH_2OH_2^+$, torsion angles were fixed at 60° (gauche, optimal GAE as defined by sp³-hybrids) and 180° (trans, no anomeric stabilization)[44]. However, a more recent study on the full rotational profile demonstrates that full anomeric stabilization is only present at a torsion angle of approx. 90° (clinal)[45]. In fact, recent definitions of the GAE implicitly state that the important lone pair electrons (on O or S) are in a π-type atomic orbital (or p-orbital, Scheme 7c)[46].

1.5. Computational Studies

The original manifestations of the anomeric effect were discovered experimentally. The increasing availability of high-powered computers and software capable of sophisticated semi-empirical and *ab initio* calculations on molecular orbital interactions has led to a burgeoning of theoretical studies directed at the GAE. Indeed, it may be argued that these calculations have eclipsed the very experimental data they were set to examine. The computational approach is ideally suited to decomposition of a conformational energy profile into its causal components. Two approaches may be taken. Firstly, one may semi-quanititatively (using a Fourier equation) decompose the energy profile into contributions from apparent individual interactions, for example: lone-pair/lone pair repulsive; steric-eclipsing repulsive; internal hydrogen bond stabilizing; electronic stabilizing. In this way, using sp³-hybrid orbitals, Deslongchamps and Grein are attempting to assign general parameters to permit summation of conformational energies and prediction of conformational preference for general application[47]. In this case, as the authors admit, the validity of using sp^3 orbitals is irrelevant, since specific atomic and molecular orbital interactions are ignored.

Secondly, one may attempt to rigorously and specifically define each effect that contributes to the overall energy profile. In this way, contributions from steric interactions, electrostatic effects and a host of electronic orbital interactions (both 4e$^-$ destabilizing and 2e$^-$ stabilizing interactions[48]) have been defined. The first method has required a very large "reverse anomeric effect" parameter (5 kcal/mol in $H_2N-CH_2NH_3^+$) ascribed to an electrostatic attraction between the positive nitrogen and a lone pair on the second heteroatom[47]. The second method in treating the reverse anomeric effect in $HO-CH_2-NH_3^+$, $HO-CH_2-OH_3^+$ and a cyclic species (Scheme 8) has shown that hyperconjugative effects are actually overwhelmed by steric and electrostatic (dipole-dipole) effects[45].

Scheme 8   X=CH$_2$:   ΔE(eq-ax)= -1.55 kcal/mol
              X=O: ΔE(eq-ax)= -2.24 kcal/mol

It is clear from the many theoretical studies that: (1) hyperconjugative effects can be important factors influencing conformation in the gas phase and; (2) that these effects seem to correlate with geometrical changes akin to those observed in crystal structures. However, hyperconjugative effects are not always the dominant factor influencing conformation, even in the gas phase. Since hyperconjugation may be overwhelmed by steric and electrostatic effects in the gas phase, is it safe to transfer these results to solution where solvation effects may dominate? The importance of anomeric effects in solution has been questioned previously[50]. Contrary to the view that contributions to the GAE should be seen in the enthalpic term, VT NMR conformational studies indicated a zero enthalpic and significant entropic term[51]. It is possible that the use of the emerging solvent approximations in molecular orbital calculations may shed light on this problem. For example, using a continuum dielectric model, solvation is shown to favor the equatorial conformer in 2-tetrahydropyranosylammonium ion[47]. Finally, molecular orbital calculations are always open to criticism on the basis of the level of approximation employed. Since hyperconjugative interactions appear important in anomeric stabilization, it may be suggested that electron correlation must be included in the computational method.

1.6. The Second Row and Below

The GAE involving sulfur was studied as early as 1971[52]. Numerous studies on second and third row effects followed, in particular by the groups of Juaristi and of Pinto[53]. The conclusion that hyperconjugation and the GAE were insignificant for these elements, owing to low π-donor ability and electronegativity, was therefore unexpected[54]. However, further theoretical studies confirmed an effect on the order of 1-2 kcal/mol for the fragments SCS, SCP and SeCSe[55]. These studies have been extended to effects at P, Si and S centres[56]. The applicability of the term anomeric effect or GAE to these centres is questionable at best, but the influence of hyperconjugation is evident. It is worthwhile considering the conformational equilibrium in pentacoordinate phosphoranes termed pseudorotation[57]:

TBP: X = equatorial   TBP: X = apical

Studies by Streitweiser, Schleyer and co-workers determined a scale of apicophilicity: that is, the energetic preference for the substituent to occupy an apical over an equatorial position in a trigonal bipyramidal phosphorane[58]. This apicophilicity scale is dominated by hyperconjugative effects. However, these calculated relative apicophilicities vary enormously from the experimentally determined scales based largely on solution NMR studies[59]. Again one asks: are hyperconjugative effects overwhelmed in solution?[60]. In addition to the influence of solvation, one must consider that the model compounds used in these calculations as many of those used in studies on the GAE are simplified to the extent that steric effects (and steric modification of hyperconjugation through bond angle changes) are all but ignored. The applicability of these computational gas phase models to large molecules in solution still provokes skepticism.

1.7 Conclusion
Substantial theoretical and experimental evidence exists for molecules that adopt conformations contrary to the GAE. In some cases, anomeric stabilization is overwhelmed by opposing effects. In others, hyperconjugative anomeric stabilization appears to be present, but is opposed by rather than reinforced by electrostatic effects. The move towards viewing the GAE as synonomous with n→σ* anomeric hyperconjugation is dangerous, since this ignores alternative contributions to anomeric stabilization, not only electrostatic, but other hyperconjugative and through-space 2e⁻ stabilizing interactions[62].

2. Transition State Structure & Reactivity
Significant contributions by Deslongchamps and Kirby have led to the widescale application of anomeric hyperconjugation as a foundation for understanding and predicting reaction pathways. The kinetic anomeric effect is conceptually simple. The no-bond↔double bond model predicts weakening of the Y-Z bond and therefore implies stereoelectronic assistance of Y-Z bond cleavage (Scheme 9a). Theoretical calculations show anomeric hyperconjugation to be sufficiently strong as to result in bond rupture, for example in $H_2NCH_2NH_3^+$[47] and $^-CH_2CH_2F$[65]. Since this anomeric hyperconjugation is stabilising, both bond forming and breaking pathways should be of lower energy if this interaction is present. Reactivity is thus controlled by conformation. Both popular versions of this theory rely on visualisation of sp³-hybrid orbitals for prediction of reaction pathways subject to stereoelectronic control. The use of sp³ orbitals and the similarity with the anti stereoelectronic requirement for β-elimination (Scheme 9b), with which most organic chemists are very comfortable, has led to the wide acceptance and application of this theory. Simply put, if an sp³-lone pair orbital is ap to a scissile bond, then that bond will be lengthened and weakened

(hence Antiperiplanar Lone Pair Hypothesis: ALPH). Or alternatively, anomeric hyperconjugation in the ground state is amplified in the transition state.

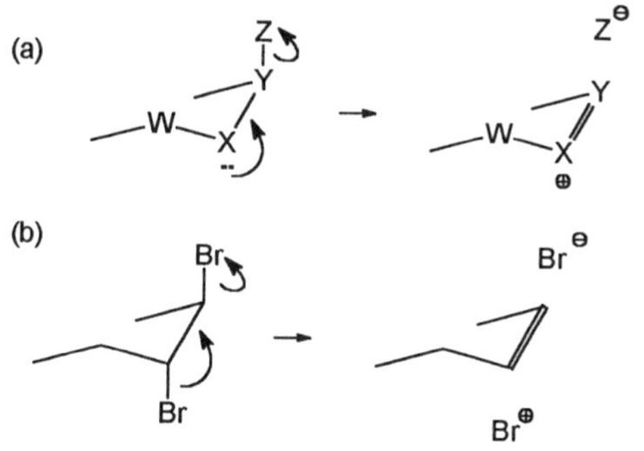

Scheme 9

2.1 ALPH and The Principle of Least Nuclear Motion in Acetals

Since the GAE emerged from studies on acetals, it is salient to initially apply the kinetic anomeric effect to acetal hydrolysis. Clearly from consideration of Schemes 1 and 9, ALPH predicts that α-glycosides will be more reactive than β-glycosides. However, the majority of studies on acetal hydrolysis, with one much-quoted exception[66], clearly show the reverse: β-glycosides are ca. 2-3 times more reactive. Very elegant experiments, which avoid the possibility of the β-glycoside reacting via anomerization, on conformationally-restricted acetals have been designed to examine this anomaly [22,29]. An alternative explanation that denies a universal, causal relationship between anomeric hyperconjugation and reactivity has been proposed, based on the little used Principle of Least Nuclear Motion (PLNM)[31]. In its fullest sense the PLNM should only be applied to elementary reaction steps and is too complex for routine application, since individual nuclear contributions should be weighted[67,68]. However it may be applied qualitatively, for example, to β-elimination. Reductive elimination of transdiaxial 1,2-dibromocyclohexane is favored over the diequatorial conformer (Scheme 10) if the PLNM is applied solely to the elementary elimination step.

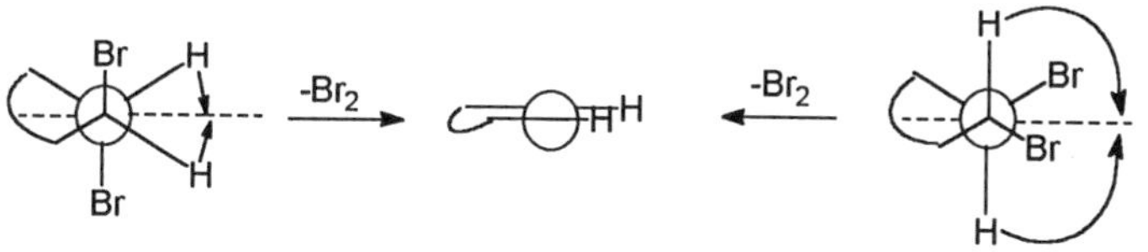

Scheme 10

Since the structure of reactant and product are dictated by electronic bonding, the PLNM implicitly includes an electronic contribution. The predictions of ALPH and the PLNM are equivalent for acetal hydrolysis. It is argued that ALPH is merely an overinterpretation of least motion effects. Furthermore, the failure of the PLNM is predicted for acetal hydrolysis, if one places more emphasis on the position of the transition state along the reaction coordinate, in analysing reaction pathways. Elegant experiments employing kinetic isotope measurements are provided as support[69].

2.2 The Synperiplanar Lone-Pair Hypothesis (SLPH)

The stereoelectronic theory of ALPH dictates that for reaction to occur, conformers not possessing ap lone-pair interactions must undergo conformational isomerization in order to acquire ap interactions. Thus β-glycosides must adopt a twist-boat conformation for reaction. In sterically unencumbered systems this does not appear to be unreasonable. However, Fraser-Reid and co-workers have estimated that for the conformationally restrained acetal (**1**) the energy cost is 6 kcal/mol[70]. Yet the rate of reaction of the β-anomer remains comparable to that of the α-anomer. As an alternative to ALPH, it is proposed that as reaction progresses, synperiplanar (sp³) lone pair interactions in the energetically accessible half-chair of the β-anomer are equivalent to ap interactions in the half-chair of the α-anomer. This is sometimes referred to as SLPH, in anology with ALPH. Further reaction co-ordinate calculations have been performed, by these workers, at the *ab initio* level[71].

2.3 ALPH and the Mechanism of Acetal Hydrolysis

A further theory, compatible with a kinetic anomeric effect in acetal hydrolysis and provided to account for the reduced relative reactivity of α-glycosyl halides, suggests that β-glycosides may react via a rapid S_N2 pathway which is dissallowed for α-glycosides which retain an S_N1 mechanism [30].   Lone pair-lone pair and other interactions are considered. Convincing experimental evidence to support this and a further pathway via radical intermediates, proposed by the same author, is required.

An alternative approach to acetal reactivity is garnered by considering the basicity and nucleophilicity of endo and exocyclic oxygens. The exo-anomeric effect is often considered to be greater in β than α-glycosides owing to the absence of "cross-hyperconjugation" in the former.   In fact, the effect has been predicted to be 1.66 times as great[37a]. The involvement of lone pair electrons in conjugative interactions should reduce their availability for donor reactions. The GAE predicts that of the four sites in α and β-glycosides, the exocyclic and endocyclic oxygen of a β-glycoside will be the least and most basic respectively. Theoretical calculation of proton affinities, in accord with the GAE, confirm the β-endocyclic oxygen as the most basic[73].   In acetal solvolysis, solvent acts as Lewis (or Bronsted) acid catalyst and nucleophile. Consideration of oxygen basicities predicts ring-opening to be prevalent for the β-anomer and competitive with exocyclic displacement for the α-anomer. This is not entirely consistent with experiment[73].

2.4. Amidine Hydrolysis

Breakdown of tetrahedral intermediates such as those invoked in hydrolysis of amidines according to ALPH requires two ap-lone pair interactions with the entering or leaving group. Thus ALPH predicts exclusively aminoamide product in hydrolysis of cyclic amidines. Lactam product can only occur via slow inversion at nitrogen (Scheme 11). Therefore the experimental observation of lactam as major product has provoked some of the hardest questioning of ALPH[74].

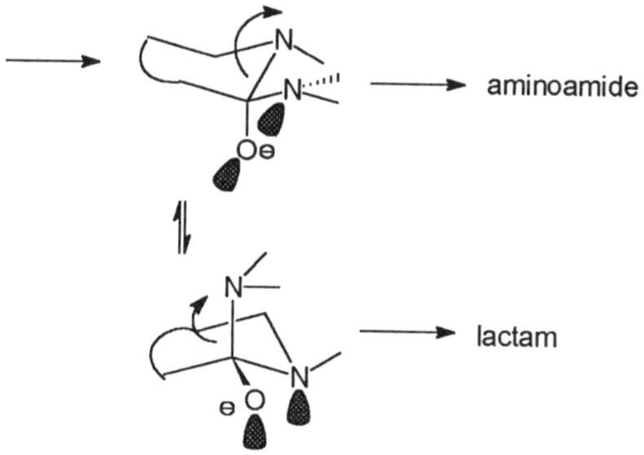

Scheme 11

A similar picture applies in amide hydrolysis in which two ap-lone pair interactions are required to expel a leaving group[8]. Thus partition of the tetrahedral intermediate is governed by the accessibility of conformations possessing such interactions. Clearly, in an unencumbered system, free rotation allows the correct orbital allignment for C-N bond cleavage (amide hydrolysis) and C-O bond cleavage (oxygen exchange) (Scheme 12). In the strained bicyclic amides studied by Brown's group, the nitrogen lone pair is unable to become antiperiplanar (or synperiplanar) to

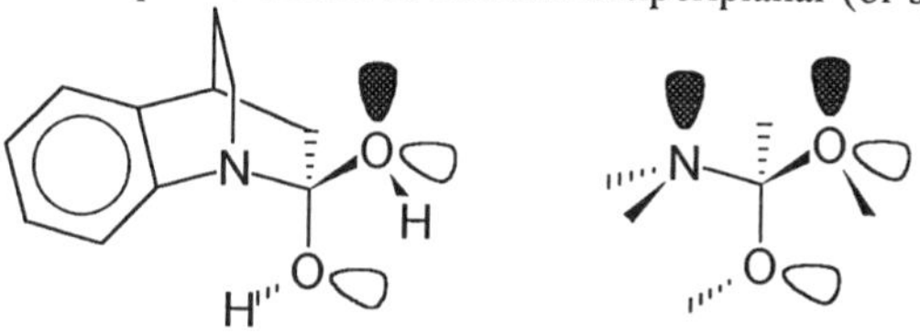

Scheme 12   open lp's ap to C-N
                filled lp's ap to C-O

either oxygen in the tetrahedral intermediate[75]. Two ap-lone pair interactions with the C-N bond promote C-N bond cleavage. This is reinforced by the relief of ring strain on ring cleavage and the good anilide leaving group. Amide hydrolysis via C-N bond cleavage is predicted by ALPH. The experimentally observed oxygen exchange at pH 4.5 requires rapid formation and breakdown of the tetrahedral intermediate via C-O bond cleavage competitive with C-N bond cleavage and hydrolysis. It is difficult therefore to resolve the stereo-electronically dissallowed breakdown of the tetrahedral intermediate via C-O bond cleavage with ALPH.

2.5. Orthoester Hydrolysis

Orthoester hydrolysis via hemiorthoesters again requires two ap-lone pair interactions with the leaving group according to ALPH. In this and related cases, discussion of stereoelectronic effects is divided into primary and secondary electronic effects. Secondary electronic effects are the n(sp^3)→σ* (C-O) interactions discussed above and commonly associated with anomeric hyperconjugation[8]. Primary electronic effects include n→π* and π(C-O)→π* interactions. Application of ALPH to predict reactivity and product distribution in such systems requires summation of primary and secondary effects in various conformers of the tetrahedral intermediate.   Again, the same experimentally observed product distributions rationalised in this way by ALPH have been accounted for using a non-stereoelectronic rationale[76].   Furthermore, a comparison of gas and condensed phase reactions of cyclic orthoesters and thioesters does not appear compatible with ALPH[77]. ALPH predicts that the axial thioester will ionize more rapidly than the equatorial owing to the presence of two ap-lone pair interactions (Scheme 13). The observed rates are identical. However, in solution a modest preference for axial attack on the carbocation is observed. It is thus argued that a medium or solvent effect may be the underlying cause of effects attributed to ALPH.

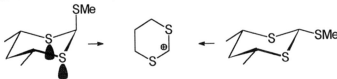

Scheme 13

2.6. Conclusion

In presenting the controversies surrounding the theory of stereoelectronic control associated with ALPH, the emphasis has been, neccessarily, on experimental exceptions and arguments against this theory.   The reader is directed to the large volume of experiments employed to support ALPH, in particular by the elegant experiments of Deslongchamps and co-workers, and to the following chapters in this volume.

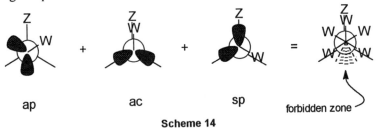

Scheme 14

ALPH explicitly requires ap-lone pair interactions with scissile bonds for bond cleavage. Although an unyielding support of ALPH neglects orbital overlap interactions other than ap-lone pair, it has been stated that two anticlinal (ac-lone pair) interactions may be better than one ap-lone pair. The proponents of SLPH, *vide supra*, see both ap- and sp-lone pair interactions as able to assist in bond cleavage.   It is therefore possible to combine these stereoelectronic lone pair interactions into one picture for the fragment W-X-Y-Z (Scheme 14: ap, sp and ac-lone pair orbitals on X interacting with the Y-Z bond). If it is approximated that adequate orbital overlap is present at the optimum angle ±30°, there remains only one sextant in which loss of leaving group Z⁻ is stereoelectronically forbidden.   Of course, two sc-lone pair interactions may be better than one sp-lone pair, in analogy with the ac-lone pair argument. The result: no "forbidden zone". This reduction *ad absurdum* illustrates the danger with dogmatic application of sp³ lone pair stereoelectronic interactions.

Although the simplicity and ease of application of a hypothesis based solely on the visualisation of sp^3 ap-lone pair interactions in the ground state remains attractive.

Several questions need to be directed at ALPH:

1. In order to accomodate experimental observations in accord with ALPH, large-scale conformational change is often required. At what stage does the molecular strain energy invoked outway the benefits of ap-orbital overlap?

2. ALPH is based upon the concept of dominant anomeric hyperconjugation. The stabilization energy resulting from hyperconjugation is proportional to the square of the orbital overlap (S) and inversely proportional to the difference in energy between electron donor and anti-bonding acceptor orbitals (ΔE). It is logical that if ΔE is small (good π-donor, polar acceptor bond), overlap (S) will become less important. It is well-accepted that donor strength follows:

$$n_{C^-} > n_N > n_O > \sigma_{C-S} > \sigma_{C-S} > \sigma_{C-H} > \sigma_{C-C} > \sigma_{C-O} > \sigma_{C-F}$$

and acceptor ability:

$$\sigma^*_{C-Cl} > \sigma^*_{C-S} > \sigma^*_{C-F} > \sigma^*_{C-O} > \sigma^*_{C-C} > \sigma^*_{C-H}$$

However, ALPH does not recognize differences between $n \rightarrow \sigma^*$ interactions and ignores all other than ap-lone pair interactions. Is one $n_O- \rightarrow \sigma^*$ interaction (oxyanion donor) worth two $n_O \rightarrow \sigma^*$ interactions? Is one $sp-n_N \rightarrow \sigma^*$ interaction (or one distorted ap-interaction with reduced overlap) equivalent to one ap-$n_O \rightarrow \sigma^*$ interaction?

3. Is it safe to apply ALPH (as it was originally and remains most frequently applied) to reactants and reactive intermediates, when the structure and position of the transition state may amplify or diminish the influence of $n \rightarrow \sigma^*$ interactions?

4. Is it safe to apply ALPH as formulated, in terms of sp^3-hybrid orbitals, when the contemporary theoretical basis of anomeric stabilization is cast largely in terms of π-type donor atomic orbitals?

5. Although there is ample support for anomeric hyperconjugation in gas phase calculations, the importance of such stereoelectronic effects in solution is uncertain. Does the experimental evidence in solution provide support for the importance of anomeric hyperconjugation when alternative rationale exist for these experimental observations?

3. ALPH and Ribonuclease: Stereoelectronic Control in Phosphoryl Transfer

Phosphoryl group transfer reactions are of profound importance in biological systems. Ribonuclease represents an important and well-studied enzyme that has been prototypical in many aspects of research on enzymic phosphoryl transfer. It has been argued that the free energy profile for ribonuclease is dominated by stereoelectronic effects based on ap-lone pair interactions[2,78]. This version of ALPH has generated more than 30 papers in the literature from the groups of Gorenstein, as summarised in a review [79], and more recently of Taira. How did Gorenstein's theory develop?

Early theoretical calculations by Lehn and Wipff indicated the influence of stereoelectronic effects on ground state tetrahedral phosphates[80]. Extension of these calculations by Gorenstein and co-workers to trigonal bipyramidal (TBP) pentacoordinate phosphoranes, putative intermediates for phosphoryl group transfer, suggested an increased role for stereoelectronic effects. This was quantified by further calculations on the free energy profile for nucleophilic substitution at phosphorus.

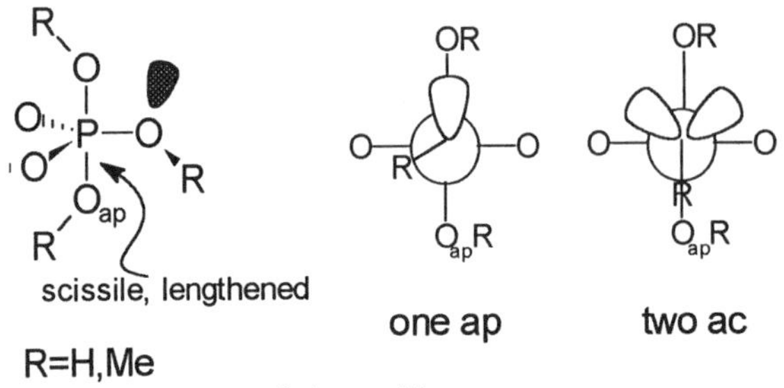

scissile, lengthened

R=H,Me

one ap   **two ac**

Scheme 15

Despite the low level of calculation and frequent use of non-relaxed geometries, this important work did provide an interesting glimpse at conformational factors influencing the energy and structure of TBP intermediates and transition states. Conformers with the equatorial substituent sc or sp to one axial P-O bond were seen to be favored. Owing to the growing awareness of the GAO and ALPH, this stereoelectronic effect was classified not in terms of these sc or sp-substituents, but ap-sp^3-hybrid lone pairs on the equatorial oxygen . Stabilization was seen to be the result of n→σ^*(P-O) orbital mixing with electron donation from one ap or two ac-sp^3-hybrid lone pairs on the equatorial oxygen (Scheme 15). The corresponding lengthening of the ap axial bond (P-O$_{ap}$) provided further similarities with the GAO and ALPH. Thus Gorenstein's theory developed as an adjunct of ALPH: a stereoelectronic requirement for ap-lone pairs on equatorial heteroatoms for expulsion of the axial leaving group from a TBP intermediate or transition state in nucleophilic substitution at phosphorus.

The first test of this theory was methyl ethylene phosphate (MEP) hydrolysis[81]. The hydrolysis of MEP and other five-membered phosphates occurs ≤10^8 faster than their acyclic counterparts. The lion's share of this rate acceleration was ascribed by Gorenstein to stereoelectronic assistance. Clearly, endocyclic cleavage of MEP is stereoelectronically favored and exocyclic cleavage disfavored (Scheme 16). The observation of up to 50% exocyclic

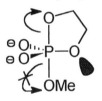

Scheme 16

cleavage in dilute acid and more modest exocyclic cleavage in strong alkali solution therefore had to be accounted for. A dipole effect on oxygen basicity was deemed to favor protonation of the exocyclic oxygen and thus exocyclic cleavage in the acid-catalysed pathway. With no explanation for the observations in strong base, the original experiments had to be proven wrong. However, subsequent work has confirmed the original experiments[82]. Conversely, Gorenstein found support for his theory in the much more negative entropy of activation observed for hydrolysis of an acyclic over a five-membered cyclic phosphate - the acyclic phosphate must freeze one more degree of rotational freedom in the transition state to obtain the correct stereoelectronic overlap. Subsequent work has shown this original experimental observation to be in error[83]. Further experimental observations used to support Gorenstein's theory have been reinterpreted and are discussed elsewhere[57].

The "textbook" mechanism of ribonuclease (Scheme 17) is well-founded in mechanistic phosphorus chemistry[84]. Two histidines are required, one as a base to deprotonate the O2 nucleophile, the second as an acid to protonate the O5 leaving group.   Stabilization of the dianionic TBP intermediate (and structurally similar transition states) is provided by electrostatic interaction with the ammonium of Lys-41. The key intermediate (**I**) bears comparison with Scheme 16.   Exocyclic P-O$_5$, bond cleavage is stereoelectronically disfavored. In order to accomodate, within Gorenstein's theory, the P-O$_5$, bond cleavage that ribonuclease must necessarily catalyse major and unprecedented conformational changes to the pentacoordinate intermediate (**I**) have been proposed[85].

Breslow, on the basis of much cited experiments on the catalysis of hydrolysis of dinucleotides by imidazole, has suggested a modification to the mechanism of ribonuclease in which the anionic phosphoryl oxygen is first protonated by histidine and in which the histidines functionally facilitate the shuttle of a proton from the 2'-hydroxyl to the phosphoryl oxygens, thence to the leaving group[86]. The advantage

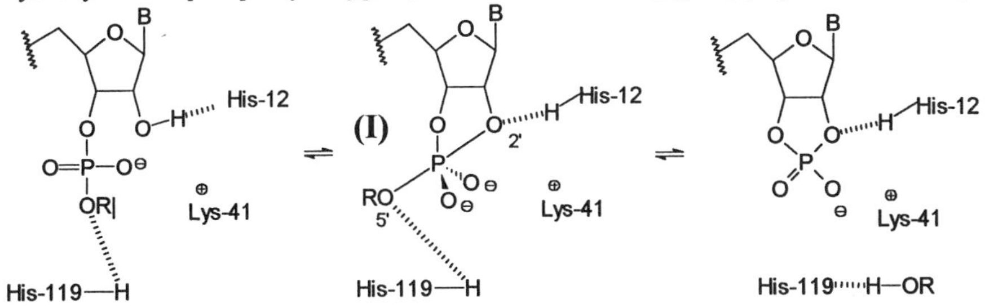

Scheme 17

of this catalytic mechanism is reasonable in that the intermediate is stabilized by protonation and the most basic oxygens are those protonated initially.   Theoretical work by Lim and Tole has provided a more detailed, but similar picture of this proton shuttle, which does differ in the order of the proton transfer steps and the active site geometry required to effect them[87].   The role of and requirement for stereoelectronic assistance in these mechanisms is not clear.   The difference in activation barriers, governed by Gorenstein's effect, for the breakdown of TBP intermediate (**I**) via exocyclic P-O$_5$, and endocyclic P-O$_2$, bond cleavage has been quoted to be as high as 29 kcal/mol in favor of the latter[78].   Yet the mechanisms of Breslow and of Lim are highly symmetrical for catalysis of breakdown of the TBP intermediate in both directions.   Acid catalysis is required for stabilization of both the endocyclic and exocyclic scissile bond transition states.

In molecular orbital calculations on pentacoordinate phosphoranes and sulfuranes, similar torsion angle correlations with conformational energy and geometry are evident. However, there is no evidence for the n→σ* interaction of the Gorenstein theory in perturbational molecular orbital analysis of sulfuranes[58] nor phosphoranes [88] and glaring anomalies with the theory exist in the calculated energies and structures of phosphoranes[89]. Furthermore, there is no general requirement for a pervading stereoelectronic effect related to ALPH to account for experimental observations on solution reactivity of phosphate esters and related compounds. However, it is unlikely that stereoelectronic effects will not play some role in the reactivity of these compounds. For example, the hydrolysis of α-carbonyl phosphonate esters proceeds a millionfold faster than that of simple alkyl phosphonates[90]. One mechanistic rationale consistent with kinetic analysis, product distribution and isotope labelling studies involves stereoelectronic assistance $\{n(O_{apical}) \rightarrow \pi^*(C=O)\}$ by the neighbouring α-carbonyl group, via a three-membered cyclic transition state (Scheme 18)[88].

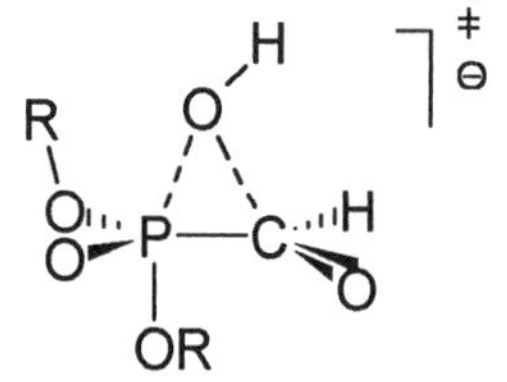

Scheme 18. Transition state (single imaginary frequency corresponding to reaction coordinate) located at HF/6-31G*: R=H,CH$_3$

Rigorous decomposition of stereoelectronic effects, observed in gas phase calculations, into individual contributions is essential. In this way, the symptomatic effects seen in the seminal work of Gorenstein's group may find a convincing rationale. Certainly, *carte-blanche* adoption of the ap-sp^3 lone pair argument of ALPH, without rigorous theoretical support, has led to erroneous conclusions in phosphorus chemistry based on the now defunct "Gorenstein stereoelectronic theory".

Note Added in Proof The publication by a Japanese physicist, Kubo, in 1936 of a dipole moment observation on dimethoxymethane and its influence on Lemieux's thinking, is discussed by Lemieux, himself, in citation 37a [91]. A reviewer has pointed out the common usage of the term *Kubo Effect* in the Japanese literature to describe an anomeric effect. The usage of the term *Edward-Lemieux Effect* suggested in this chapter is limited to carbohydrates and is not equivalent to the Japanese term. The term *Edward-Lemieux Effect* has previously been proposed by Wolfe and co-workers [92].

Acknowledgement
The Natural Sciences and Engineering Research Council of Canada is thanked for financial support.

Literature Cited
1.   Reed A.E., Schleyer PvR., *Inorg. Chem.* **1988**, *27*, 3969.
2.   Taira K., *Bull. Chem. Soc. Jpn.*,**1987**, *60*, 1903; Taira K., Uebayasi M., Maeda H., Furukawa K., *Protein Engineering,* **1990**, *3*, 691; Storer J.W., Uchimara T., Tanabe K., Uebajasi M., Nishikawa S., Taira K., *J. Am. Chem. Soc.*, **1991**, *113*, 5216.
3.   Bizzozero S.A., Dutler H., *Bioorg. Chem.* **1981**, *10*, 46.
4.   Post C.B., Karplus M., *J. Am. Chem. Soc.*, **1986, *108*,** 1317.
5.   Makinen M.W., Kukuyama J.M., *J. Biol. Chem.*, **1982**, *257*, 24.
6.   Ninibiar K.P., Stauffer D.M., Kolodzies P.A., Benner S.A., *J. Am. Chem. Soc.,* **1983**, *105*, 5886.
7.   Dugas H., *Bioorganic Chemistry: A Chemical Approach to Enzyme Action*, 2nd edn, Springer-Verlag, New York, 1989.
8.   Deslongchamps P., *Stereoelectronic Effects in Organic Chemistry*, Pergamon, Oxford, 1983.
9.   Lemieux R.U., *Abstr. Pap. Am. Chem. Soc.*, **1959**, *135*, 5E.
10. Juaristi E., Cuevas G., *Tetrahedron*, **1992**, *48*, 5019
11. Hirsch J.A., *Concepts in Theoretical Organic Chemistry*, Allyn & Bacon, Boston, 1985.
12. IUPAC Physical Organic Glossary Review, *Pre Appl. Chem.*, **1983**, *55*, 1281.

13. Lowry T.H., Richardson K.S., *Mechanism and Theory in Organic Chemistry*, 3rd edn., Harper & Row, New York, 1974.
14. Edward J.T., *Chem. Ind. (London)* **1955**, 1107; Lemieux R.U., Chu N.J., *Abst. Pap. Am. Chem. Soc. Meeting*, **1958**, *33*, 31N; Lemieux R.U., in *Molecular Rearrangements*. Pt. 2 ed. de Mayo P. Interscience Publishers, **1964**; Lemieux R.U., Kullnig R.K., Bernstein H.J., Schneider W.G., *J. Am. Chem. Soc.*, **1958**, *80*, 6093.
15. For example, the conformational preference of the phosphorane BH_3PH_4 ascribed to $\sigma(P-H) \rightarrow \sigma^*(P-B)$ hyperconjugation has been described as an anomeric effect[16].
16. Wang P., Zhang Y., Glaser R., Reed A.E., Schleyer PvR., Streiweiser A., *J. Am. Chem. Soc.*, **1991**, *113*, 55..
17. Lemieux R.U., Pavia A.A., Martin J.C., Watanabe K.A., *Can. J. Chem.* **1969**, *47*, 4427.
18. This definition is adapted from the original: Praly J-P, Lemieux R.U., *Can. J. Chem.* **1987**, *65*, 213.
19. Franck R.W., *Tetrahedron*, **1983**, *39*, 3251; Juaristi E., Lopez-Nunez N.A., Glass R.S., Petsom A., Hutchins R.O., Stercho J.P., *J. Org. Chem.*, **1986**, *51*, 1357.
20. Wolfe S., *Acc. Chem. Res.* **1972**, *5*, 102.
21. Epiotis N.D., *J. Am. Chem. Soc.*, **1977**, *99*, 8379.
22. Kirby A.J., *The Anomeric Effect and Related Stereoelectronic Effects at Oxygen*, **1983**, Springer, Verlag, Berlin.
23. Wolfe S., Whangbo M-H, Mitchell D.J., *Carbohydr. Res.*, **1979**, *69*, 1.
24. Taira K., Fanni T., Gorenstein D.G., *J. Am. Chem. Soc.*, **1984**, *106*, 1521; Taira K., Fanni T., Gorenstein D.G., *J. Org. Chem.*, **1984**, *49*, 4531.
25. Lemieux R.U., *Pure Appl. Chem.* **1971**, *25*, 527.
26. Lowe G., Thatcher G.R.J., Turner J.C.G., Waller A., Watkin D.J., *J. Am. Chem. Soc.*, **1988**, *110*, 8512.
27. For example, Jeffrey, G.A., Yates J.H., *J. Am. Chem. Soc.*, **1979**, *101*, 820.
28. (a) Fuchs B., Schleifer L., Tartakovsky E., *New J. Chem.*, **1984**, *8*, 275; (b) Cosse-Barbi A., Dubois J.E., *J. Am. Chem. Soc.*, **1987**, *109*, 1503; (c) Schleifer L., Senderowitz H., Aped P., Tartakovsky E., Fuchs B., *Carbohyd. Res.* **1990**, *206*, 21; (d) Aped P., Fuchs B., Goldberg I., Senderowitz H., Tartakovsky E., Weinman S., *J. Am. Chem. Soc.*, **1992**, *114*, 5585.
29. Edwards M.R., Jones P.G., Kirby A.J., *J. Am. Chem. Soc.*, **1986**, *108*, 7067 and references therein.
30. Box V.G.S., *Heterocycles*, **1984**, *22*, 891; **1990**, *31*, 1157; **1991**, *32*, 795.
31. Sinnott M.L., *Adv. Phys. Org. Chem.* **1988**, *24*, 113.
32. Booth H., Dixon J.M., Khedhair K.A., Readshaw S.A., *Tetrahedron*, **1990**, *46*, 1625.
33. Babirad S.A., Wang Y., Goekjian P.G., Kishi Y., *J. Org. Chem.*, **1987**, *52*, 4825 and references therein.
34. Deslongchamps P., Pothier N., *Can. J. Chem.* **1990**, *68*, 597.
35. Paulsen H., Gyorgdeak, Z., Friedmann M., *Chem. Ber.*, **1974**, *107*, 1590.
36. Perrin C.L., Nunez O., *J. Chem. Soc. Chem. Commun.*, **1984**, 333; Ouedrago A., Lessard J., *Can. J. Chem.* **1991**, *69*, 474.
37. (a) Lemieux R.U., *Explorations with Sugars: How Sweet It Was*, Profiles, Pathways and Dreams, **1990**, Seeman J.I., ed., ACS, Washington; (b) Chu, N-J., PhD Thesis, University of Ottawa, **1959**.
38. Altona C., Romers C., Havinga E., *Tetrahedron Lett.* **1959**, *8*, 16; Altona C., PhD Thesis, University of Leiden, **1964**.
39. Hutchins R.O., Kopp L.D., Eliel E.L., *J. Am. Chem. Soc.*, **1968**, *90*, 7174.
40. Romers C., Altona C., Buys H.R., Havinga E., *Top. Stereochem.*, **1969**, *4*, 39.
41. Foster J.P., Weinhold F., *J. Am. Chem. Soc.*, **1980**, *102*, 7211.
42. This remains a deficiency of the most recent championing of $n \rightarrow n$ $4e^-$ destabilization [30]

43. (a) Wiberg K., *J. Am. Chem. Soc.*, **1992**, *115*, 614;   (b) Krol M.C., Huige C.J.M., Altona C., *J. Comp. Chem.* 1990, *11*, 765.

44. Woods R.J., Szarek W.A., Smith V.H., *J. Chem. Soc. Chem. Commun.*, **1991**, 334.

45. Cramer C.J., *J. Org. Chem.*, **1992**, 57, *7034*.

46. For example, Reed A.E., Schleyer PvR., *Inorg. Chem.*, **1988**, *27*, 3969.

47. Grein F., Deslongchamps P., *Can. J. Chem.* **1992**, *70*, 1562, and references therein.

48. A host of specific through-bond hyperconjugative and through-space interactions have been proposed including $n_\sigma \rightarrow \sigma^*$, $n_\pi \rightarrow \sigma^*$, $\pi \rightarrow \sigma^*$, $\sigma \rightarrow \sigma^*$. In addition, PMO analysis has been used to extremely good effect in defining stabilizing and destabilizing interactions between molecular fragments[49]

49. Pinto B.M., Schlegel H.B., Wolfe S., *Can. J. Chem.*, **1987**, *65*, 1658; Epiotis N.D., Cherry W.R., Shaik S., Yates R.L., Bernardi F., *Structural Theory of Organic Chemistry*, Topics in Current Chemistry 70, **1977**, Springer-Verlag, Berlin.

50. For example, Lipkind G.M.,Verovsky V.E.,Kochetkov N.K., *Carbohydr. Res.*,**1984**, *133*,1.

51. Booth H., Grindley T.B., Khedhair, K., *J. Chem. Soc. Chem. Commun.*, **1982**, 1047 .

52. Zefirov N.S., Blagoveschchenski V.S., Kazmirchik I.V., Yakovleva O.P., *J. Org. Chem.*, **1971**, *7*, 599.

53. Juaristi E., Gonzalez, Pinto B.M., Johnston B.D., Nagelkerke R., *J. Am. Chem. Soc.*, **1989**, *111*, 6745; Juaristi E., Aguilar M.A., *J. Org. Chem.*, **1991**, *56*, 5919; Pinto B.M., Johnston B.D., Nagelkerke R., *Heterocycles* , **1989**, *28*, 389..

54. Schleyer PvR., Jemmis E.D., Spitznagel G.W., *J. Am. Chem. Soc.*, **1985**, *107*, 6393.

55. Sabzner U., Schleyer PvR., *J. Chem. Soc. Chem. Commun.*, **1990**, 190.

56. For example: Pikies J., Wojnowski W., *J. Organomet. Chem.*, **1990**, *393*, 187; Apeliog Y., Stango A., *J. Organomet. Chem.* **1988**, *346*, 305; Reed A.E., Schade C., Schleyer PvR., Kamath P.V., Chandrasekhar J., *J. Chem. Soc. Chem Commun.*, **1988**, 67.

57. Thatcher G.R.J., Kluger R., *Adv. Phys. Org. Chem.* **1989**, *25*, 99.

58. Wang P., Zhang Y., Glaser R., Reed A.E., Schleyer PvR, Streiweiser A., *J. Am. Chem. Soc.*, **1991**, *113*, 55; McDowell R.S., Streiweiser A., *J. Am. Chem. Soc.*, **1985**, *107*, 5849.

59. Holmes R.R., *J. Am. Chem. Soc.*, **1978**, *100*, 433; , Trippett S., *Pure Appl. Chem.* **1974**, *40*, 595.

60.   For example, CH_3 is more apicophilic than OH by 2 kcal/mol from calculation, but 9 kcal/mol less apicophilic in Holmes's empirical scale. For a full discussion and evidence for the influence of solvent see ref. [61].

61. Thatcher G.R.J., Campbell, A.S., *J. Org. Chem.*, **1993**, *53*, 2272.

62. The simplest of these are through space charge transfer interactions comparable to internal hydrogen bonding[63] and putative π-bonding interactions. For alternative views on the significance of hyperconjugation as the basis of the GAE, see refs [64].

63.   Cameron D.R, Thatcher G.R.J., *The Anomeric Effect and Associated Stereoelectonic Effects*, ACS Symposium Series, **1993**, ACS, Washington, this issue.

64. Dewar M.J.S., *J. Am. Chem. Soc.*, **1984**, *106*, 669; Epiotis N.D., *THEOCHEM*, **1991**, *229*, 205; Smits G.F., Altona C., *Theor. Chim. Acta* **1985**, *67*, 461.

65. Bach R.D., Badger R.C., Lang T.L., *J. Am. Chem. Soc.*, **1979**, *101*, 2845.

66. Eikeren P. v., *J. Org. Chem.* , **1980**, *45*, 4641.

67. Tee O.S., Altmann J.A., Yates K., *J. Am. Chem. Soc.*, **1974**, *96*, 3141..

68. Hine J., *Adv. Phys. Org. Chem.*, **1977**, *15*, 1.

69. For example, Bennett A.J., Sinnott M.L., *J. Am. Chem. Soc.*, **1986**, *108*, 7287.

70. Ratcliffe A.J., Mootoo D.R., Andrews C.W., Fraser-Reid B., *J. Am. Chem. Soc.*, **1989**, *111*, 7661.

71. Andrews C.W., Fraser-Reid B., Bowen J.P., *J. Am. Chem. Soc.*, **1991**, *113*, 8293.

72. Andrews C.W., Bowen J.P., Fraser-Reid B., *J. Chem. Soc. Chem. Commun.*, **198**, 1913.

73. McPhail D.R., Lee J.R., Fraser-Reid B., *J. Am. Chem. Soc.*, **1992**, *114*, 1905.
74. Perrin C.L., Nunez O., *J. Am. Chem. Soc.*, **1986**, *108*, 5997; Perrin C.L., Nunez O., *J. Am. Chem. Soc.*, **1986**, *108*, 5997.
75. Somayaji V., Brown R.S., *J. Org. Chem.* **1986**, *51*, 2676.
76. Perrin C.L., Arrhenius G. M. L., *J. Am. Chem. Soc.*, **1982**, *104*, 2839.
77. Caserio M.C., Shih P., Fisher C.L., *J. Org. Chem.*, **1991**, *56*, 5517.
78. Taira K., Uchimaru, T., Tanabe K., Uebayasi M., Nishikawa S., *Nucl. Acids Res.*, **1991**, *19*, 2747.
79. Gorenstein D.G., *Chem. Rev.*, **1987**, 1047.
80. Lehn J.M., Wipff G.H., *J. Chem. Soc. Chem. Commun.*, **1975**, *99*, 800.
81. Kluger R., Westheimer F.H., *J. Am. Chem. Soc.*, **1969**, *91*, 4143.
82. Kluger R., Thatcher G.R.J., *J. Am. Chem. Soc.*, **1985**, *107*, 6006; Kluger R., Thatcher G.R.J., *J. Org. Chem.*, **1986**, *51*, 207
83. Kluger R., Taylor S., *J. Am. Chem. Soc.*, **1990**, *112*, 6669.
84. Fersht A.J., *Enzyme Structure and Mechanism* 2nd edn. **1984**, Freeman, New York.
85. Gorenstein D.G., Luxon B.A., Findlay J.B., Momii R., *J. Am. Chem. Soc.*, **1977**, *99*, 4170.
86. Breslow R., *Acc. Chem. Res.*, **1991**, *24*, 317, and references therein.
87. Lim C., Tole P., *J. Am. Chem. Soc.*, **1992**, *114*, 7245.
88. Thatcher G.R.J., Cameron D.R., Krol E.S., unpublished results.
89. Tole P., Lim C., *The Anomeric Effect and Associated Stereoelectonic Effects*, ACS Symposium Series, **1993**, ACS, Washington, this issue.
90. Krol E.S., Davis J.M., Thatcher G.R.J., *J. Chem. Soc. Chem. Commun.*, **1991**, 118; Krol E.S., Thatcher G.R.J., *J. Chem. Soc. Perkin 2*, **1993**, 793.
91. Kubo M., *Papers Inst. Phys. Chem. Res. (Tokyo)* **1936**, *29*, 179.
92. Wolfe S., Rauk A., Tel L.M., Csizmadia I.G., *J. Chem. Soc. (B)*, **1971**, 136.

RECEIVED July 6, 1993

Chapter 3

Intramolecular Strategies and Stereoelectronic Effects

Glycosides and Orthoesters Hydrolysis Revisited

P. Deslongchamps

Département de chimie, Faculté des sciences, Université de Sherbrooke, Sherbrooke, Québec J1K 2R1, Canada

It has been generally accepted that stereoelectronic effects play an important role in hydrolytic processes. However, the stereoelectronic theory has been criticized because the relative rates of glycosides hydrolysis could not be explained. A brief survey of the evidence provided by us and others which support stereoelectronic principles is presented. More recent works on the kinetically controlled spiro acetalization of hydroxy enol ethers show that spiroacetal formation takes place via an early transition state while following the antiperiplanar pathway. This approach corresponds closely to the Bürgi-Dunitz angle of attack of a nucleophile on a π-system. As a result, transition states having a geometry which corresponds to the beginning of a chair form are preferred over those corresponding to a boat form. On that basis, reactions occurring at the anomeric center in α and β-glycosides, including kinetic data for hydrolysis are readily explained. The results previously reported on the acid hydrolysis of cyclic orthoesters have also been reexamined on the basis of a late transition state while following the antiperiplanar hypothesis. Again, these results are well rationalized.

Evidence from our laboratory that stereoelectronic effects are a key element in the understanding of the chemical reactivity of organic reactions started with the discovery of ozonolysis of acetals. In this work, it was first discovered (1-3) that in order to observe an oxidation, an acetal must take a conformation where each oxygen can have an electron lone pair antiperiplanar to the C-H bond. For instance, it was found (Scheme 1) that the β-glycosides 1 were selectively oxidized to the corresponding hydroxy ester 3 whereas the corresponding α-isomers 5 were found to be unreactive under the same conditions.

0097–6156/93/0539–0026$08.25/0
© 1993 American Chemical Society

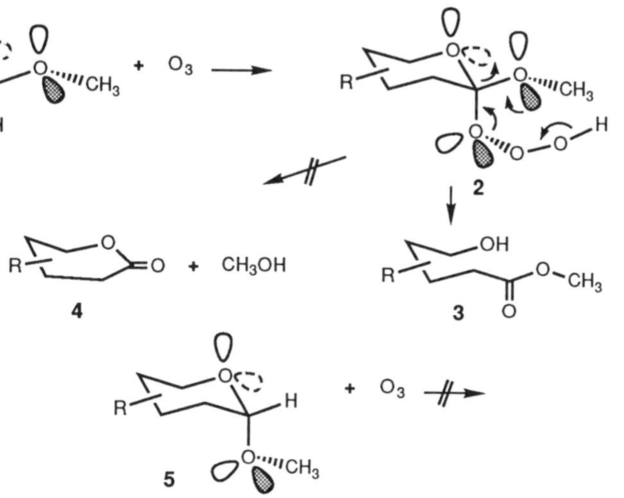

SCHEME 1

The preferential formation of hydroxy-ester **3** from the hydrotrioxide tetrahedral intermediate **2** which is formed during the ozonolysis of the acetal led us to further postulate that stereoelectronic effects might be the main driving force in hydrolytic processes. In this case, it was assumed that intermediate **2** is equivalent to a tetrahedral intermediate and we postulated that specific cleavages can take place when two oxygen atoms of such a tetrahedral intermediate can have each an electron lone pair antiperiplanar to the leaving group. Under such conditions, intermediate **2**, in the chair form, can only give the hydroxy methyl ester **3**, none of the corresponding lactone **4** and methanol can be produced.

More direct experimental evidence that stereoelectronic effects (4) control hydrolytic processes were then obtained by studying the behavior of cyclic orthoesters (5-8). For instance, the conformationally rigid mixed orthoester **6** was shown to yield the hydroxy-methyl ester **9** following the pathway described in Scheme 2. Indeed the antiperiplanar hypothesis predicts preferential loss of the protonated axial deuterated methoxy group from **6** to yield the cyclic dioxocarbonium ion **7** which must be hydrated to give the corresponding tetrahedral intermediate **8** having an axial hydroxyl group. This intermediate can then only lead to the hydroxy methyl ester **9**. [Compound **6** is also formed stereoselectively by the addition of CD_3O^- to the cation **7** under aprotic conditions (6). It is also produced stereospecifically when the corresponding dimethyl orthoester undergoes acid-catalyzed exchange with CD_3OH in methanol-d_4-dichloromethane (9-10). The exchange of the equatorial OCH_3 group is much slower (~100 times)].

We have subsequently obtained experimental evidence that there is stereoelectronic control in the acetal formation by studying the mild acid cyclization of bicyclic hydroxypropyl acetal **10** (Scheme 3) under kinetic and thermodynamic conditions (11-12). Upon acid treatment (PTSA-

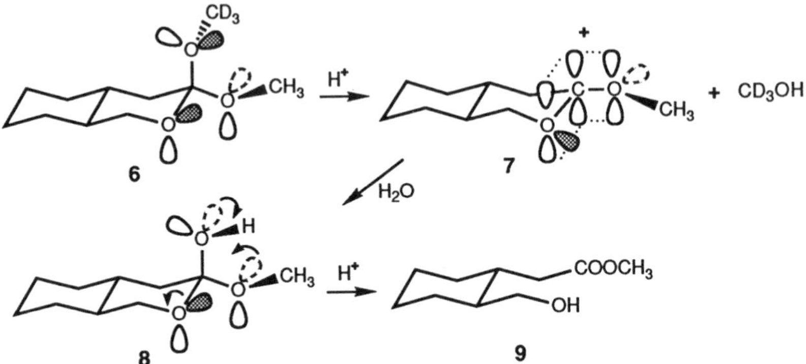

SCHEME 2

MeOH) at room temperature, **10** gave the *cis* tricyclic acetal **13** whereas an equilibrium mixture (45:55 ratio) of *cis* and *trans* acetals **13** and **15** was obtained after refluxing conditions. The kinetically controlled formation of **13** was then explained in the following way. Upon acid treatment, it was assumed that **10** gave first the cyclic oxocarbonium ion **11** which underwent a stereoelectronically controlled cyclization via an antiperiplanar attack to give the *cis* acetal **13** via a chair-like pathway (**11→12→13**). The formation of the *trans* acetal **15** was not observed under kinetic conditions because this compound can be produced only via a high energy twist-boat pathway (**11→14→15**) which is the result of an antiperiplanar attack of the incoming hydroxyl group.

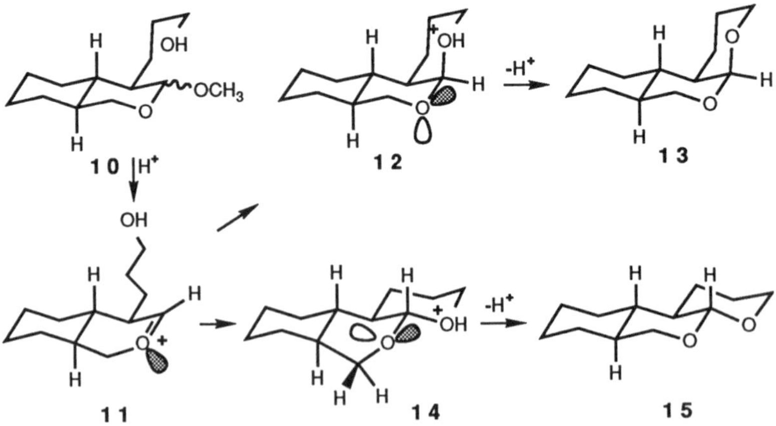

SCHEME 3

On that basis, we further postulated (4) that α-glycosides (axial anomer) must hydrolyze via their ground state chair conformation whereas β-glycosides (equatorial anomer) must first assume a twist-boat conformation in order to fulfill the stereoelectronic requirement of the antiperiplanar hypothesis as shown in Scheme 4. It follows that if one <u>assumes</u>

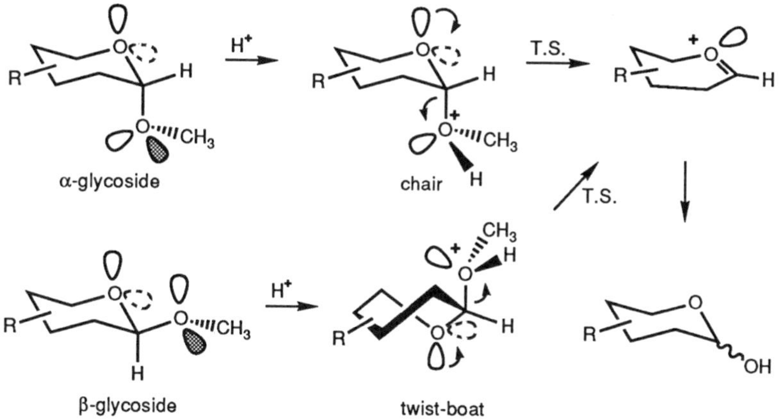

SCHEME 4

that the hydrolysis takes place via an early transition state which resembles the reactive conformation of the protonated acetal rather than the cyclic oxocarbonium ion, the hydrolysis of α-glycosides via a chair-like transition state should be faster than that of β-glycosides which must proceed via a twist-boat transition state.

The rates of hydrolysis of α and β-glycosides have been measured and they do not agree with the above hypothesis, the β-anomers being generally hydrolyzed at a slightly faster rate (up to 2 or 3 times). Eikeren (13) has measured the relative rate of hydrolysis of the conformationally rigid bicyclic acetals **16** and **17** (Scheme 5) and found that the axial anomer **16** is hydrolyzed slightly faster by a factor of 1.5. Since α and β-glycosides bear additional hydroxyl groups which are likely to influence the rate of hydrolysis, compounds **16** and **17** are better models for comparing the reactivity of α and β-anomers. [By comparison, D-glucopyranosides and 2-deoxy-D-arabino-hexapyranosides are hydrolyzed at 10^{-7} and 10^{-3} times the rate of tetrahydropyranyl ethers respectively (14-16)]. Eikeren measured the activation parameters for the hydrolysis of **16** and **17** and found a slightly larger entropy term for the axial

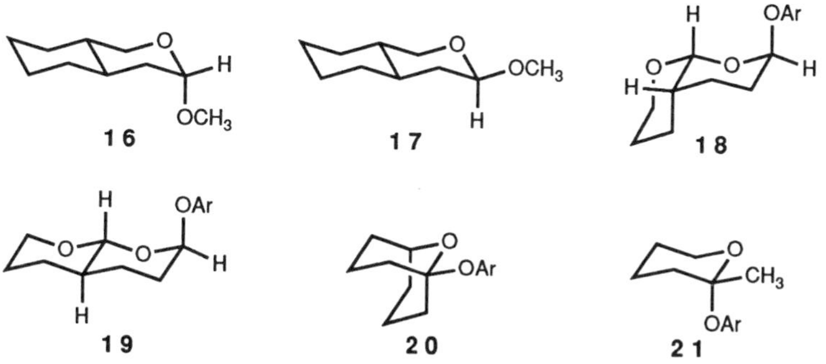

SCHEME 5

anomer and this observation led him to postulate an early transition state for the equatorial anomer with less C-O bond cleavage, and a slightly later transition state for the axial anomer with more extensive C-O bond cleavage. However, since the difference in rate between model compounds **16** and **17** (or between α and β-glycosides) is so small, it is clear that these results cannot be explained by the antiperiplanar hypothesis while assuming an early transition state along the reaction coordinates. This topic will be rediscussed later.

A large body of experimental evidence showing that the hydrolysis of acetals is governed by stereoelectronic effects has also been reported by Kirby and his co-workers. This work has been explained in details in his book (*17*) and in two recent reviews (*9-10*). The most convincing results were obtained by studying the spontaneous hydrolysis of conformationally restricted aryloxytetrahydropyranyl acetals **18** and **19** (Scheme 5). Isomer **18** underwent hydrolysis 200 times faster than 19. More strikingly, the completely rigid bicyclic acetal **20** was found to hydrolyze 1.2 x 10^{13} times less rapidly than the tetrahydropyranyl acetal **21**. In the case of **20**, the oxygen atom can only destabilize the cation through an inductive effect, since its lone pair electrons are not properly aligned to provide the stabilization which is normally observed by electronic delocalization.

Kirby and his collaborators (*9-10*) have also obtained rigorous evidence by making accurate crystal-structure determination of a series of aryloxytetrahydropyranyl acetals which revealed a striking and systematic patterns of changes in the bond lengths at the anomeric center. They found that in axially oriented aryloxytetrahydropyrans, the endocyclic C-O bond is significantly shortened and the C-OAr bond lengthened by an amount which depends on the electronegativity of the leaving group. They further noticed that the variation in bond length is related to the rate of hydrolysis of these acetals in a simple manner. Indeed, the rate of hydrolysis shows a linear variation with the pKa of the leaving group when the C-O cleavage is rate determining.

Since the relative rate of hydrolysis of α and β-glycosides could not be explained by the antiperiplanar hypothesis along with an early transition state, alternative proposals have been put forward. Fraser-Reid and his collaborators (*18*) have recently proposed that the hydrolysis (or formation) of glycosides could take place in some cases by a synperiplanar rather than an antiperiplanar lone pair pathway. Another pathway which is based on the principle of least motion but completely ignores stereoelectronic principles has also been strongly advocated by Sinnott (*19*) in recent years. We will examine these alternative rationalizations in detail toward the end of this article. However, the yet non totally general acceptance of the stereoelectronic theory has convinced us to reinvestigate the mechanism of acetal formation and hydrolysis. It appeared to us that it was necessary to establish firmly the position of the transition state along the reaction coordinate in order to better understand the mechanism of acetal hydrolytic processes. This work is now described.

We have started looking for experiments where acetals could be produced in specific configurations under kinetically controlled conditions. For that we have reexamined the formation of 1,7-dioxaspiro[5.5]-undecanes, *i.e.*, spiroacetals.

We have previously reported (20) a study which revealed that the unsubstituted 1,7-dioxaspiro[5.5]undecane exists exclusively in conformation **22a** (Scheme 6) even at room temperature. This experimental observation was explained by the fact that conformation **22a** is stereoelectronically and sterically more stable than conformations **22b** and **22c** which were estimated to be less stable respectively by a value of 2.4 and 4.8 kcal/mol. [The relative energy of 0, 2.4 and 4.8 kcal/mol for **22a, 22b**, and **22c** was estimated by using the following values: anomeric effect (e) = -1.4 kcal/mol; steric effect: gauche form of *n*-butane (CC) = 0.9 kcal/mol, gauche form of CH_2-CH_2-CH_2-O (CO) = 0.4 kcal/mol and gauche form of CH_2-O-CH_2-O (CO) = 0.4 kcal/mol. Conformer **22a** = 2e + 4 CO = -1.2 kcal/mol, conformer **22b** = 1e + 2 CC + 2 CO = 1.2 kcal/mol and conformer **22c** = 4 CC = 3.6 kcal/mol]. This result was further confirmed experimentally by comparing the behavior of 1-oxaspiro[5.5]undecane which was shown to exist as an equilibrium mixture of two conformers, **23a** and **23b** in a 4:1 ratio (21).

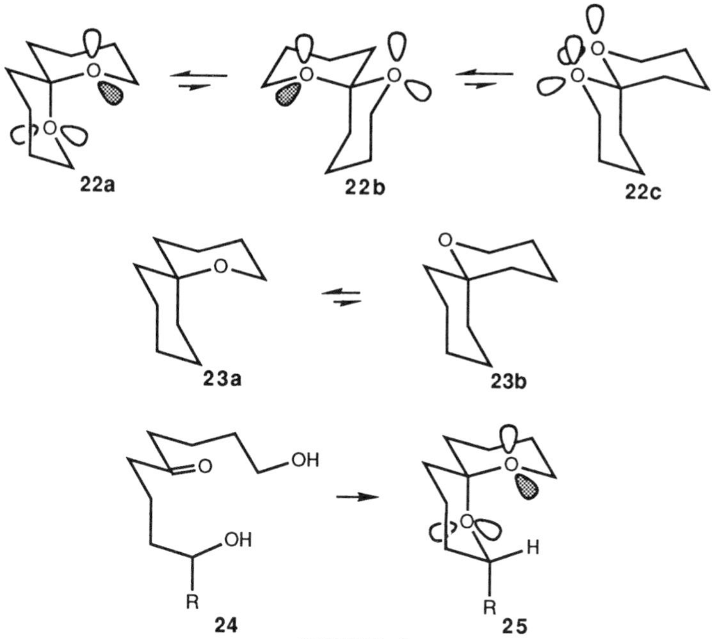

SCHEME 6

We have also showed (20) that the acid cyclization of ketodiols **24** led only to thermodynamically controlled conditions producing substituted 1,7-dioxaspiro[5.5]undecanes **25** having a conformation corresponding to that of **22a**.   For instance, 2-methyl-1,7-dioxaspiro[5.5]undecane was formed as isomer **30** (Scheme 7), none of isomer **31** (which corresponds to conformation **22b**) was observed from the acid cyclization of the keto-diol precursor.   Similarly, the tricyclic spiroacetal isomer **35** was produced exclusively under similar conditions, none of isomer **36** being formed.

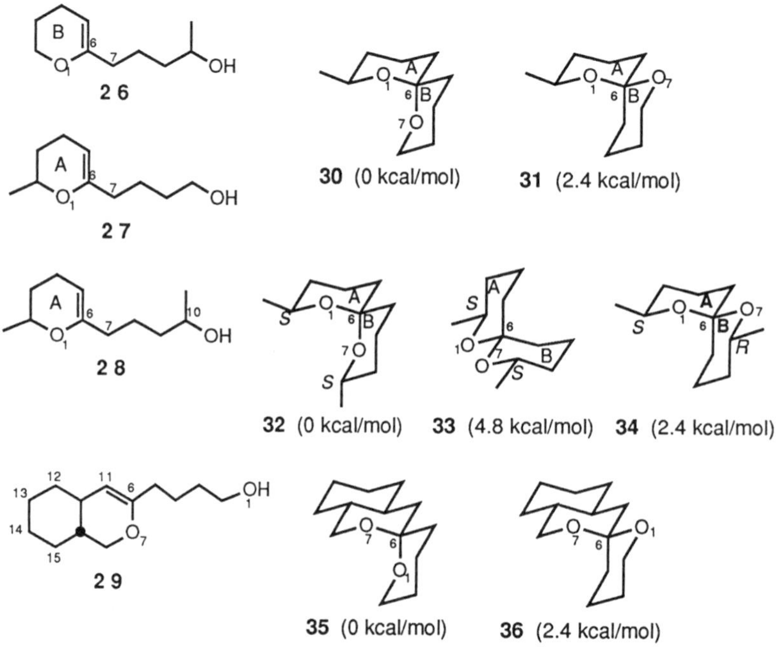

30 (0 kcal/mol)   **31** (2.4 kcal/mol)

32 (0 kcal/mol)   **33** (4.8 kcal/mol)   **34** (2.4 kcal/mol)

35 (0 kcal/mol)   **36** (2.4 kcal/mol)

SCHEME 7. (Reproduced with permission from references 22 and 41. Copyright 1992, 1993).

We have recently found a method to produce spiroacetals under kinetically or thermodynamically controlled conditions (22) (Deslongchamps, P. *Pure Appl. Chem.*, in press.)   This method involves the acid cyclization of hydroxy-enol ethers.

Cyclization of hydroxy-enol ether **26** (Scheme 7) with trifluoroacetic acid / benzene was complete in two hours and gave the known spiroacetal **30** (20) in quantitative yield.   On the other hand, treatment of hydroxy-enol ether **26** with acetic acid / benzene during 19 hours gave a 1:1 mixture of spiroacetals **30** and **31**.   This ratio was shown to remain unchanged under these mild acidic conditions.   It was also observed that the mixture of spiroacetals **30** and **31** was equilibrated (<2 h) upon treatment with trifluroacetic acid / benzene to give only spiroacetal **30**.   These results show rigorously that acetic acid / benzene and trifluoroacetic acid / benzene provide respectively kinetically and thermodynamically controlled cyclization conditions.   Repeating similar experiments with hydroxy-enol ether **27** gave again under thermodynamic control (TFA/benzene, 2 h) only spiroacetal **30**.   Under kinetic control

(AcOH/benzene, 19 h), compound 27 provided a 3:2 ratio of spiroacetals 30 and 31.

Analogous results were obtained with bicyclic hydroxy-enol ether 29. Under thermodynamic conditions (TFA/benzene, 2 h), the known (11) tricyclic spiroacetal 35 was formed exclusively whereas under kinetically controlled conditions (AcOH/benzene, 10 h), a 3:2 ratio of isomeric spiroacetals 35 and 36 was observed. Again, upon treatment with trifluoroacetic acid / benzene (<2 h), the mixture 35 and 36 underwent equilibration to give exclusively spiroacetal 35.

The acid cyclization of hydroxy-enol ether 28 which is a 1:1 mixture of diastereoisomeric racemic pairs due to the presence of the two secondary methyl groups, was next examined. Cyclization of the diastereoisomeric mixture 28 with trifluoroacetic acid / benzene gave a 1:1 mixture of spiroacetals 32 and 34. On the other hand, cyclization of 28 with acetic acid / benzene provided a mixture of three spiroacetals 32, 33, and 34 in a relative ratio of 3:2:5. Upon treatment with trifluoroacetic acid / benzene, this mixture was converted into a mixture of spiroacetals 32 and 34 in a 1:1 ratio.

As previously discussed, a spiroacetal exists in conformation 22a unless there is a severe 1,3-diaxial steric interaction between the substituents and the ring skeletons. In such a case, the compound will normally adopt conformation 22b unless there is again severe 1,3-diaxial steric interaction which will force the compound to adopt conformation 22c.

The results obtained under thermodynamically controlled conditions can be easily rationalized because we can evaluate the relative energy of the various possible conformations of the spiroacetal isomers, as well as the relative stability of the spiroacetals isomers which can be inter-converted under acid conditions. Thus, since isomer 31 is estimated to be 2.4 kcal/mol less stable than isomer 30, the exclusive formation of 30 when the cyclization is carried out with trifluoroacetic acid / benzene is readily understood. Similarly, only isomer 35 was observed starting with 29 under thermodynamic conditions, because this isomer is more stable (~2.4 kcal/mol) than isomer 36. As previously discussed, hydroxy-enol ether 28 is a 1:1 mixture of two racemic diastereoisomers. One of them (racemic mixture of SS and RR) can give racemic spiroacetals 32 and 33 whereas the other (racemic mixture of SR and RS) can only lead to racemic spiroacetal 34. However, since we know that 32 and 33 are inter-convertible under acid conditions (but not 34) and that 33 is estimated to be less stable than 32 by 4.8 kcal/mol, it follows that the cyclization of the racemic mixture 28 under thermodynamically controlled conditions should lead to a 1:1 mixture of 32 and 34 in complete agreement with the experimental results.

It remains to explain the results under kinetically controlled conditions. Under such conditions, the reaction products are independent of the relative stability of the various spiroacetal isomers, but rather depend upon the relative energy of the transition states leading to the formation of the various spiroacetal isomers.

The formation of **30** and **31** in a 1:1 ratio by the kinetic cyclization of hydroxy-enol ether **26** will first be examined. It is reasonable to assume that **26** will be protonated (*14-16*) to give an oxocarbonium ion (Scheme 8) which can exist in two rapidly equilibrating conformations **37a** and **37b**. This would then be followed by a stereoelectronically controlled reaction assuming an antiperiplanar attack (*4*). There are four possibilities, considering reactions on each face of the oxocarbonium ring leading either to a chair-like or a twist-boat-like transition state. Thus, an α-attack on conformation **37a** (**40→41**) and a β-attack on conformation **37b** (**42→43**) lead to chair-like transition states while a β-attack on **37a** (**38→39**) and an α-attack on **37b** (**44→45**) leads to twist-boat transition states. If it is assumed that transition states are late, resembling the protonated spiroacetals, the sterically disfavored twist-boats (**39** and **45**) are eliminated and it appears possible to understand the 1:1 ratio of **30** and **31** from the chair-like transition states. Indeed, stereoelectronic effects are

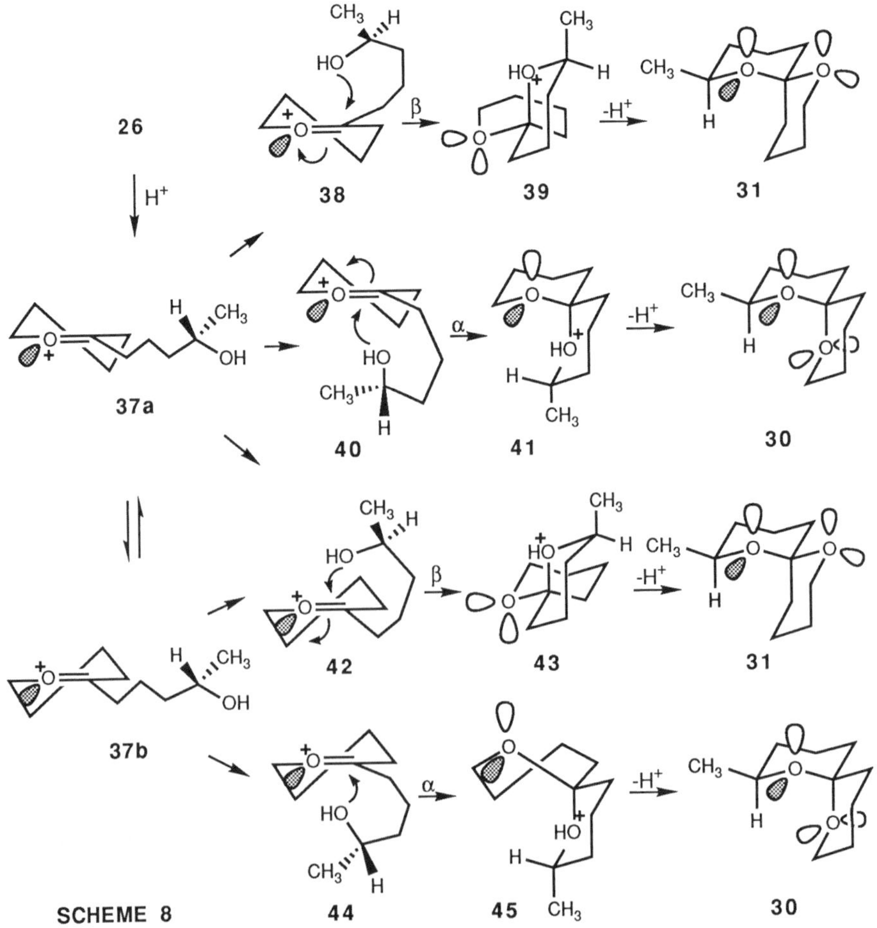

SCHEME 8

equivalent and steric effects are relatively similar in **40→41** and **42→43**, especially if it is considered that the formation of the C-O bond is not yet completed at the transition state. If it is assumed however that the transition state is very early, resembling **38, 40, 42,** and **44,** steric effects appear to be close in all cases and the 1:1 ratio of isomers could be explained on that basis as well. Thus, the experimental results described so far cannot distinguish between an early or a late transition state.

Kinetic cyclization of hydroxy-enol ether **27** giving a 3:2 ratio of spiro-acetals **30** and **31** will now be examined. Protonation of **27** will produce an oxocarbonium ion which can have conformation **46a** or **46b** (Scheme 9) where the former having a pseudo equatorial methyl group is more stable than the latter. In this case, there are again 4 possible modes of cyclization, since there are two chair-like (**49→50** and **51→52**) and two twist-boat-like (**47→48** and **53→54**) transition states.

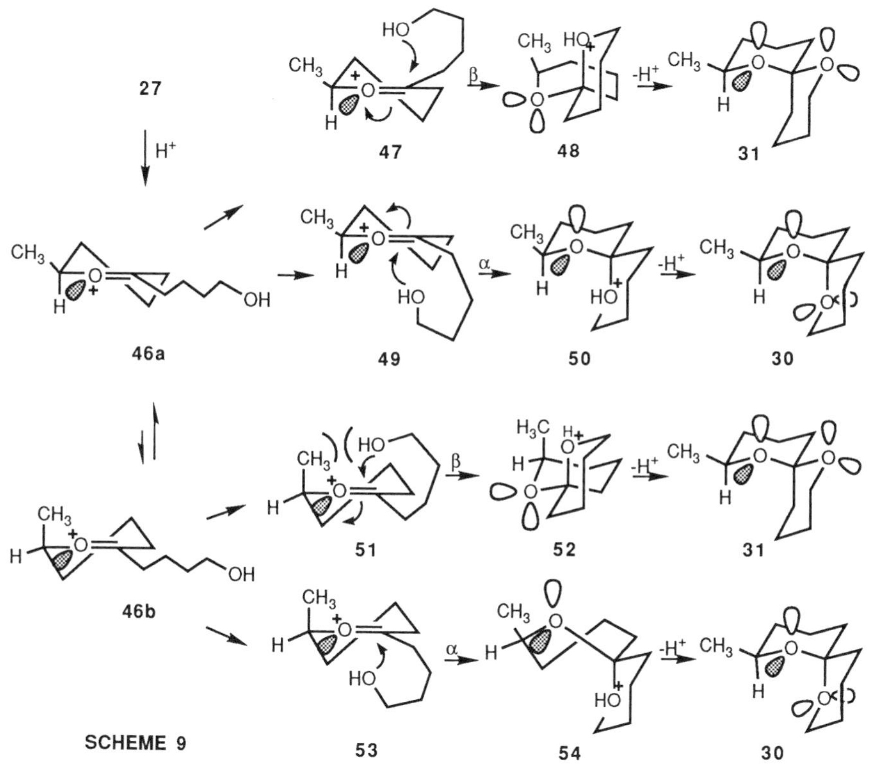

SCHEME 9

We will consider first the two processes having a chair-like transition state. The first one (**49→50**) should produce (after loss of a proton) the more stable spiroacetal **30** whereas the second one (**51→52**) should give (after loss of a proton and conformational inversion of both rings) the less stable spiroacetal **31**. The first process (**49→50**) is essentially devoid of severe steric interactions but the second one (51→52) is severely hindered since the methyl group in 52 is in a 1,3-diaxial disposition relative to the protonated oxygen.

The two pathways which involve an antiperiplanar attack leading to a twist-boat intermediate are now considered. The first one is the result of a β-attack on 46a (*i.e.*, 47→48) while the second one comes from an α-attack on 46b (*i.e.*, 53→54). These two pathways are stereoelectronically equivalent (they are mirror images) except for the fact that the methyl group is in a pseudo equatorial orientation in the process 47→48 and in a pseudo axial orientation in the process 53→54.

If the possibility of a late transition state is first considered for this cyclization, it becomes impossible to rationalize the rather close ratio of 3:2 in favor of spiroacetal 30. The two twist-boat like transition states (48 and 54) are readily eliminated as well as the chair-like transition state 52 which experience a severe 1,3-diaxial steric interaction by comparison with 50 which is essentially sterically free. Indeed, on that basis, only spiroacetal 30 should have been produced (via 49→50) under kinetically controlled conditions and this possibility is thus eliminated.

The possibility of an early transition state must next be considered. The two pathways involving the oxocarbonium ion 46b could be disfavored because this ion is sterically less favored than ion 46a. Furthermore, the pathways 51→52 must experience some steric hindrance between the pseudoaxial methyl group and the incoming OH group. On the other hand, both processes taking place on 46a are relatively sterically free. It is therefore possible that the cyclization would take place only via 46a where the chair-like process 49→50 would be very slightly favored over the twist-boat process 47→48 because of the very early nature of the transition state. Indeed, on that basis, the 3:2 ratio of 30 and 31 is readily explained.

Examination of the results obtained with bicyclic hydroxy-enol ether 29 confirms this conclusion. Protonation of 29 produces oxocarbonium ion 55 (Scheme 10) which has a conformation (55a) essentially identical to that of 46a (or 37a) but with the difference that ring inversion is no more

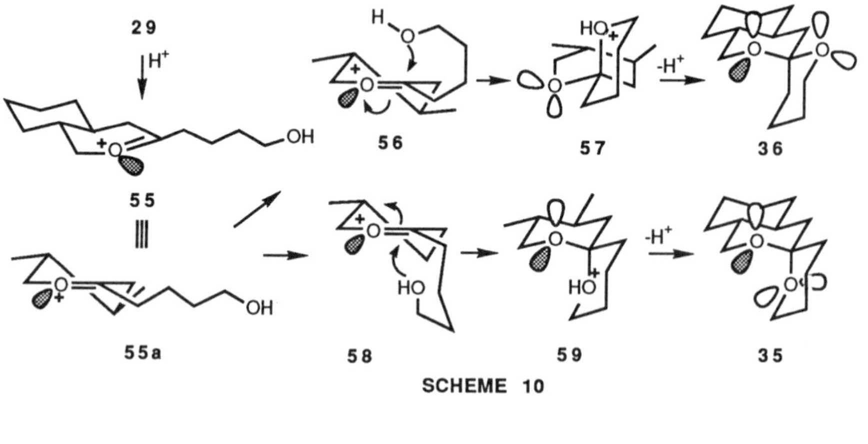

SCHEME 10

possible due to the *trans*-junction of the bicyclic skeleton. There are therefore only two modes of attack which respect the antiperiplanar hypothesis, the first one takes place via a twist-boat process (56→57) to give the less stable spiroacetal **36**, whereas the second one occurs via a chair-like process (58→59) to give after deprotonation the more stable spiroacetal **35**.

Again, a late transition state cannot explain the experimental results since a geometry close to **57** is energetically too high by comparison with **59**. On the other hand, an early transition state can explain the fact that **56** (beginning of a twist-boat) should be slightly higher in energy than **58**. Since the 3:2 ratio is the same for the kinetic cyclization of hydroxy-enol ethers **27** and **29**, it can be concluded that the cyclization of oxocarbonium **46** probably takes place only from the ion **46a** (via the pathways 47→48→31 and 49→50→30 in a 2:3 ratio). On that basis, the 1:1 ratio of **30** and **31** observed from hydroxy-enol ether **26** is also readily explained. Indeed, with an early transition state situation, the processes 38→39 and 40→41 from oxocarbonium ion **37a** should lead to a 2:3 ratio of **31** and **30**. Similarly, the processes 42→43 and 44→45 from oxocarbonium ion **37b** should take place with a 3:2 ratio of **31** and **30**. Since cyclization from **37a** and **37b** should take place with equal ease, the net result should be a 1:1 ratio of **30** and **31** in complete accord with experiments.

It remains to examine the cyclization of the diastereoisomeric mixture of hydroxy-enol ether **28** which gave a 3:2:5 ratio of spiroacetals **32, 33**, and **34**. We will first discuss the formation of racemic spiroacetals **32** and **33** which come from the cyclization of the racemic SS and RR diastereoisomers of hydroxy-enol ether **28**. The formation of **32** and **33** in a 3:2 ratio is now easily explained. Taking the SS enantiomer of **28** as an example, upon protonation, it will give an oxocarbonium ion (equivalent to **46a**) which can cyclize either via a twist-boat mode or a chair-like mode where the latter would predominate over the former in a 3:2 ratio via an early transition state.

The amount of racemic spiroacetal **34** in both the kinetic and thermodynamic mixtures can be easily rationalized from the fact that in this case only one such racemic product might be obtained from the cyclization of racemic diastereoisomer RS/SR **28**. Indeed, taking the SR isomer of **28** as an example, cyclization on one face of the ring yields the enantiomer corresponding to **34** while the cyclization on the other side provides the enantiomeric form of **34**. In the light of the preceding discussion, one enantiomer of **34** must be formed preferentially over the other. Of course, a racemic mixture of **34** is finally obtained because the starting hydroxy-enol ether **28** was racemic (SR and RS mixture).

In conclusion, the spiro acetalization of the four enol ethers **26-30** under kinetically controlled conditions can be explained on the basis of the antiperiplanar hypothesis while postulating an early transition state, with the early chair-like being slightly favored over the early boat-like transition state. Indeed, it explains the 1:1 ratio of spiroacetals **30** and **31** from enol ether **26**. It also provides a simple explanation for the same

3:2 ratio obtained from the spiro acetalization of the other three enol ethers 27, 28, and 29.

In more polar solvents, it is anticipated that the transition state will be more advanced, closer to the tetrahedral intermediate, the net results being a greater energy difference between the chair and the boat-like transition states leading to a larger preference for the formation of one spiroacetal. This is observed experimentally (cf. Table I). For example, the acetic acid catalyzed cyclization of bicyclic - enol ether 29 in aceto-nitrile produced a 2.6:1 (rather than the 3:2) ratio of spiroacetals 35 and 36 (Li, S. Université de Sherbrooke, unpublished data).

Table I. The Influence of Solvent on the Ratio of Spiroacetals 35 and 36

Solvent (ε)	C_6H_6 (2.3)	$CDCl_3$ (4.8)	CD_2Cl_2 (9.1)	CD_3CN (37.5)
35/36	1.5	1.6	1.7	2.6

ε = Dielectric constant.

It is now pertinent to reanalyze the previously reported mild acid cycliza-tion of bicyclic hydroxypropyl acetal 10 (11-12) (Scheme 11). At room temperature, 10 gives only the *cis* tricyclic acetal 13 upon acid treatment (PTSA-MeOH). An equilibrium mixture (45:55 ratio) of *cis* and *trans* acetals 13 and 15 was obtained after a reflux under the same conditions.

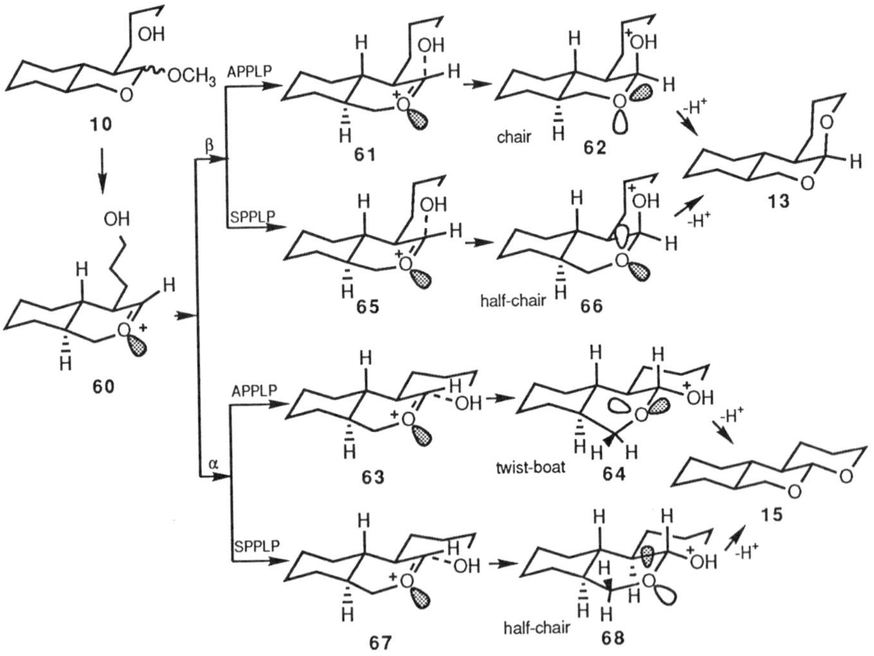

SCHEME 11.   *ANTI* VERSUS *SYN* LONE PAIRS HYPOTHESIS.

The kinetically controlled formation of **13** was explained in the following way. Upon acid treatment, it was assumed that **10** gave first the oxocarbonium ion **60** which can undergo a stereoelectronically controlled cyclization to give either the *cis* acetal **13** via a chair-like pathway (**61**→**62**) or the *trans* acetal **15** via a twist-boat pathway (**63**→**64**). Since, the formation of the *cis* acetal was exclusive, it was believed (*11-12*) that the transition state must be late resembling **62** rather than the less stable **64**. Nevertheless, in the light of the preceding results and other experimental evidence (*23-24*), it is unlikely that the exclusive formation of *cis* acetal **13** comes from a late transition state. Interestingly, it is relatively easy however to understand the exclusive formation of *cis* acetal **13** via an early transition state provided that there is a proper alignment of the incoming hydroxyl group with the π-orbital of the oxocarbonium ion. This stereoelectronic parameter corresponds approximately to the Bürgi-Dunitz angle of attack of a nucleophile on a π-system (*25-26*). This alignment is readily achieved in **61** but not in **63**. Modeling studies on oxocarbonium species **60** also support this argument. Indeed a conformational analysis (30° steps) on all exocyclic bonds of MINDO-3 minimized **60** shows that the vast majority of the 1103 allowed conformers has the hydroxyl above the plane of the oxocarbonium ion (like in **61**) rather than underneath that plane (like in **63**). The distance from the hydroxyl oxygen to the oxocarbonium carbon covers a range between 2.63 to 6.33 Å (0.1 Å grid). An examination of those conformers having shortest distances should give an indication about the easiest path of approach between the two centers (hydroxyl oxygen and oxocarbonium carbon). Table II lists salient parameters for representative lowest energy conformers of both types (*e.g.*, **61**-like and **63**-like), together with their heat of formation obtained from MINDO-3 semi-empirical calculations. It is easily observed that the shortest O•••C=O distance (2.65 Å) is obtained for a **61**-like conformer (entry 1, O•••C=O angle = 106°, heat of

Table II. Representative 61-like and 63-like Conformers

		O•••C=O Distance (Å)	O•••C=O Angle (deg.)	Heat of formation (kcal/mol)
61-like	1	2.65	106	50
	2	2.75	103	51
	3	2.85	101	51
	4	2.85	111	49
	5	2.95	101	51
63-like	6	2.75	142	54
	7	2.95	144	55
	8	3.15	152	50
	9	3.25	148	50
	10	3.35	146	48

formation = 50 kcal/mol). The lowest energy **61**-like conformer (entry 4) has an O•••C=O distance of 2.85 Å, and an O•••C=O angle of 111°. Shortest O•••C=O distances in **63**-like conformers are obtained at 2.75Å and 2.95 Å. However in this case heats of formation are about 3 to 4 kcal/mol higher than in the corresponding **61**-like conformers (compare entries 2 and 6, 5 and 7). The lowest energy conformer (entry 10) has an O•••C=O distance of 3.35 Å. Another point of interest is the fact that whereas O•••C=O angles of 101-111° may be readily achieved in **61**-like conformers, this is not the case for **63**-like conformers where such angles are bigger than 142°. Therefore since **61**-like conformers have shortest O•••C=O distances, lower energies and Bürgi-Dunitz O•••C=O angles, this approach path of the nucleophile (OH) should be preferred over the one starting from **63**-like conformers. In conclusion, an early transition state situation readily explains the exclusive formation of *cis* acetal **13**.

We will now examine the two alternative mechanisms proposed respectively by Fraser-Reid (*18*) and Sinnott (*19*). Taking the bicyclic hydroxyacetal **10** as an example, the synperiplanar hypothesis suggested by Fraser-Reid means that the cyclization of **60** could take place via a β-attack leading to a half-chair (**65→66**) in order to give *cis* acetal **13** or via an α-attack leading to a half-chair (**67→68**) before giving *trans* acetal **15**. Now, if these reactions take place via a <u>very early</u> transition state, it can be seen that **61** and **65** on one hand and **63** and **67** on the other hand are respectively virtually identical! Kinetically, they are equivalent and there is no need to discuss further the *anti* versus the *syn* mechanisms on that basis. Consequently, it should be noted that the experimental results on <u>acetal cleavages</u> reported by Fraser-Reid (*18*) cannot be considered as evidence in favor of the synperiplanar hypothesis because there is no reason to believe that these processes do not take place via a <u>late transition state</u>. On that basis, and as previously discussed in this article, these results are equally well explained by the antiperiplanar hypothesis.

The structure of the transition state for the spontaneous hydrolysis of axial tetrahydropyranyl acetals has been estimated from experimental structural and kinetic data by Bürgi and Dubler-Steudle (*27*). This analysis indicates a late transition state. It follows from this work that the transition state for the proton catalyzed cleavage must also be late and therefore, it must be early for the reverse process which is consistent with the results given in the present work. Very recently, Andrews, Fraser-Reid and Bowen (*28*) have carried out an *"ab initio"* study of transition states in glycoside hydrolysis based on axial and equatorial 2-methoxytetrahydropyrans. This theoretical work also indicates that acetal hydrolysis takes place via a late transition state. Indeed, for the protonated axial anomer (α-glycoside), cleavage occurs via a half-chair transition state which gives a half-chair oxocarbonium ion. In the case of the protonated equatorial anomer (β-glycoside), cleavage takes place via an ^4E *endo sofa* transition state and hence via oxocarbonium ion having the same geometry. It should be pointed out that the half-chair and the sofa oxocarbonium ions have a very close geometry and similar energy, the half-chair being slightly more stable (0.15 kcal/mol). The next question which can be asked however, concerns what happens just after the transition state; which pathway is preferred? Is the chair pathway 61→62 preferred over the half-chair 65→66 in the formation of *cis* acetal 13? Similarly, is the *trans* acetal 15 more easily produced via the twist-boat 63→64 than the half-chair 67→68? These questions are important, especially when the reverse process which takes place via a late transition state is considered. Indeed, in the reverse process, it becomes pertinent to know precisely which conformational change occurs in compounds 13 and 15 prior to the cleavage step.

The antiperiplanar hypothesis has received support both theoretically (29-31) and experimentally (4-12, 17) which indicates that it is a lower energy pathway than the synperiplanar one. It should also be pointed out that there are yet no rigid model compound supporting the synperiplanar hypothesis. Consequently, it appears safe to conclude that the antiperiplanar process is normally favored over the synperiplanar, unless unusual steric effects would prevent the former over the latter. On that basis, the chair process 61→62 would be definitely preferred over the half-chair 65→66 based on steric and electronic reasons. For similar reasons, the process 63→64 should also be preferred over 67→68. Indeed, recent calculations (Grein, F. In *On the Nature of the Anomeric and Reverse Anomeric Effect in Acetals and Related Functions*. This book) indicate that although the *syn* and antiperiplanar electronic effect are energetically similar, the former process is sterically disfavored over the latter, eclipsed (*syn* process) being higher energetically than gauche conformer (*anti* process).

Sinnott's opposition to the stereoelectronic theory and his attempt to rationalize the reactivity of acetals on the basis of the principle of least motion will now be examined. His conclusions were reached on the following basis. In pyridinium glycosides and due to a phenomenon called the reverse anomeric effect (32), the α-isomer is known to exist in the unusual twist-boat conformation 69b (Scheme 12) rather than the usually more stable chair conformation 69a, whereas the β-isomer remains in the usual chair conformation 70a. It was then postulated by Sinnott (19) that the α and β-pyridinium glycosides undergo hydrolysis directly from their ground state conformations 69b and 70a respectively to yield the corresponding oxocarbonium ion 71 by simply following the principle of least nuclear motion (33) while completely neglecting the importance of stereoelectronic effects (or orbital overlap) during these processes. On that basis, Sinnott concluded that the acid hydrolysis of α and β-glycosides follows similar pathways because on protonation of the OR side chain, the β-glycosides would remain in the chair conformation 73a whereas the α-glycosides would change from the chair 72a to the twist-boat 72b. Then, the formation of the oxocarbonium ion 71 would

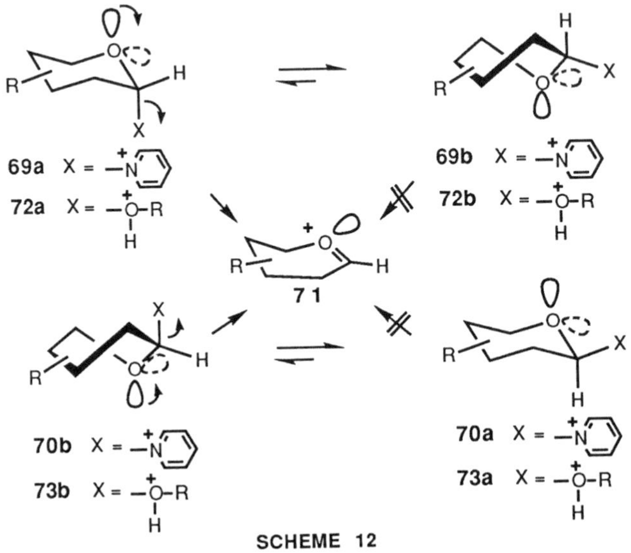

SCHEME 12

come directly from **72b** and **73a** following the principle of least nuclear motion.

This hypothesis, based on the principle of least nuclear motion, provides no driving force for the cleavage to take place, and we are convinced that it is false on the following basis. Keeping the bicyclic hydroxyacetal **10** as an example, the pathway proposed by Sinnott predicts that **10** would form *cis* acetal **13** via the twist-boat process **74→75** (Scheme 13), whereas *trans* acetal **15** would be formed from the chair pathway **76→77**. Following this hypothesis, the formation of tricyclic *trans* acetal **15** from the oxocarbonium ion **60** should take place with equal ease as the *cis* acetal **13** under kinetically controlled conditions, and hence the specific formation of *cis* acetal **13** cannot be explained.

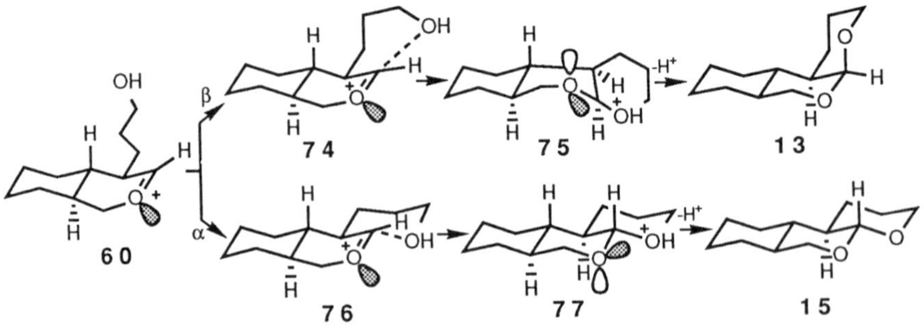

SCHEME 13.   PRINCIPLE OF LEAST MOTION HYPOTHESIS.

Furthermore, *ab initio* calculations have been carried out recently (*30-31, see also F. Grein, this book*) on protonated species $H_nX-CH_2-Y^+H_n$ (X and Y=O and/or N). It was found that when a lone pair of X is antiperiplanar to the $C-Y^+$ bond, the C-X bond shortens and the C-Y bond becomes much longer. In some cases, the tetrahedral species switch to a π-complex ($H_nX^+=CH_2 \bullet\bullet\bullet YH_n$). On the other hand, when there are no lone pair of X antiperiplanar to the $C-Y^+$ bond, the $C-Y^+$ bond does not become longer. These calculations revealed also that positively charged tetrahedral intermediates having no antiperiplanar lone pairs are energetically quite stable species, indicating that the reverse anomeric effect is a stabilizing electronic effect. Furthermore, these calculations strongly suggest that the reverse anomeric effect is the result of an electrostatic attraction between the electron lone pairs of atom X and the positive charge of atom Y. Thus, in α-pyridinium glycosides, the twist-boat **69b** (Scheme 12) is more stable than the chair conformation **69a** because the N^+ is gauche to the two lone pairs of the ring oxygen. In the case of the β-isomer, it stays in the same chair conformation **70a**, because the N^+ is already gauche to the two lone pairs of the ring oxygen.

On that basis, one can postulate that although β-pyridinium glycosides exist in the ground state chair conformation **70a**, they will then undergo a conformational change to the twist-boat **70b** before reaching the transition state which will produce the oxocarbonium ion **71**. On the other hand, α-pyridinium glycosides which exist in the ground state twist-boat conformation **69b** will undergo a conformational change to the chair conformation **69a** before reaching the transition state which will eventually produce the cyclic oxocarbonium ion **71**. α and β-Glycosides would behave similarly. Upon protonation of the OR group, β-glycosides would remain in the chair conformation **73a**, but would have to undergo a conformational change to the twist-boat **73b** prior to reaching the transition state required for the formation of the oxocarbonium ion **71**. On the other hand, upon protonation, the α-glycoside would undergo a conformational change from the chair **72a** to the now more stable twist-boat **72b**. However, **72b** cannot undergo a cleavage with stereoelectronic control, and would therefore undergo a conformational change back to the chair **72a** in order to eventually reach the transition state leading to the half-chair oxocarbonium ion **71**. In other words, a cleavage with stereoelectronic control corresponds to a cleavage following the principle of least electronic motion, which contrary to the principle of least nuclear motion, has the advantage of providing the necessary chemical driving force for the ejection of the leaving group (*33*).

The overall reaction coordinates and conformational changes occurring in the hydrolysis of glycosides are summarized in Scheme 14. Upon protonation a β-glycoside **82** would remain in the chair conformation **83** further stabilized by the reverse anomeric effect. Protonation of an α-glycoside **78** would produce the protonated α-chair **80** which is much less stable than the protonated β-chair **83** as it is not stabilized by the reverse

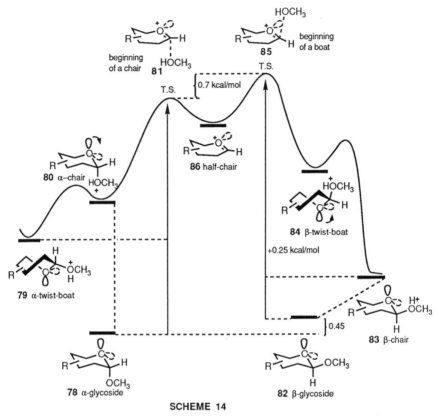

SCHEME 14

Reaction coordinate and conformational change in hydrolysis of α and β-methoxy tetrahydropyrans. (Reproduced with permission from references 22 and 41. Copyright 1993 Publishers).

anomeric effect. The protonated α-chair form **80** could however undergo a conformational change into the more stable protonated α-twist boat **79** which is stabilized by the reverse anomeric effect. The protonated α-twist-boat **79** is however much less stable than the protonated β-chair **83** due to steric effects (twist-boat versus chair), both species having the same stabilizing reverse anomeric effect. The protonated α-chair **80** and β-twist-boat **84** are stereoelectronically equivalent, however the former is more stable than the latter for steric reasons. Finally, protonated β-twist-boat **84** is much higher in energy than the protonated β-chair **83** being disfavored by steric effect and the reverse anomeric effect.

The cleavage steps for α and β-glycosides would thus start from the protonated α-chair **80** and β-twist-boat **84**, each species having an electron pair antiperiplanar to the leaving group. In each case, the position of the transition state along the reaction coordinate would be late, having a geometry close to the half-chair oxocarbonium ion **86**. Looking at the reverse process, the net result would be that starting from the half chair oxocarbonium ion **86**, the transition state for an α and a β-attack of methanol would have a geometry corresponding to the beginning of a chair (81) and the beginning of a twist-boat (85) respectively. The relative rate of hydrolysis of O-alkyl α and β-glycosides are explained by the difference in energy between their ground state conformation (78 and 82) and their respective transition state (81 and 85). The axial and equatorial bicyclic acetals 16 and 17 can be taken as models for α and β-glycosides. The α-anomer 16 is more stable than the β-anomer 17 by 0.45 kcal/mol, on the other hand, 16 is hydrolysed faster by a factor of 1.5 (~0.25 kcal/mol) (13), the difference in energy between the two transition states should be approximately 0.7 kcal/mol. [Andrews, Fraser-Reid and Bowen have obtained a value of the same order ($\Delta E_{T.S.}(\beta-\alpha)$ = 1.05 kcal/mol) in their 6-31G *ab initio* studies (28)]. The beginning of a chair form is at a lower energy level than the beginning of a twist-boat, it is therefore normal that the α is lower than the β-transition state. Then, when the difference in energy between the α and β-ground state conformations are taken into account, the relative rate of hydrolysis (α faster than β by 1.5) is readily understood.

The relative rate of hydrolysis of α and β-pyridinium glycosides which differ considerably from the corresponding methoxy glycosides can be also easily explained. α and β-Pyridinium glycosides exist respectively in the chair and the twist-boat conformations. Their ground state conformation correspond to that of the protonated α-twist-boat 79 and β-chair 83 of O-alkyl glycosides. Their energy difference should therefore be similar to that of 79 and 83, i.e., much larger (difference between a twist-boat and a chair, i.e., 3 to 5 kcal/mol) than that for the non-protonated α and β-glycosides which exist in their respective ground state chair forms 78 and 82 (0.45 kcal/mol for 16 and 17). On the other hand, taking into consideration a late transition state, the relative energy of the α and β-transition states should be relatively similar to those of O-alkyl glycosides (cf. 81 and 85). On that basis, α-pyridinium isomers are expected to hydrolyze at a much faster rate. This is indeed the case, α-pyridinium glucoside is hydrolyzed 80 times more rapidly than the β-anomer (34). Thus, even if an antiperiplanar lone pair requirement is necessary for cleavage, it is not essential that it must be satisfied in the ground state conformation as long as some accessible appropriate conformations are available. The only criteria is simply that the conformational barrier involved should be smaller (usually the case) than the activation energy required for the cleavage reaction. Thus, the kinetic data for the hydrolysis of glycosides cannot be used against a rationalization based on stereoelectronic principles. On the contrary, the kinetic results of hydrolysis are readily understood taking into account an appropriate modification of the position of the transition state along the reaction coordinates (transition state near the oxocarbonium ion) while respecting the principle of stereoelectronic control. [In this respect, the oxidative hydrolysis of conformationally restrained 4-pentenyl glycosides reported by Fraser-Reid and co-workers (18) may not represent experimental evidence for a the synperiplanar hypothesis. Indeed, molecular models of the 2,3 and

4,6 bis-isopropylidene derivative of β O-4-pentenyl glucose indicates that the sugar ring can take the twist-boat conformation required for the antiperiplanar hypothesis. Furthermore, this process must take place via a late transition state having a geometry close to that of the cyclic half-chair oxocarbonium ion. A similar conclusion can be reached for amidines hydrolysis reported by Perrin and Nunez (35-36) and the gas phase ionization of cyclic orthoester as reported by Caserio and co-workers (37). It should be emphasized that stereoelectronic control (or the antiperiplanar hypothesis) does not mean that this electronic effect controls product formation. In fact, experiments designed to demonstrate stereoelectronic control are very difficult to discover because this principle applies in probably all cases and therefore other factors decide which reaction pathway will be favored (relative energy barrier for a conformational change, best leaving group, etc)].

There are other well known organic reactions which are believed to proceed via a conformational change prior to cleavage in order to undergo stereoelectronically controlled processes. A very well known case is the reductive elimination of *trans* 1,2-dibromocyclohexane. This compound exists in the diequatorial conformation in the ground state, but it must undergo a conformational change to the less stable diaxial orientation in order to produce cyclohexene. Note that the formation of cyclohexene does not occur directly from the diequatorial conformer although this would follow the principle of least nuclear motion! Similarly, in the reverse process, the addition of bromine to cyclohexene will produce first *trans* 1,2-dibromocyclohexane in the diaxial conformation. Indeed, these reactions are taking place with stereoelectronic control thus following the principle of least electronic motion at the transition state level.

It is well known that reactions at the anomeric center in α and β-glycosides proceed in some cases with retention and in others with inversion of configuration. These reactions are explained on the basis of an SN_1 and an SN_2 process respectively. When the displacement reaction takes place via an SN_1 mechanism, it is definitely a process with a very late transition state, near the oxocarbonium ion which is a discrete species in these conditions (cf. **87** and **88** in Scheme 15). The results described in this article indicate that starting from the oxocarbonium ion, the geometry of the transition state which corresponds to the beginning of a chair is slightly preferred over that of a boat form for steric reasons. When the process takes place via an SN_2 mechanism, it is again a late transition state operation, but in this case, the attacking species is nucle-ophilic enough to start reacting before the leaving group is completely ejected (SN_1 mechanism). Recent studies by Banait and Jencks (38) on the reactivity of α-D-glucopyranosyl fluoride are in complete accord with this conclusion. The SN_2 displacement reaction must therefore have a geometry at the transition state where the C_1 and O_5 atom of the glyco-sides must be sp^2 hybridized (cf. **89**). So, these processes are also con-trolled stereoelectronically at the transition state level (39). In other

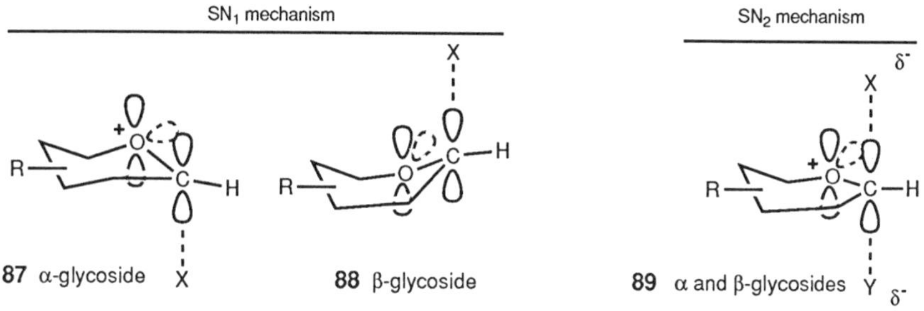

SCHEME 15. TRANSITION STATE.

words, the theory of stereoelectronic control (4) does not represent "an over interpretation of small and elusive least motion effects" (19), but it predicts the stereochemistry of the overall process including the transition state, although, it cannot pinpoint the exact position of the transition state along the reaction coordinates. The position of the transition state will of course vary depending on the nature of the substrate and the reaction conditions (solvent, nucleophile, catalyst, etc).

In conclusion, the rate of hydrolysis of α and β-glycosides are explained while assuming a late transition state following the antiperiplanar hypothesis. The geometry of transition states for α and β-glycosides corresponds to the end of a chair and of a twist-boat respectively in order to produce the half-chair cyclic oxocarbonium ion.

Hydrolysis of Cyclic Orthoesters

In order to have a complete discussion on tetrahedral intermediates derived from esters, it is now pertinent to consider the primary and secondary stereoelectronic effects which are operative in their breakdown. As previously reported (4, 5-8), these intermediates should always undergo cleavage with the help of primary stereoelectronic effects, and the most favored one should have the maximum number of secondary stereoelectronic effects.

At the tetrahedral intermediate stage, the primary and secondary stereoelectronic effects are identical, corresponding to the classic anomeric effect of acetals, *i.e.*, one electron lone pair of an oxygen atom being antiperiplanar to a carbon-oxygen bond, (a n_O/σ^*_{C-O} interaction). The primary stereoelectronic effect is the result of the two oxygen lone pairs being antiperiplanar to the leaving group whereas the secondary ones are simply due to the other oxygen lone pairs which are antiperiplanar to other C-O σ bonds. At the transition state, the two lone pairs antiperiplanar to the leaving group are in the process of forming the conjugated system of the ester product. The other oxygen lone pairs oriented antiperiplanar to a C-O σ bond will remain in the same relative orientation at the transition state and in the ester product.

Taking intermediate **90** and **92** as examples (Scheme 16), each of them has O_1 and O_2 with a lone pair oriented antiperiplanar to the leaving group Y. In addition, O_1 and O_2 have each a lone pair antiperiplanar to a C-O bond (dotted arrows) in **90** whereas only O_1 has such an arrangement with the C-O_2 bond in **92**. In other words, **90** has four n_O/σ^*_{C-O} interactions whereas **92** has three. By ejection of Y in **90** and **92**, the two n_O/σ^*_{C-O} interactions antiperiplanar to the leaving group are converted into a π_{CO} and a n_O/π^*_{C-O} interaction in the resulting ester products **91** and **93**. The remaining n_O/σ^*_{C-O} interactions are simply transferred in the reaction products, two in the case of **91** which is a Z ester and one in the case of **93** which is an E ester. In fact, comparing a Z and an E ester, both have the primary stereoelectronic effects, *i.e.*, a π_{CO} and a n_O/π^*_{C-O} interaction which corresponds to the two resonance form of an ester. In addition, a Z ester has two secondary stereoelectronic effects, *i.e.*, two n_O/σ^*_{C-O} interactions whereas an E ester has only one. These different streoelectronic effects are clearly illustrated in Scheme 17.

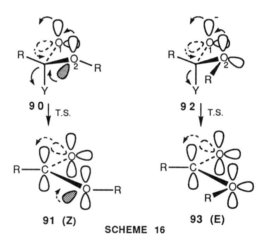

SCHEME 16

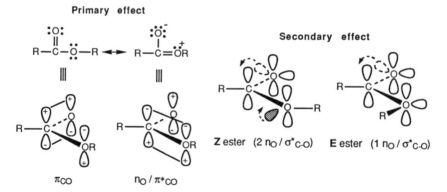

SCHEME 17.   RESONANCE STRUCTURE AND STEREOELECTRONIC EFFECT IN ESTER.

If conformational change between **90** and **92** is allowed (usually the case), than the cleavage **90→91** leading to a Z ester should be stereoelectronically favored over the cleavage **92→93** which is producing an E ester. This is of course due to the additional stabilizing secondary stereoelectronic effect present in the cleavage **90→91**. These effects are consequently also present in the ester function, and the greater stability of the Z over the E form in ester is primarily due to the presence of this additional secondary stereoelectronic effect.

In the light of the conclusion reached for glycosides hydrolysis, it is likely that the position of the transition state will vary depending on the nature of the substrate. For instance, a dioxocarbonium ion like **94** (Scheme 18) is more stable, thus less reactive than the corresponding oxocarbonium ion **86**. The hydration of **94** should therefore take place via a more advanced transition state along the reaction coordinates than that of **86**. The transition state should move closer to the tetrahedral intermediate geometry, it is therefore expected that steric effect should play a greater discriminating role. For example, the α-attack leading to a chair-like (**94→95**) should now be highly preferred over the β-attack (**94→96**) leading to a twist-boat transition state.

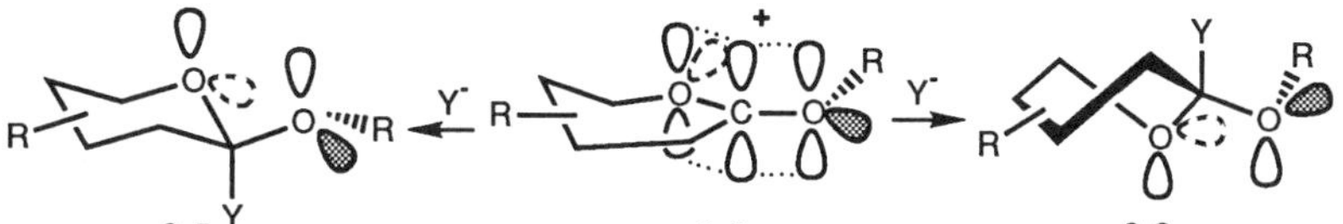

SCHEME 18

A reexamination of the results previously reported (*5-8*) on the acid hydrolysis of cyclic orthoesters supports entirely this conclusion. In fact, these results are now more completely rationalized by taking into account that the position of the transition state is not as near the tetrahedral intermediate as previously anticipated. Indeed, the relative importance of steric effects due to chair-like and twist-boat like processes is better evaluated. Also, an entropy factor favoring the breakdown of an intermediate into two different products rather than to a single one, is better understood.

For example, the loss of an axial or an equatorial methoxy group in bicyclic mixed orthoester **6** (Scheme 2) should take place via a chair or a twist-boat like transition state respectively. On that basis, the loss of the axial methoxy group should be highly preferred. This is confirmed experimentally as the rate of exchange of the equatorial methoxy group upon treatment of non-deuterated **6** with acid was shown to be much slower than the axial one by a factor close to 100 (*9-10*). By comparison, this ratio is much greater than the 1.5:1 ratio observed for α (axial) and β (equatorial) glycosides. In the reverse process, the formation of the orthoester **6** from the reaction of deuterated sodium methoxide on the bicyclic dioxocarbonium ion **7** should also be highly favored. This is in complete agreement with the experimental result (6) which gave only orthoester **6** with the axial deuterated methoxy group.

The mild acid hydrolysis of cyclic orthoester **6** has been reported (7) to give the hydroxy methyl ester **9** (94%) along with a small amount of the corresponding lactone (6%). These products must come from the competition between the cleavages **97→98→99** and **100→101→102** of tetrahedral intermediate **8** as illustrated in Scheme 19. Interestingly, with transition states having a geometry further away from a real chair and twist-boat (closer to the cyclic dioxocarbonium ion with a longer bond for the leaving group, cf. **98** and **101**), it becomes possible to understand why the latter process ion can compete to some extent with the former. Indeed, with such a case, the steric energy difference between a chair-like and a twist-boat-like becomes less important than normally expected.

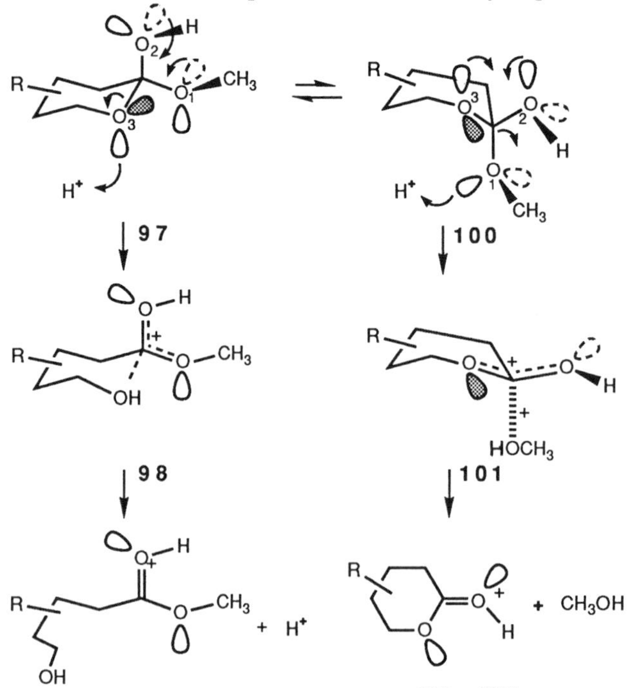

SCHEME 19

We have also previously proposed that the formation of lactone and alcohol should be more favored entropically than the formation of hydroxy-ester. This entropy phenomena is now better understood with a transition state having a longer weak bond for the leaving group. Indeed, any molecular vibration due to heat should favor more the reverse of a bimolecular process (**101→102**) over that of an intramolecular one (**98→99**).

It can also be seen that the cleavage $97 \rightarrow 98 \rightarrow 99$ yields the protonated hydroxy-ester in the ZZ configuration having two secondary stereoelectronic effects whereas the cleavage $100 \rightarrow 101 \rightarrow 102$ produced the protonated lactone in the EZ configuration which has only one secondary effect. The former process is thus sterically and stereoelectronically favored but it is disfavored entropically. Since the major product observed is the hydroxy-ester **9**, the combination of stereoelectronic and steric effects overweighs almost completely the entropy factor.

Conformationally mobile cyclic orthoesters which behave differently than the rigid orthoester **6** can now be reexamined. These orthoesters were found to give a larger amount of lactone, the percentage being different above and below pH 3 (7). For example, 2,2-dimethoxytetrahydropyran gave a 30:70 ratio of lactone and hydroxy-ester at pH<3 and a 10:90 ratio at pH>3. With 2,2-diethoxytetrahydropyran these values changed to 20:80 and 5:95 respectively indicating that ethanol, being slightly less acidic, is not as good a leaving group as methanol. Consequently, the percentage of lactone is smaller. These orthoesters (**103**) (Scheme 20) will produce first tetrahedral intermediate **104** with the hydroxyl group in the axial orientation, but since these compounds are conformationally mobile, they are capable of ring inversion to produce tetrahedral intermediate **105** having the hydroxyl in the equatorial orientation.

Cleavage of **104** will take place as illustrated in Scheme 19 giving mainly the hydroxy-ester by the process $97 \rightarrow 98 \rightarrow 99$ along with a small percentage of lactone via $100 \rightarrow 101 \rightarrow 102$. On the other hand, cleavage at **105** can also give hydroxy-ester by the process $106 \rightarrow 107 \rightarrow 108$ or produce the lactone via the chair-like transition state $109 \rightarrow 110 \rightarrow 111$. Thus, the main difference between conformationally mobile and rigid orthoesters is that lactone can be produced by a chair-like (**110**) in the former and only by a boat-like (**101**) in the latter. Larger proportion of lactone in orthoesters which can undergo ring inversion is thus clearly understood.

The difference in product ratio as a function of pH can now be considered. The observation that the hydroxy-ester/lactone product ratio varies as a function of pH was first observed by McClelland and Alibkai (*40*). They showed that this behavior was due to a difference in mechanism corresponding to the acid (pH<3) and the base-catalyzed (pH>3) breakdown of the tetrahedral intermediate. In other words, at pH<3, it is the protonated form T$^+$ which breaks down whereas at pH>3, it is the hydroxyde-catalyzed cleavage of T$^\circ$ (Scheme 21).

The cleavage of T$^+$ will produce hydroxy-ester or lactone in their protonated form whereas that of T$^\circ$ plus OH$^-$ (or T$^-$) will yield the same products in their neutral form. However, the position of the transition state will not be the same in both cases. If one makes the reasonable assumption that the position of the transition state should be closer to the protonated ester than to the ester (or protonated lactone than to the lactone), the breakdown of T$^+$ by comparison with T$^-$, will take place with a longer and weaker bond for the leaving group. On that basis, the

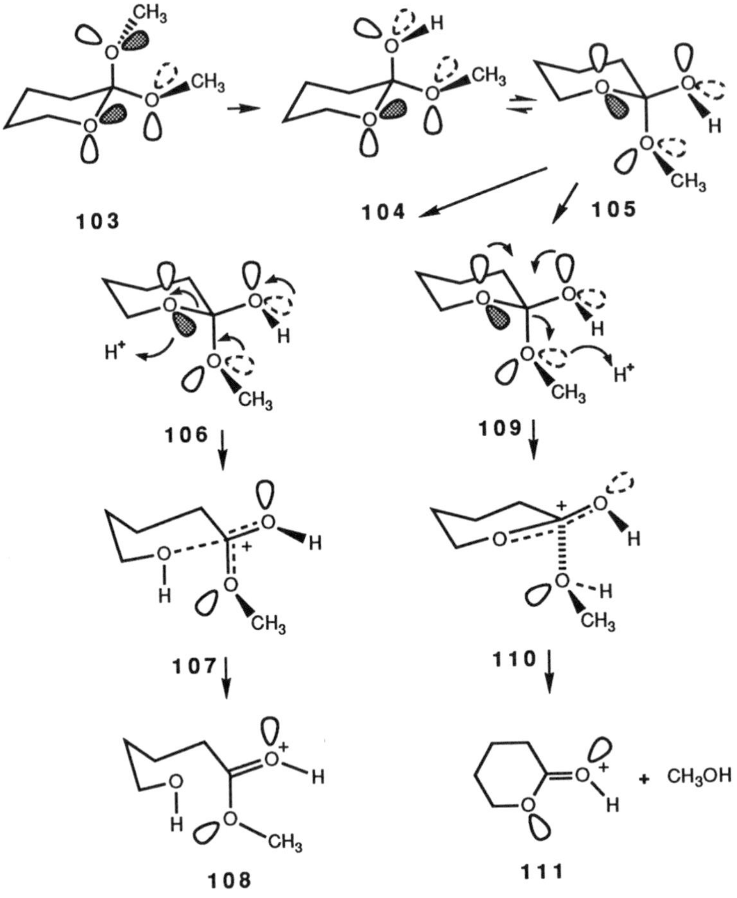

SCHEME 20

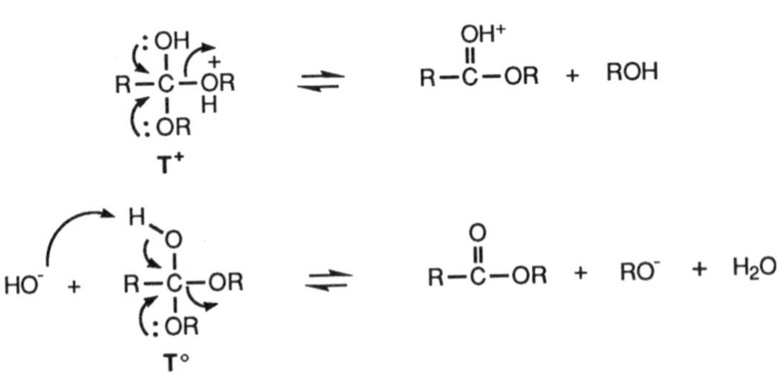

SCHEME 21

entropy factor should play a more important role in the breakdown of T⁺. The formation of a lactone which is the reverse of a bimolecular process should be more favored than the formation of a hydroxy-ester which is the reverse of an intramolecular process. Thus, more lactone should be produced in the breakdown of T⁺ as observed experimentally. Results previously reported on *cis* and *trans* tricyclic orthoesters (8) can be explained on the same basis.

In conclusion, while respecting the antiperiplanar hypothesis but moving the geometry of the transition states away from that of the tetrahedral intermediate and closer to the carbonyl species, the observed hydrolytic behavior of acetals and orthoesters is clearly understood.

Acknowledgments

The excellent contribution of my collaborators Mr Shigui Li, Dr Normand Pothier and Dr Solo Goldstein is deeply appreciated. This work was supported financially by NSERCC (Ottawa) and FCAR (Québec).

Literature Cited

1   Deslongchamps, P.; Moreau, C. *Can. J. Chem.* **1971**, *49*, 2465.
2   Deslongchamps, P.; Moreau, C.; Fréhel, D.; Atlani, P. *Can. J. Chem.* **1972**, *50*, 3402.
3   Deslongchamps, P.; Atlani, P.; Fréhel, D.; Malaval, A.; Moreau, C. *Can. J. Chem.* **1974**, *52*, 3651.
4   Deslongchamps, P. In *Stereoelectronic Effects in Organic Chemistry*, Baldwin, J. E., Ed.; Organic Chemistry Series, Vol. I; Pergamon Press: Oxford, England, 1983.
5   Deslongchamps, P.; Atlani, P.; Fréhel, D.; Malaval, A. *Can. J. Chem.* **1972**, *50*, 3405.
6   Deslongchamps, P.; Chênevert, R.; Taillefer, R. J.; Moreau, C.; Saunders, J. K. *Can. J. Chem.* **1975**, *53*, 1601.
7   Deslongchamps, P.; Lessard, J.; Nadeau, Y. *Can. J. Chem.* **1975**, *63*, 2485.
8   Deslongchamps, P.; Guay, D.; Chênevert, R. *Can. J. Chem.* **1985**, *63*, 2493.
9   Kirby, A. J. *Acc. Chem. Res.* **1984**, *17*, 305.
10   Kirby, A.J. *C.R.C. Critical Reviews in Biochemistry* **1987**, *22*, 283.
11   Beaulieu, N.; Dickinson, R.A.; Deslongchamps, P. *Can. J. Chem.* **1980**, *58*, 2531.
12   Deslongchamps, P.; Guay, D. *Can. J. Chem.* **1985**, *63*, 2757.
13   Eikeren, P. v. *J. Org. Chem.* **1980**, *45*, 4641.
14   BeMiller, J. N. *Adv. Carbohydr. Chem.* **1967**, *22*, 25-108.
15   BeMiller, J. N. *Adv. Carbohydr. Chem.* **1970**, *25*, 544.
16   Legler, G. *Adv. Carbohydr. Chem.* **1970**, *48*, 319.

17 Kirby, A. J. In *The Anomeric Effect and Related Stereoelectronic Effects at Oxygen;* Springer-Verlag: Berlin, 1983.

18 Ratcliffe, A.J. ; Mootoo, D. R.; Andrews, C. W.; Fraser-Reid, B. *J. Am. Chem. Soc.* **1989**, *111*, 7661.

19 Sinnott, M. L. *Chem. Rev.* **1990**, *90*, 1171, and references quoted therein.

20 Deslongchamps, P.; Rowan, D. D.; Pothier, N.; Sauvé, G.; Saunders, J. K. *Can. J. Chem.* **1981**, *59*, 1105.

21 Deslongchamps, P.; Pothier, N. *Can. J. Chem.* **1987**, *68*, 597.

22 Pothier, N.; Goldstein, S.; Deslongchamps, P. *Helv. Chim. Acta* **1992**, *75*, 604.

23 Young, P. R.; Jencks, W. P. *J. Am. Chem. Soc.* **1977**, *99*, 8238.

24 Bennett, A. J.; Sinnott, M. L. *J. Am. Chem. Soc.* **1986**, *108*, 7287.

25 Bürgi, H. B.; Dunitz, J. D.; Shefter, E. *J. Am. Chem. Soc.* **1973**, *95*, 5065.

26 Bürgi, H. B.; Dunitz, J. D. *Acc. Chem. Res.* **1983**, *16*, 153.

27 Bürgi, H. B.; Dubler-Steudle, K. C. *J. Am. Chem. Soc.* **1988**, *110*, 7291.

28 Andrews, C. W.; Fraser-Reid, B.; Bowen, J. P. *J. Am. Chem. Soc.* **1991**, *113*, 8293.

29 Irwin, J. J.; Ha, T. K.; Dunitz, J. D. *Helv. Chim. Acta* **1990**, *73*, 1805.

30 Grein, F.; Deslongchamps, P. *Can. J. Chem.* **1992**, *70*, 604.

31 Grein, F.; Deslongchamps, P. *Can. J. Chem.* **1992**, *70*, 1562.

32 Lemieux, R. U.; Morgan, A. R. *Can. J. Chem.* **1965**, *43*, 2205.

33 Hine, H. *Adv. Phys. Org. Chem.* **1977**, *15*, 1.

34 Hosie, L.; Marshall, P. J.; Sinnott, M. L. *J. Chem. Soc. Perkin Trans.* **1984**, *2*, 1121.

35 Perrin, C. L.; Arrhenius, G. M. L. *J. Am. Chem. Soc.* **1982**, *104*, 2839.

36 Perrin, C. L.; Nunez, O. J. *J. Chem. Soc., Chem. Commun.* **1984**, 333.

37 Caserio, M. C.; Souma, Y.; Kim, J. K. *J. Am. Chem. Soc.* **1981**, *103*, 6712.

38 Banait, N. S.; Jencks, W. P. *J. Am. Chem. Soc.* **1991**, *113*, 7951, 7958.

39 Berson, J. A. *Acc. Chem. Res.* 1991, *24*, 215.

40 McClelland, R. A.; Alibkai, M. *Can. J. Chem.* **1981**, *59*, 1169.

41 Deslongchamps, P.; *Pure Appl. Chem.* **1993**, *65* (6), 1161.

RECEIVED June 16, 1993

Chapter 4

Anomeric and Gauche Effects
Some Basic Stereoelectronics

Anthony J. Kirby and Nicholas H. Williams

University Chemical Laboratory, Cambridge CB2 1EW, England

The anomeric effect and its extensions - the generalised anomeric effect and the gauche effect - are neatly interpreted in terms of competing bonding interactions - n–σ* and σ–σ* - between filled and vacant orbitals. (The reverse anomeric effect is an exception, but the evidence for its existence is not unequivocal.) In a given system this represents just one of several factors controlling conformation, and perhaps reactivity. Its importance is specifically indicated by the changing pattern of bond lengths at the centres concerned. New data - both experimental and calculational - on the structure and conformation of very simple systems - aryl tetrahydropyranyl acetals and 2-substituted ethanol derivatives - are consistent with the dominant role of these stereoelectronic effects.

After 15 years we have a fair understanding of the basis of the anomeric effect. Because of its practical importance and fundamental interest it remains a very active area of research. I will discuss recent work on three different but related aspects where basic questions remain to be answered. In all three cases the immediate interest is in conformation. What we are really interested in though is reactivity, and possible applications to reactivity are never far from our thinking.

In what terms do we currently understand the basis of the anomeric effect? The first thing to realise - and it applies to almost any problem in Chemistry - is that explanations in terms of a single factor very often have more or less serious limitations. In this case we can manage pretty well with just two, at least as long as we restrict ourselves to structural

0097–6156/93/0539–0055$06.00/0

properties.   The two are: *bonding n–σ* interactions,* and *electrostatic effects.*

The idea of a stabilising interaction between a non-bonding electron-pair on oxygen and the low-energy antibonding orbital of a C—X bond is familiar, and of general importance in determining structure (*1*).

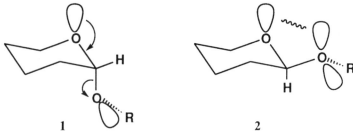

The most direct evidence for this interaction is the systematic variation in the pattern of bond lengths at the carbon centre of such systems when the electronegativity of the group X is changed (*2*). The geometrical requirement for optimum n_O–σ^*_{C-X} overlap is that the orbitals should be antiperiplanar, which means that the substituent R on the donor oxygen must be gauche to the C—X bond.   In pyranoses, and in tetrahydropyranyl acetals in general, this requirement is fulfilled only for the axial anomer (**1**) when the ring oxygen is the donor.   (A similar geometrical requirement for the exocyclic oxygen as donor is of course the basis of the exo-anomeric effect, which operates equally well in the equatorial (**2**) or the axial conformation (**1a**)).

The sterically favored equatorial conformation **2** has a larger nett dipole moment, and also suffers to some extent from electrostatic (lone pair-lone pair) repulsion, as shown. These factors will be strongest for single molecules in the gas phase, and thus show up - like all electrostatic effects - as significant in *ab initio* calculations. A destabilizing interaction cannot of course be the sole origin of the anomeric effect, which involves net stabilization of the acetal structure, compared with comparable systems having two isolated C—O bonds.

The relative importance to the anomeric effect of these two factors - the stabilisation of axial conformation **1** by the $n_O-\sigma^*_{C-X}$ interaction shown, and electrostatic destabilisation of the equatorial conformation **2** - is an area of strong and continuing interest. Calculations are a good starting point, but we have to bear in mind that electrostatic effects will be reduced in the condensed phase, particularly in polar solvents; while $n_O-\sigma^*_{C-X}$ interactions, insofar as they result in charge separation, will be strengthened. The three cases I want to discuss all bear on this question.

1. The Anomeric Effects of OAr and OAc Groups.

Ouedraogo and Lessard (3) recently measured conformational equilibria for a series of 2-(4-substituted-phenoxy) tetrahydropyrans (**1** $\rightleftarrows$ **2**, R = Ar), and showed that the axial preference of the OAr group (resulting from the anomeric effect) increases with increasing electron-withdrawal by the *para*-substituent. This is the expected effect, referred to above, of increasing the electronegativity of the OAr group, and can be explained in terms of an increasingly significant $n_O-\sigma^*_{C-X}$ interaction. Consistent with this explanation, the endocyclic C—O bond is shortened, and the C—OAr bond lengthened over this series of compounds (2).

The alternative explanation, in terms of the electrostatic interaction, is that increasing electron withdrawal from the exocyclic oxygen increases the dipole moment of the exocyclic C—O bond, and thus the unfavorable dipole-dipole interaction in the equatorial isomer. Note that it *reduces* the size of the lone pairs on the exocyclic oxygen, and thus the magnitude of the lone pair-lone pair interaction (**2**), which would have therefore to be a smaller effect. (An explanation based on this latter effect has been used previously to rationalise the unexpectedly small apparent size of acetoxy groups in sugars (*1c*). The question is raised again, in a slightly different form, by another result of Ouedraogo and Lessard (3), for 2-acetoxytetrahydropyran. The Canadian authors found that the axial preference of the acetoxy group in CF_2Br_2 is no larger - perhaps even slightly smaller - than that of the phenoxy group, even though it is considerably more electron-withdrawing.)

The data of Ouedraogo and Lessard (3) do not in fact define the absolute magnitude of the anomeric effect for aryloxy groups, because the steric preference (A-value) of an OAr group for the equatorial position on a cyclohexane has not been accurately measured. The steric effect of a *meta* or *para*-substituted aryloxy group is expected to be constant, but it seemed clear to us that the A-value should depend to some extent on the electronic effect of the substituent: because the axial conformation of an aryloxycyclohexane (**3**) would be expected to show two relatively stabilising $\sigma_{C-H}-\sigma^*_{C-OAr}$ interactions, similar to the $n_O-\sigma^*_{C-X}$ interaction of a tetrahydropyranyl acetal; though smaller because a C—H bond is a weaker σ-donor than a lone pair, and thus less different from the corresponding $\sigma_{C-C}-\sigma^*_{C-OAr}$ interaction of the equatorial isomer. To test this idea we have measured (4,5) conformational preferences for four cyclohexyl 4-substituted-phenyl ethers (**3**, Ar = phenyl, *p*-methoxy-, *p*-chloro- and *p*-nitrophenyl).

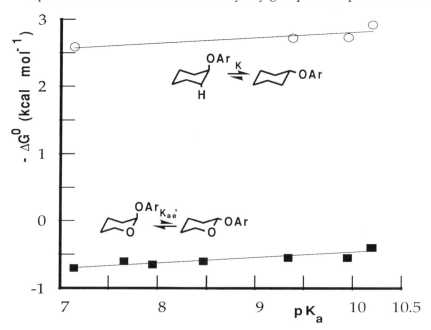

3

Our results are summarised in Figure 1.   There is indeed a weak dependence of the A-value for the aryloxy-group on the para substituent

Figure 1.   Axial preferences of 2-aryloxytetrahydropyrans (data taken from Ouedraogo & Lessard (3)) compared with equatorial preferences (A-values) for the same groups in aryl cyclohexyl ethers (5).

(corresponding to a Hammett ρ-value of -0.29). This has to be set against the observed axial preference in a tetrahydropyranyl acetal, where electron-withdrawal favours the axial conformation. The comparison is illustrated in the Figure, where the axial preferences are also plotted, most conveniently against the pK_a of ArOH. The anomeric effect of an aryloxy group is thus a function of its pK_a, and is given by (7)

Anomeric effect for ArO (kcal mol^{-1}) $= 1.7\pm0.2 - (0.05\pm0.02)\,pK_{ArOH}$

For the phenoxy group this gives a best value of 1.2 kcal mol^{-1} (8).

This work also led us to look at the question of the apparent size of the acetoxy group, suggested by various authorities to be anomalous. The anomeric effect as properly defined (1d) involves the sum of the axial preference in a tetrahydropyran acetal plus the steric factor favoring the equatorial position, as measured by the A-value. An anomaly in the apparent anomeric effect could therefore be due to a genuinely anomalous anomeric affect, an anomalous A-value, or even a combination of both.

It is possible to extend the correlation of the A-value with the pK_a of (the conjugate acid of) the group concerned, as shown in Figure 1, to other derivatives $C_6H_{11}OX$ of cyclohexanol. This extended correlation is shown in Figure 2, where our new data for aryloxycyclohexanes (open circles) are combined with data from the literature (6) for the OTs, OMs, OCOCF$_3$ and OCHO groups (squares); the filled circle is Jensen and Bushweller's (6) 'best value' for acetoxy. There does indeed seem to be a deviation from what looks like a sensible correlation of A-value with pK_a, and thus a possible anomaly. The deviation is, however, positive, consistent with the acetoxy group being unexpectedly large, rather than small. If genuine it would not explain the unexpectedly small apparent size of acetoxy groups in sugars referred to above. It could however explain an unexpectedly small axial preference for the acetoxy group in 2-acetoxytetrahydropyran.

One's first reaction is that the result shown for acetoxy in Figure 2 looks wrong. The plot includes data for two other esters, the formate and trifluoroacetate, both of which conform to what we consider expected behaviour. But we need to be sure, so we have measured the thermodynamic parameters for the axial-equatorial equilibrium of cyclohexyl acetate at a series of temperatures between 170 - 200 K: and find a result which is interesting, if only preliminary.

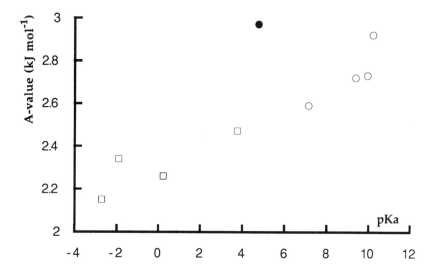

Figure 2. Plot of the equatorial preferences for cyclohexanol derivatives *vs.* the pK_a of the conjugate acid (ArOH, XOH) of the leaving group. Data points, in increasing order of pK_a, are for OTs, OMs, OCOCF$_3$ and OCHO (*6*), and (open circles) for OAr (*5*). The closed circle represents Jensen and Bushweller's 'best value' for acetoxy (*6*).

Unlike most cyclohexyl derivatives, including the aryloxycyclohexanols we measured previously, the equatorial preference of the acetoxy group appears to be temperature-dependent. The entropy change for the equilibrium:

is not zero but 2.15 ± 0.35 cal mol^{-1} K^{-1}, consistent with some minor conformational restriction of the acetoxy group in the axial position. (An almost identical result was found by Squillacote for the isopropyl group (*9*). The equatorial preference, measured in the usual way as $-\Delta G°$, would thus be made up at 298 K of contributions of 0.42 kcal (1.76 kJ) mol^{-1} from $\Delta H°$ and 0.64 kcal mol^{-1} from the entropy term (*10*). This is some evidence that acetoxy is indeed relatively small for an oxygen-based group in enthalpy terms. (More recent evidence (*10*) shows a $\Delta S°$ close to zero for the equatorial-axial equilibrium for cyclohexyl trifluoroacetate.) However, these measurements are notoriously imprecise, and we prefer to wait for more results before attaching too much credence to this set (*11*).

2. The Reverse Anomeric Effect.

This effect, the apparently extra-strong *equatorial* preference of positively-charged electronegative substituents at the anomeric position of sugars and 2-substituted tetrahydropyrans, (*1e*) is of particular interest in that it is the opposite of the effect predicted by the $n_O-\sigma^*_{C-X}$ interaction, which should still operate with $X = R_3N^+$, as in **4a**. It is however consistent with electrostatic stabilisation of the equatorial form, **4e**, where N^+ is close, because gauche, to both lone pairs on the ring oxygen.

4a   4e

An obvious complicating factor is that any trialkylammonium group will also be large (Me_3N^+ being sterically equivalent to *t*-butyl), and I know of no case where a trialkylammonium group has been shown to be axial. However, no quantitative measurements of the reverse anomeric effect are available, and it is clearly time that this omission was remedied. I cannot help thinking that the most desirable (perhaps even the most likely) outcome would be for the effect to disappear on closer examination, and I know that Dr. Perrin will present evidence pointing in that direction.

There is (just) one detailed collection of ^1H NMR data, on pentopyranoses, from Paulsen *et al.* (*12*) which supports the existence of the reverse anomeric affect. For example, pyridinium α-D-xylopyranoside clearly exists in the $^1C_4$ conformation, **6**, with the three acetoxy groups axial:

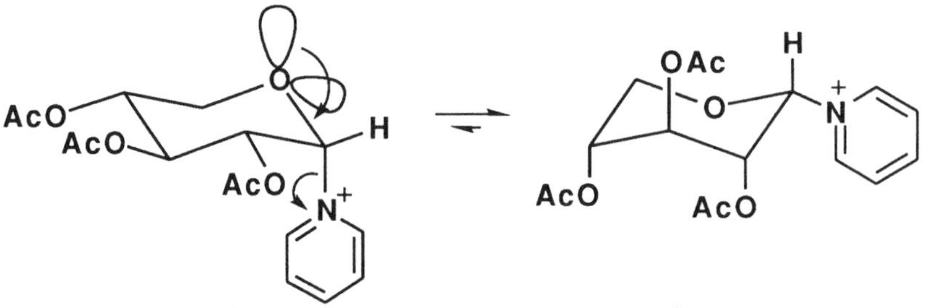

5   6

This is true also for the corresponding imidazolium glycoside, though this already exists to a considerable extent in the $^1C_4$ conformation even in the neutral form.

There is on the other hand a growing body of evidence that the reverse anomeric effect, if it exists, cannot be generalised as can the anomeric effect itself. Thus phosphonium substituents in the anomeric position of dithianes apear to show normal anomeric effects. The (base catalysed) equilibrium shown, for example, clearly favors the axial conformation, by about 3:1 (13a):

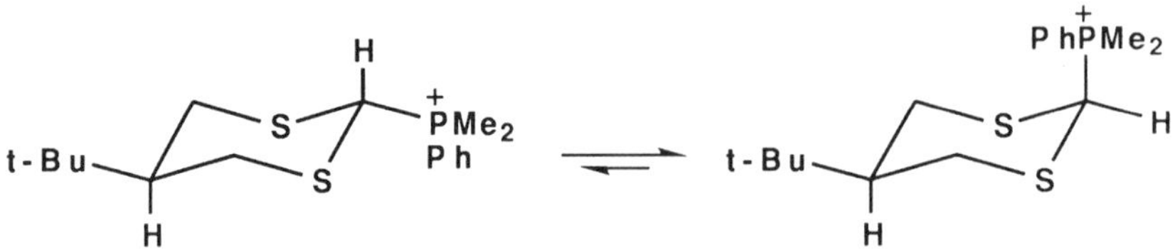

The evidence for the reverse anomeric effect is not therefore unequivocal, and we have made occasional abortive attempts over the years to make compounds that would be definitive. We were very interested therefore in another recent report (13b) from Lodz, á propos of measurements of the large anomeric effect of the $P(O)Ph_2$ group, in which Graczyk and Mikolaczyk made 7, as a the minor component in a mixture of cis and trans diastereoisomers.

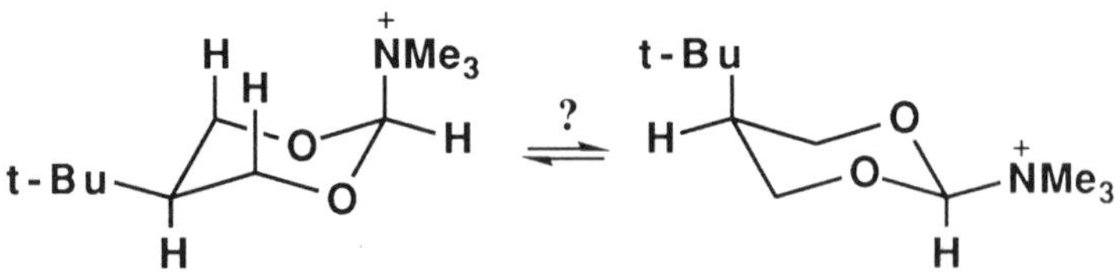

7a   7e

This compound suggests a possible definitive experiment: if t-butyl and 2-Me$_3$N$^+$ groups are in balanced conformational opposition a significant reverse anomeric effect should drive the t-butyl group axial, whereas if the $^+$NMe$_3$ group shows an ordinary anomeric effect it would itself prefer to be axial. The *cis*-2,5 isomer **7** is not in fact suitable for this experiment: it has a powerful built-in conformational bias favoring **7e**, because of the absence of axial hydrogens 1,3 to the t-butyl group. (An effect strong enough for even a 2-methyl group to drive a 5-t-butyl into the axial position (*14*)). We are currently making the related *trans*-2,4 compounds **8a** and **8e** (*10*) in the hope that they will be stable enough at least for NMR investigation.

3. The Gauche Effect.

The gauche effect (*1f*) is the preference of systems with *vicinal* electronegative groups for a conformation, **8**, in which they adopt a gauche, rather than the usual antiperiplanar relationship, **9**.

The accepted explanation is in terms of the σ-σ^* orbital interactions shown. These are qualitatively similar to those which contribute to the anomeric effect. A C—H bond provides a better donor orbital in the σ-σ^*_{C-X} interaction shown than the C—Y bond when Y is more electronegative than H, thus stabilising the gauche conformation selectively. The analysis does not suffer from the ambiguities which complicate the discussion of the related anomeric and reverse anomeric effects, because steric and electrostatic effects both favour the alternative antiperiplanar conformation.

We have looked at a large series of compounds having the general structure Y—C—C—OX, using the method of crystal structure correlation (*15*). We established earlier (*2*) that the length of the C—OX bond is a criterion of its ease of cleavage. We have found linear relationships between bond length and reactivity in more than a dozen series of acetals and other alcohol derivatives R—OX: in general, in a given system, the longer the bond, the faster it breaks, and the more sensitive is the length of the bond to changing X (2). The relevant correlations for primary and secondary alkyl (here all cyclohexyl) derivatives are shown in Figure 3. (Note that the plots use pK_a as convenient measure of reactivity. The free energy of activation for the cleavage of the C—OX bond in a system ROX is expected be a linear function of the pK_a of HOX.)

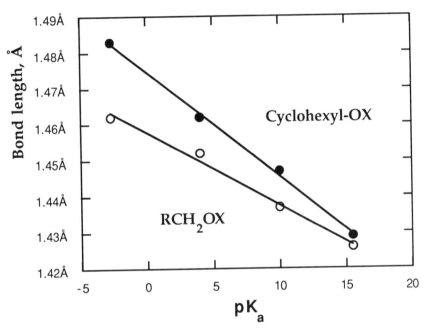

Figure 3. Bond length-pK_a correlations for primary and secondary (cyclohexyl) alkyl derivatives. The mean C—OX bond length of a large number of systems R—OX is plotted against the pK_a of the conjugate acid, HOX, of the leaving group.

It is known that a vicinal substituent Y can have a profound effect on reactivity . Rates of C–OX cleavage can be enormously enhanced by antiperiplanar silicon (Y = R_3Si) (16), whereas electron-withdrawing 2-substituents (Y = OX' or halogen) reduce reactivity. This latter effect has not in fact been studied systematically, except in some early work by Hine (17) who found 7 to 8-fold retardations in the rates of substitution reactions of 2-alkoxy, fluoro, chloro and bromo-ethyl halides by PhS^-. Such factors are already large enough to be useful in synthesis. They account, for example, for the regioselective ring opening of some epoxides, such as those derived from allylic alcohols (*18*); and for the extra stability of the bis-*p*-toluenesulphonate and even the bis-perchlorate (*19*) esters of 1,2-diols.

The cumulative effect of more than one electron-withdrawing group can be very large. S_N2 reactions of 2,2,2-trifluoroethyl halides are at least 10^4 times slower than those of parent, non-fluorinated, compounds(*20*); and the stabilising effect of multiple hydroxyl substitution on derivatives such as sulfonate esters of sugars is well-known (*21*).

We find good evidence that there are parallel and substantial effects on bond lengths. The C—OX bond lengths (a) in two compounds (**10** and **11**) with silicon antiperiplanar and gauche to ester groups are both significantly longer than found for the parent (H) compounds (*15*), but we do not have sufficient data to make judgements about the relative magnitude of these effects. Conversely, though the C—OS bond length in the monofluoro-derivative **12** (Y = H; 1.463(3)Å) is not significantly different from that in 2-phenylethyl tosylate (1.459(4)Å), in the trifluoroethyl ester (**12**, Y = F) this bond is clearly shortened, at 1.435(3)Å.

This latter result does not tell us whether the cumulative effect of three fluorines is independent of geometry, or derives disproportionately from the fluorine antiperiplanar to the C—O bond. So we have looked at crystal structures of a large set of 2-fluorocyclohexanol derivatives of fixed geometry, with fluorine either gauche or antiperiplanar to the C—OX bond (*15*).

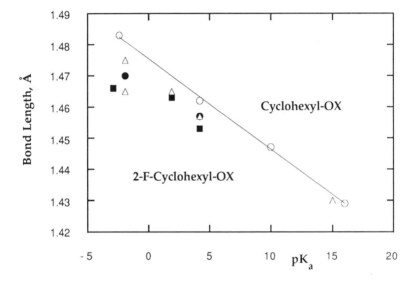

In every case the introduction of vicinal fluorine is accompanied by a shortening of the C—OX bond. However, the effects of gauche and antiperiplanar fluorine are not significantly different, as shown in Figure 4.

Figure 4.   Bond length-pK_a correlation for 2-fluorocyclohexanol derivatives compared with the same correlation for the parent cyclohexyl systems taken from Figure 3.   Triangles represent data points for axial OX compounds, filled symbols for equatorial derivatives with fluorine axial or equatorial. (From reference 15.)

This could be because the effects of gauche and antiperiplanar fluorine really are not different, or because they differ too little to be significant at the level of accuracy of crystal structure determinations at room temperature (as is the case for axial and equatorial sulfonate and carboxylate esters of cyclohexanol (15)). So we have complemented these measurements with a large set of *ab initio* calculations, in collaboration with Professor N.C. Handy of our department.   Calculations were carried out for the 20 structures shown, to compare systems with antiperiplanar and gauche fluorine and silicon with each other and with the parent, unsubstituted compounds.

Y = H, F, SiH$_3$ Y = F, SiH$_3$

X = Me, CHO (E and Z), NO$_2$

The resulting very substantial data set confirms that the effects of antiperiplanar fluorine and silicon are in fact greater than those in the gauche systems, at least in the gas phase, and shows that the differences are indeed of the magnitude of the experimental errors in crystal structure determinations. The work has been published (*15*), so will not be discussed in detail. But one set of results of special interest for this symposium is shown in Figure 5.

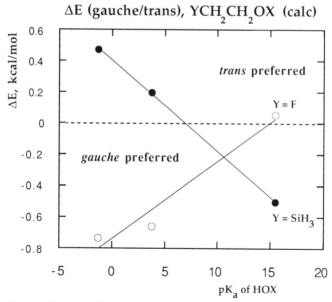

Figure 5. Calculated energy differences between gauche and anti conformations. Taken from reference 15.

Here the calculated energy differences between gauche and antiperiplanar conformations are plotted against the pK_a of the conjugate acid of the 'leaving group' OX. Note how the preference for the gauche conformation increases for the 2-fluoro compounds as the group OX becomes more electronegative. For the 2-SiH_3 compounds, on the other hand, the same change of leaving group increases the preference for the antiperiplanar conformation. In these systems at least, we can see a clear-cut manifestation of the stabilising effect of σ-σ^* interactions.

Acknowledgements. We are grateful to the Science and Engineering Research Council of Great Britain for support of this work, and to Professor Peter G. Jones (Braunschweig) for the crystal structure determinations.

References.
1. A. J. Kirby, in *'The Anomeric Effect and Related Stereoelectronic Effects at Oxygen,'* Springer Verlag, Berlin, 1983; (a) p. 62 *ff.* (b) p. 38 (c) p. 13 (d) p. 7 (e) p. 15 (f) p. 32
2. A.J. Kirby, *Accts. Chem. Res.*, 1984, **17**, 305 .
3. A. Ouedraogo and J. Lessard, *Can. J. Chem.*, 1991, **69**, 474.
4. By ^{1}H NMR measurements at low temperatures (5). Spectra were obtained in the slow exchange region at 400 MHz (solvent 9:1 $CFCl_3$:CD_2Cl_2), at 5° intervals from 200K to 170K.
5. A. J. Kirby and N. H. Williams, *J. Chem. Soc., Chem. Commun.*, 1992, 1285.
6. F. R. Jensen and C. H. Bushweller, *Adv. Alicyclic Chem.*, 1971, **3**, 140.
7. A. J. Kirby and N. H. Williams, *J. Chem. Soc., Chem. Commun.*, 1992, 1286. Note that because equatorial preferences are larger in tetrahydropyrans compared with cyclohexanes (because of the shorter in-ring C—O bonds) the absolute values quoted here and elsewhere for the anomeric effect are in fact minimum values. The required correction factor is not accurately known, but is likely to be on the order of 1-2 kJ mol^{-1}. (The equatorial preference of the considerably larger 2-methyl group is 12.0 kJ mol^{-1} in tetrahydropyran, compared with 7.3 kJ mol^{-1} in cyclohexane. [E. L. Eliel, K. D. Hargrave, K. M. Pietrusiewicz and M. Manoharan, *J. Am. Chem. Soc.*, **1982**, *104*, 3635.] A factor of this size would add 1.7 kJ mol^{-1} to the values given for the anomeric effect. The true correction factor will be smaller.
8. At low temperature in 9:1 $CFCl_3$:CD_2Cl_2, so not necessarily significantly different from the approximate value of 1.6 kcal mol^{-1} obtained for the neat liquid by G. O. Pierson and O. A. Runquist, *J. Org. Chem.*, 1968, **33**, 2572.
9. M. E. Squillacote, *J. Chem. Soc., Chem. Commun.*, **1986**, 1406.

10. A. J. Kirby and P. D. Wothers, unpublished results.

11. Our results for cyclohexyl acetate (*10*), obtained from ^{1}H NMR data, differ to a minor extent from those obtained by natural abundance ^{13}C measurements at similar low temperatures by E. A. Jordan and M. P. Thorne, *Tetrahedron*, 1986, **42**, 93. However, a plot, corresponding to Figure 2, of the available (*5,10*) $\Delta H°$ data *vs* the pK_a of the 'leaving group' shows the expected good linear relationship ($r = 0.970$ for 5 data points).

12. H. Paulsen, Z. Györgydeák and M. Friedmann, *Chem. Ber.*, 1974, **107**, 1590.

13. P.P. Graczyk and M. Mikolaczyk, (a) *Angew. Chem., Intl. Ed. Engl.*, 1991, **30**, 578. (b) *Tetrahedron*, 1992, **48**, 4209.

14. E. L. Eliel and M. C. Knoeber, *J. Am. Chem. Soc.*, **1966**, *88*, 5347: **1968**, *88*, 3444.

15. R. D. Amos, N. C. Handy, P. G. Jones, A. J. Kirby, J. K. Parker, J. M. Percy and M. D. Su, *J. Chem. Soc., Perkin Trans. 2*, **1992**, 549-558.

16. J.B. Lambert, G.-T. Wang, R.B. Finzel, and D.H. Teramura, *J. Am. Chem. Soc.*, 1987, **109**, 7838.

17. J. Hine, and W. H. Brader, *J. Am. Chem. Soc.*, 1953, **75**, 3964.

18 C. H. Behrens, and K. B. Sharpless, *Aldrichim. Acta*, 1983, **16**, 67.

19. J. B. Lambert, H.W. Mark, A.G. Holcomb and E.S. Magyar, *Accts. Chem. Res.*, 1979, **12**, 317.

20. N.S. Zefirov, and A.S. Koz'min, *Accts. Chem. Res.*, 1985, **18**, 154. F. G. Bordwell, and W. T. Brannen, *J. Am. Chem. Soc.*, 1964, **86**, 4645.

21. A. C. Richardson, *Carbohydrate Res.*, 1969, **10**, 395.

RECEIVED June 21, 1993

Chapter 5

Anomeric Effects

An Iconoclastic View

Charles L. Perrin

Department of Chemistry, University of California—San Diego, La Jolla, CA 92093-0506

We have investigated experimentally the reverse anomeric effect (AE), the origin of the AE, and stereoelectronic control. (1) To probe the reverse AE claimed for cationic substituents, the proportions of axial anomers of N-alkylglucopyranosylamine derivatives were determined by ^{1}H NMR in a variety of solvents, including acidic media. The change upon N-protonation is small and can be accounted for on the basis of steric effects and a small normal AE. Thus we conclude that the reverse AE does not exist. (2) To determine whether MO or electrostatic interactions are more important for the AE, the conformational equilibrium of 2-methoxy-1,3-dimethylhexahydropyrimidine was studied by ^{1}H and ^{13}C NMR. The proportion of axial conformer is almost the same as in 2-methoxy-1,3-dioxane, but steric corrections indicate that the AE is weaker in the former. Since nitrogen is less electronegative than oxygen, its stronger MO interactions ought to have increased the AE. We therefore conclude that in nonpolar solvents electrostatic interactions predominate. Moreover, it is shown that dipole-dipole interactions can account for structural changes considered to be evidence for MO interactions. (3) To evaluate stereoelectronic control in cleavage of tetrahedral intermediates, product ratios from hydrolyses of cyclic amidinium ions and rates of methoxy exchange in 2-methoxy-4,6-dimethyl-1,3-dioxanes were determined. The results indicate a weak preference for antiperiplanar lone pairs in six-membered rings, but not in five- or seven-membered ones. The amidinium results cannot be rationalized by assuming reaction via a boat conformer.

Reverse Anomeric Effect

Introduction. The anomeric effect (*1*) is the preference for the axial conformer **1A** shown by tetrahydropyranyl derivatives (Y = O) with electronegative X at C1. It opposes the steric repulsion, which prefers the equatorial conformer **1E**. The quantitative relation is given in eq 1, where E_{An} is the preference (energy > 0) for the

$$E_{An} = \Delta G_{1A \rightarrow 1E} + A_X \qquad (1)$$

0097–6156/93/0539–0070$07.75/0

axial position due to the anomeric effect, $\Delta G_{1A \to 1E}$ is the observed free-energy change, $-RT\ln([1E]/[1A])_{obs}$, and A_X is $RT\ln([E]/[A])_{model}$, the steric preference for equatorial X as measured in a model compound such as a cyclohexane. Values of E_{An} in O-C-O systems are 5-10 kJ mol⁻¹, which is small but significant. However, when X is a positively charged nitrogen substituent, the equilibrium lies toward **1E**. This change in conformational preference has been attributed to a so-called reverse anomeric effect.

The reverse anomeric effect represents a significant puzzle regarding molecular structure. It is important for our understanding of the conformations of carbohydrates, simple organic heterocycles, and nucleosides. It is also necessary for understanding the reactivity of such molecules, which often react via their protonated forms, and the reverse anomeric effect is sometimes invoked (2) to account for stereoselectivity when there is a cationic leaving group.

The first examples of a reverse anomeric effect were in N-(α-glucopyranosyl)-4-methylpyridinium ion and N-(α-D-glucopyranosyl)- and N-(α-D-mannopyranosyl)-imidazolium ions, where the cationic heterocycle is equatorial despite the consequence that other substituent groups must be axial (3). Similar behavior is shown by many such ions, although some may take a twist-boat conformation (4). However, an aromatic heterocycle is quite bulky, especially with a positive charge that must be solvated, and all the results could be due simply to avoidance of prohibitive steric repulsions associated with placing that group in the axial position.

Yet it is claimed (5) that the conformational preference in protonated N-(tri-O-acetyl-α-D-xylopyranosyl)imidazole **2** is greater than can be attributed to steric factors.

In CDCl₃ the steric bulk of the imidazolyl group favors conformer **2E** by 65%, but with trifluoroacetic acid the proportion increases to >95%. Since protonation is considered not to change the size of the substituent, the shift of the equilibrium is attributed to the positive charge. However, protonation might change the bond length or the solvation, so that the effective size of the substituent may increase by more than just the size of a distant hydrogen.

The anomeric effect itself has been explained as the result of either dipole-dipole interactions or molecular-orbital interactions (1). Orbital overlap between an oxygen lone pair (n) and the C-X antibonding (σ*) orbital is a stabilizing interaction and is most effective in axial conformer **1A**, where a lone pair is antiperiplanar to the C-X bond. This is the MO equivalent of including the resonance form **1A'**, which is not possible for **1E**, where the p orbital on oxygen is orthogonal to the C-X bond. Alternatively, according to the electrostatic interpretation, repulsion between the dipoles of the oxygen lone pairs and of the C-X bond is minimized in the axial conformer. Both of these kinds of interactions are probably operative, but the molecular-orbital interpretation is currently favored.

Yet the reverse anomeric effect is better accounted for by electrostatic forces, which reverse upon introduction of a positive charge. This has been attributed (3) to a reversal of the dipole moment of the C-N bond, but it must be remembered that dipole moment is undefined when there is net charge, since it is not invariant to a change of origin. Instead, it is proper to recognize that monopole-dipole attraction is maximal with the substituent in an equatorial position, where the positive charge is closer to the negative end of the dipole.

The inability of the molecular-orbital or resonance interpretation to account for the reverse anomeric effect has been noted (1a). The introduction of a positive charge makes a nitrogen substituent even more electronegative. Then resonance form $1A'$ ought to contribute more and the anomeric effect ought to increase. Alternatively, the energy of the σ^* orbital is lowered so that it interacts more strongly with the n orbital. Since n-σ^* interaction is stabilizing, the proportion of axial conformer ought to increase, but that is not what is seen experimentally.

The conformational behavior claimed for bulky cationic substituents is therefore suspect. Indeed, there is considerable theoretical and experimental evidence counter to the reverse anomeric effect (6). To investigate this, it is essential to quantify steric effects. Pyridinium and imidazolium rings are too bulky to provide a reliable measure. We need a substituent whose steric sizes, both unprotonated and protonated, are known.

Proposal. Such a substituent is NH_2. From the conformational equilibria of cyclohexylamine and of its conjugate acid both A_{NH_2} and $A_{NH_3^+}$ have been determined (7). Moreover, the increase of A_{NH_2} upon N-protonation provides an estimate of the steric contribution to the shift in the anomeric equilibrium of **1** (X = NH_2) upon N-protonation. If a reverse anomeric effect exists, then N-protonation should increase the proportion of **1E** by more than what the increase in $A_{NH_3^+}$ would suggest.

Therefore we have studied the anomeric equilibrium in various glucopyranosyl-amines, with R' = H = R" (**3**) or R' = Ac = R" (**4**) or R' = H and (R")$_2$ = PhCH (**5**)

β   R = H, Me, Et, Bu (±H⁺)   α

3, R' = H = R"

4, R' = Ac = R"

5, R' = H, (R")$_2$ = PhCH

and with R = H, Me, Et, and Bu, both unprotonated and protonated. Glucose derivatives were chosen because these are most readily available in pure form. Further advantages are that the hydroxymethyl and three hydroxy substituents help maintain a chair conformation with all these groups equatorial and that all the ring protons are axial, with ^1H NMR chemical shifts consequently well upfield of H1, which is nicely isolated in the spectrum. The tetraacetate was chosen to permit solubility in nonaqueous solvents, and the benzylidene acetal fixes the ring against ring inversion. In contrast to the previous studies of the reverse anomeric effect, this equilibrium involves only the conversion of the amino substituent betweem equatorial and axial while all other substituents remain equatorial. It is surprising that such a study had never been done before. The difficulties are the sensitivity of an aminal (**1**, X = NR$_2$) to hydrolysis and the low concentration of its axial stereoisomer.

Results. The ^{1}H NMR spectrum of glucose (**8**) guided signal assignments in glucopyranosylamines. Since the equilibria lie far toward the β anomers (**3-5β**), their signals were easy to identify. Representative spectra are shown in Figure 1. In the free amine there is a distinctive 8.7-Hz doublet at δ3.92, assignable to H1, and there are other characteristic peaks assignable to other hydrogens around the ring, with chemical shifts quite similar to those of β-D-xylopyranosylamine (**9**).

The signals of the α anomer are weaker and less apparent, even its H1. Except for tetra-*O*-acetylglucopyranosylamine, α H1 appears as a characteristic doublet well downfield of all other peaks. Figure 1a shows that weak doublet 0.5 ppm downfield of the prominent β H1 doublet and with a 4.8-Hz coupling constant.

Tables I-III contain the relevant spectral parameters of all the glucopyranosylamines. Similar to glucose itself, α H1 is found an average of 0.65±0.2 ppm downfield of β H1. Also α H1 exhibits an average J_{12} of 5.1±0.4 Hz typical for gauche coupling constants and smaller than the J_{anti} of 8.7±0.2 Hz for β H1. Although

Table I. ^{1}H NMR parameters of *N*-alkylglucopyranosylamines (**3**)

R	solvent	$\delta_1{}^\beta$	$J_{12}{}^\beta$	$\delta_2{}^\beta$	$\delta_1{}^\alpha$	$J_{12}{}^\alpha$	$\delta_2{}^\alpha$	$J_{23}{}^\alpha$
H	D$_2$O	3.85	8.7	2.91	4.59	-	3.39	-
H	Py-d_5/D$_2$O	4.2	8.7	3.40	5.02	5.1	-	-
H	Py-d_5	4.49	8.6	3.75	5.41	5.3	4.42	9.4
H	DMSO-d_6	3.73	8.5	2.77	4.63	4.8	3.20	9.5
H$_2{}^+$Cl$^-$	D$_2$O	4.37	8.7	3.27	5.05	5.4	3.89	9.8
H$_2{}^+$Cl$^-$	DMSO-d_6	4.24	8.6	3.14	4.88	5.0	3.49	9.6
Me	D$_2$O	3.92	8.7	3.21	4.43	4.8	3.60	9.8
Me	DMSO-d_6	3.57	8.8	2.86	4.17	4.8	3.27	9.5
Me	Py-d_5	4.26	8.5	3.80	4.85	5.0	4.24	9.2
MeH$^+$Cl$^-$	D$_2$O	4.30	8.7	3.34	4.88	5.4	3.93	9.3
MeH$^+$Cl$^-$	DMSO-d_6	4.18	-	2.65	4.71	4.8	3.56	9.3
Et	D$_2$O	3.81	8.8	3.00	4.44	4.6	3.46	9.0
Et	DMSO-d_6	3.64	8.8	2.84	4.30	4.8	3.25	9.3
EtH$^+$Cl$^-$	D$_2$O	4.32	8.9	3.35	4.90	5.6	3.83	10.0
EtH$^+$Cl$^-$	DMSO-d_6	4.19	8.6	2.92	4.75	5.4	3.56	9.7
Bu	D$_2$O	3.84	8.7	3.04	4.48	-	3.52	-
Bu	DMSO-d_6	3.63	8.5	2.65	4.28	4.9	3.25	9.5
Bu	Py-d_5	4.35	8.8	3.80	4.99	4.8	4.22	9.3
Bu	Py-d_5/D$_2$O	4.05	8.7	3.42	4.72	5.0	-	-
BuH$^+$Cl$^-$	D$_2$O	4.32	8.7	3.05	4.92	5.5	4.30	9.7

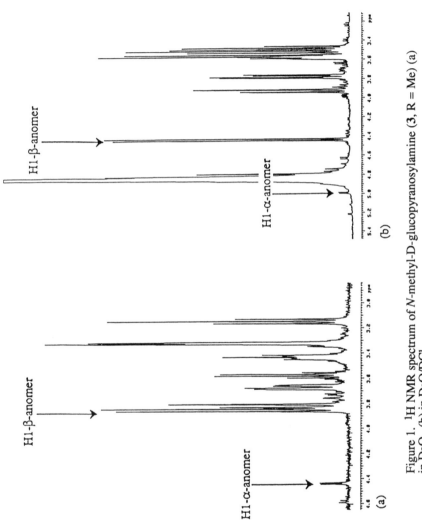

Figure 1. ^{1}H NMR spectrum of N-methyl-D-glucopyranosylamine (**3**, R = Me) (a) in D_2O. (b) in D_2O/DCl.

Table II. ^{1}H NMR Parameters of Tetra-*O*-acetylglucopyranosylamine (**4**)

R	solvent	$\delta_1{}^\beta$	$J_{12}{}^\beta$	$\delta_2{}^\beta$	$\delta_1{}^\alpha$	$J_{12}{}^\alpha$	$\delta_2{}^\alpha$	$J_{23}{}^\alpha$
H	Py-d_5	4.54	8.6	3.80	5.25	5.3	4.24	9.4
H	DMSO-d_6	4.28	8.9	4.62	4.85	5.0	4.90	10.1
H	CDCl$_3$	4.19	9.2	4.82	5.13	5.1	4.99	10.0
H	C$_6$D$_6$	3.65	-	4.90	4.85	4.9	5.08	10.1
H$_2{}^+$Cl$^-$	DMSO-d_6	4.98	8.8	4.92	5.18	5.4	5.07	9.9
H$_2{}^+$CF$_3$CO$_2{}^-$	DMSO-d_6	5.03	9.0	-	5.34	5.5	5.15	9.8
H$_2{}^+$CF$_3$CO$_2{}^-$	CDCl$_3$	4.93	9.0	5.15	5.52	4.6	5.33	9.6
H$_2{}^+$CF$_3$CO$_2{}^-$	C$_6$D$_6$	4.18	-	4.80	4.96	4.5	5.04	-

Table III. ^{1}H NMR Parameters of *N*-Alkyl-4,6-*O*-benzylideneglucosylamines (**5**)

R	solvent	$\delta_1{}^\beta$	$J_{12}{}^\beta$	$\delta_2{}^\beta$	$\delta_1{}^\alpha$	$J_{12}{}^\alpha$	$\delta_2{}^\alpha$	$J_{23}{}^\alpha$
H	Py-d_5/D$_2$O	4.38	8.6	3.68	5.17	5.2	4.07	9.4
H	DMSO-d_6	3.95	8.5	2.94	4.70	5.2	3.37	9.2
H	Py-d_5	4.52	8.7	4.15	5.40	5.1	4.70	9.5
H$_2{}^+$CF$_3$CO$_2{}^-$	DMSO-d_6	4.46	8.6	3.33	4.99	5.4	3.75	9.2
Bu	Py-d_5/D$_2$O	4.27	8.7	3.72	4.83	5.5	4.02	9.5
Bu	DMSO-d_6	3.88	8.7	3.02	4.40	5.0	4.70	9.9
Bu	CDCl$_3$	3.97	8.8	3.21	4.60	5.3	3.76	10.0
BuH$^+$CF$_3$CO$_2{}^-$	DMSO-d_6	4.48	8.7	3.43	4.94	5.4	3.75	9.2

α H2 was obscured by the intense signals of the predominant β anomer, J_{23} could be determined by decoupling difference spectra (*10a*).

Comparison of a ^{13}C NMR spectrum of glucopyranosylamine-1-^{13}C with that of glucose (*11*) substantiates the ^{1}H NMR assignments. The ^{13}C spectrum is dominated by the β C1 signal at δ85.7. Moreover, there is a small additional signal at δ81.9, which is assigned to α C1. *N*-Butylglucopyranosylamine likewise shows C1 signals separated by 3.9 ppm. Two-dimensional ^{13}C-^{1}H correlation spectra (*10b*) of gluco-pyranosylamine-1-^{13}C and of its conjugate acid show crosspeaks that correlate these ^{13}C peaks with the corresponding β and α H1 signals.

The chemical shifts and coupling constants in Tables I-III are consistent with the assignments and with the conclusion that these amines have been converted to their conjugate acids under the acidic conditions used. Chemical shifts and coupling constants match those expected from model compounds. In every sample a prominent 9-Hz doublet was clearly assignable as H1 of the β anomer, and a weaker 5-Hz doublet 0.5-0.8 ppm downfield, with a J_{23} of 9-10 Hz, was then assignable as H1 of the α anomer. The downfield shifts of α H1 and α C1 are clearly inconsistent with a 2-keto derivative, the product of Amadori rearrangement. The large J_{23} is inconsistent with an inverted chair or with a furanose derivative. Moreover, the uniformity of spectral characteristics across the entire series of glucopyranosylamines corroborates the assignments. Yet the only resolved resonances of the α anomer are those listed in

Tables I-III. Since these signals are weak and therefore conceivably due to impurities, further evidence to support their assignments was sought.

The best evidence for the assignment of the weak signals to an α-glucopyranosylamine is the observation that it is in rapid equilibrium with the dominant β anomer. Under conditions of acid catalysis, transfer of magnetization could be detected from the β H1 signal to the α H1 for nearly all of the protonated glucopyranosylamines. The extents of saturation transfer correspond to lifetimes for chemical interconversion on the order of one second.

Anomerization in basic media is too slow to lead to magnetization transfer. An indirect method was successful in establishing a relation between the β H1 and α H1 signals in aqueous base. N-Methylglucopyranosylammonium ion, whose α H1 signal at δ4.88 was identified by saturation transfer, shows 4.8% of the α anomer in D_2O/DCl. A ^1H spectrum acquired immediately after addition of enough methylamine to deprotonate the ion shows that same signal, shifted upfield but still present at the 4.8% characteristic of the anomeric equilibrium in acid. After an hour the amine equilibrium, with 8.5% α anomer, was established. This observation supports the assignment of the signal at δ4.43 in basic D_2O to α H1, since it exchanges with H1 of β-N-methylglucopyranosylamine.

According to the values in Tables I-III, the coupling constant J_{12} is 5.1±0.4 Hz for all the α anomers and 8.7±0.2 Hz for all the β anomers. The coupling constant J_{23} is 9-10 Hz not only for the β anomers but also 9.6±0.3 Hz for the α. These coupling constants then confirm that the α and β anomers are the chair conformers depicted as **3-5**. It is entirely expected that a β-glucopyranosylamine would be a chair conformer, since this permits every substituent to be equatorial. To verify that the α-glucopyranosylamines are also chair conformers, 4,6-O-benzylideneglucopyranosylamine and its N-butyl derivative (**5**) were investigated. The data in Table III show that not only the β anomers but also the α display the same coupling constants as **3-4**. Therefore we may be confident that none of the α-glucopyranosylamines distorts significantly from the chair conformation.

Relative concentrations of the two anomers were measured by integration of representative ^1H NMR signals not only under basic conditions but also under conditions acidic enough that the amino substituent was entirely protonated. The percentages are listed in Table IV. The key result is that there is appreciable α anomer for both amine and its conjugate acid. The corresponding values of $\Delta G°_{\beta \to \alpha}$, the free-energy change for conversion of β anomer to α, are listed in Table V.

Discussion. According to those values, the β anomer of the primary glucopyranosylamines (**3-5**, R = H) is favored by an average of 6.6±1.5 kJ mol^{-1} across a wide range of solvents. To gauge the steric effect, the simplest model is cyclohexylamine (**7a**), for which A_{NH_2} is 6.7±0.3 kJ mol^{-1} in D_2O and 5.0±0.3 or 6.15±0.1 kJ mol^{-1} in nonpolar solvents (average 5.6±0.6 kJ mol^{-1}). These are underestimates, since the C-O bond is shorter than a C-C bond, leading to more severe steric repulsions in a tetrahydropyran. Therefore A_{NH_2} from cyclohexanes should be corrected by using eq 2 (*12*), with A_{Me}^{cHx} = 7.3±0.3 kJ mol^{-1} and A_{Me}^{THP} = 12.0±0.8 kJ mol^{-1} (*13*). Then A_{NH_2} on a

$$A_{NH_2}^{Het} = A_{NH_2}^{cHx} (A_{Me}^{Het}/A_{Me}^{cHx}) \qquad (2)$$

pyranose ring can be estimated as 9-11 kJ mol^{-1}, depending on solvent. The observed $\Delta G°_{\beta \to \alpha}$ is close enough to this estimate that we may conclude that the preference for equatorial NH_2 is predominantly due to its steric bulk.

According to the data in Table IV, the N-alkylglucopyranosylamines (**3,5**, R = Me, Et, Bu) exhibit an average proportion of α anomer 2.4-fold that of the primary

Table IV. Equilibrium percentage of α-glucopyranosylamines (25 °C) [a]

solvent	3,R=H	3,R=Me	3,R=Et	3,R=Bu	4,R=H	5,R=H	5,R=Bu
D_2O	2.7(1)	8.5(13)	7.2	6.1(4)			
DMSO-d_6	7.0(5)	14.4(10)	21	21.4(15)	10.9(4)	11.5(10)	18.4(9)
Py-d_5	8.4	18(1)	19(1)	21(1)	10.7	10.5	23
Py/D_2O	2.9(5)[b]			8.7		3.0	7.1
$CDCl_3$					7.3(5)		11.5(10)
C_6D_6					7.4(10)		
D_2O/DCl[c]	3.2(4)	4.8(5)	3.0	3.2			
DMSO/DCl	6.5[b]	12.1(7)	7.1	6.3(4)	7.1(5)	8.1(8)	9.5(10)
DMSO/TFA					7.4(5)	6.9	7.2
$CDCl_3$/TFA					6.5		
C_6D_6/TFA					5.9(7)		

[a]Standard deviation of last digit in parentheses. [b]At 40 °C.

Table V. $\Delta G°_{\beta \to \alpha}$ (kJ mol[-1]) for anomerization of glucopyranosylamines [a]

solvent	3,R=H	3,R=Me	3,R=Et	3,R=Bu	4,R=H	5,R=H	5,R=Bu
D_2O	8.9(1)	5.9(4)	6.3	6.8(2)			
DMSO-d_6	6.4(2)	4.4(2)	3.3	3.2(2)	5.2(1)	5.1(2)	3.7(1.5)
Py-d_5	5.9	3.8(2)	3.6(2)	3.3(1.5)	5.3	5.3	2.95
Py/D_2O	9.1(5)			5.8		8.6	6.4
$CDCl_3$					6.3(2)		5.1(2)
C_6D_6					6.3(4)		
D_2O/DCl[c]	8.5(3)	7.4(3)	8.6	8.5			
DMSO/DCl	6.9	5.2(2)[b]	6.4	6.7(2)	6.4(2)	6.0(3)	5.6(3)
DMSO/TFA					6.3(2)	6.5	6.3
$CDCl_3$/TFA					6.6		
C_6D_6/TFA					6.9(3)		

[a]Standard deviation of last digit in parentheses.

glucopyranosylamines (**3,5**, R = H). The factor near two may be accounted for through consideration of the rotational conformers about the C-N bond. According to observed $^3J_{NH-H1}$s of 4.8 Hz and 8.5 Hz, respectively, in α- and β- tetra-*O*-acetyl-glucopyranosylamine (**4**), there are two α conformers available to the molecule when R = alkyl but only one when R = H or for the β conformers. If those two α conformers are equally populated, the resulting entropy of mixing for the *N*-alkyl-α-glucopyrano-sylamines is then *R*ln2. If the free-energy changes for the *N*-alkylglucopyranosyl-amines are corrected by this entropy term, the β anomer is favored by an average of

6.3 ± 1.3 kJ mol^{-1}. This is again close to A_{NH_2} in cyclohexanes. Therefore the preference for equatorial NH_2 or NHR in glucopyranosylamines is largely due to steric bulk. Actually, with the correction for the larger steric effects in a tetrahydropyran, there is a small additional preference for axial NH_2 or NHR that has been noted before and attributed to a normal anomeric effect (*12*).

The real question is the anomeric equilibrium of the glucopyranosylammonium ions. The data in Table IV show that N-protonation does not greatly reduce the proportion of α anomer of a glucopyranosylamine. Even in acid there is a substantial proportion, up to 12%. Were a reverse anomeric effect operative, there would have been very little.

According to the values in Table V, the β anomer of the N-protonated glucopyranosylamines is preferred by an average of 8.2 ± 0.5 kJ mol^{-1} in D_2O and 6.3 ± 0.5 kJ mol^{-1} in other solvents. This preference is not much different from that of the neutral glucopyranosylamines. It is also close to $A_{NH_3^+}$ in cyclohexanes (*7a*). As in the glucopyranosylamines the anomeric equilibrium of their conjugate acids can be accounted for almost entirely by steric effects. However, with the correction (*12*) for the shorter C-O bond of a tetrahydropyran $A_{NH_3^+}$ becomes 12 kJ mol^{-1} in aqueous solution and ca. 10 kJ mol^{-1} in nonaqueous. Therefore the preference of the NH_3^+ or NH_2R^+ group for the equatorial position is *less* than can be expected on the basis of steric bulk. According to eq 1, E_{An}, representing the extra preference for the axial position, is a small but significant 4 kJ mol^{-1}. This represents an anomeric effect, albeit weak, but not a reverse anomeric effect!

Table VI lists $\Delta\Delta G°_{N\rightarrow N^+} = \Delta G°_{\beta\rightarrow\alpha}(NH^+) - \Delta G°_{\beta\rightarrow\alpha}(N)$, evaluated from the data in Tables IV and V, as well as the values at 25 °C when N-alkylglucopyranosylamines are corrected by $RT\ln2$. These are measures of the extent to which N-protonation increases the preference of an amino substituent for the equatorial position. Even the uncorrected values are quite small, averaging 1.4 ± 1.1 kJ mol^{-1}. For comparison, the A values (*7a*) for NH_2 and NH_3^+ on a tetrahydropyran (*12*) predict a

Table VI. Effect of N-Protonation on Anomeric Ratio, $\Delta\Delta G°_{N\rightarrow N^+}$, kJ mol^{-1}

amine	R	solvent	$\Delta\Delta G°_{N\rightarrow N^+}$	$\Delta\Delta G°_{N\rightarrow N^+}$[a]
3	H	D_2O	-0.4 ± 0.3	-0.4 ± 0.3
3	H	DMSO-d_6	0.5	0.5
3	Me	D_2O	1.5 ± 0.5	-0.2 ± 0.5
3	Me	DMSO-d_6	0.8 ± 0.3	-0.9 ± 0.3
3	Et	D_2O	2.3	0.6
3	Et	DMSO-d_6	3.1	1.4
3	Bu	D_2O	1.7	0.0
3	Bu	DMSO-d_6	3.5 ± 0.3	1.8 ± 0.3
4	H	DMSO-d_6	1.2 ± 0.2	1.2 ± 0.2
4	H	CDCl$_3$	0.3	0.3
4	H	C$_6$D$_6$	0.6 ± 0.5	0.6 ± 0.5
5	H	DMSO-d_6	1.1 ± 0.4	1.1 ± 0.4
5	Bu	DMSO-d_6	1.9 ± 0.3	0.2 ± 0.3

[a]Corrected by $RT\ln2$ for entropy contribution to N-alkylglucopyranosylamines.

$\Delta\Delta G°_{N \to N^+}$ of 2.0±0.5 kJ mol^{-1}. This means that the shift in the position of the anomeric equilibrium upon N-protonation can be accounted for simply by the increased steric demands of a protonated amino group. There is no need to invoke a reverse anomeric effect.

Alternatively, if the entropy correction is applied, the average $\Delta\Delta G°_{N \to N^+}$ is only 0.5±0.7 kJ mol^{-1}, which is not significantly different from zero. N-Protonation leads to *no* increase in the proportion of equatorial isomer, despite the increased size of the protonated amino group. This is opposite to te reverse anomeric effect, which therefore is not operative. The observed $\Delta\Delta G°_{N \to N^+}$ is so small that this conclusion holds independently of any uncertainty about A values.

The crucial result is that the average $\Delta\Delta G°_{N \to N^+}$, corrected for entropy, is less than that expected from the increase in steric bulk. Although this increase is uncertain because A values are uncertain, NH_3^+ is certainly bulkier than NH_2. Yet the proportion of axial isomer does not decrease on N-protonation. This represents a slight tendency for cationic nitrogen to be axial, not equatorial. Rather than a reverse anomeric effect there appears to be a small normal anomeric effect!

Since these findings repudiate the reverse anomeric effect, a closer look at the previous evidence is warranted. In many cases the preference for the equatorial conformer could be due simply to the steric bulk of a heterocyclic substituent. In the study of xylopyranosylimidazole (**5**) the conformer ratio was not based on direct observation of the separate conformers but on small changes in coupling constants, determined from NMR spectra where near coincidences of chemical shifts led to second-order behavior. Yet coupling constants are sensitive to substituent and to small conformational deviations, and a cationic nitrogen is quite different from an uncharged one. For example, the data in Table III show small changes in J_{12} and J_{23} upon N-protonation of a single conformer of the 4,6-O-benzylidene derivatives (**5**) even though no conformational change can occur, and there are larger variations of J_{12} and J_{23} with solvent for the glucopyranosyl derivatives (**3**) in Table I. We therefore conclude that reliable equilibrium constants cannot be obtained from such coupling constants.

How general is the conclusion that there is no reverse anomeric effect? Our results for N-protonated glucopyranosylamines cannot necessarily be extrapolated to the glycosides with quaternary ammonium substituents, which are the original examples of the effect. The major difference between the two kinds of substituents is the possibility of hydrogen bonding with the N-protonated ones. However, there is no strong solvent dependence, and there is no difference associated with the tetra-O-acetyl derivative (**4**), which would have different hydrogen-bonding properties. Therefore the anomeric equilibrium is not being affected by intramolecular hydrogen bonding between the NH_2 or NH_3^+ substituent and the OH or OAc substituent on C2.

Anomeric Effect and Its Origin

Introduction. We next turn to a more general question, namely, the origin of the anomeric effect itself. Two explanations have been proposed, electrostatic interactions and molecular-orbital interactions. The electrostatic theory (*14*) invokes the destabilizing interaction between μ_{CX}, the dipole moment of the C-X bond of **1**, and μ_O, the dipole moment that is the resultant of individual dipole moments from C-Y bonds and the lone pairs of Y (Y = O). Such a dipolar interaction is minimized when X is axial, so that conformer **1A** is preferred over conformer **1E**. The alternative molecular-orbital explanation (*15*) for the anomeric effect considers the n-σ^* overlap between a filled nonbonding electron pair n on Y and the vacant σ^* orbital of the C-X bond. This is a stabilizing interaction that is more effective when X is axial, and so conformer **1A** is favored. In valence-bond terms this corresponds to a contribution of the double-bond/no-bond resonance form **1A'**. The stabilization is much weaker in conformer **1E**, where the high-energy p atomic orbital on Y is orthogonal to the C-X bond. We next consider the evidence for each of these possible explanations.

The n-σ^* interaction or the involvement of **1A'** is expected to produce certain structural changes. Indeed, crystallographic data (*15b,16*) show that species with Y-C-X fragments as in **1A** have shorter C1-Y bonds and longer C1-X bonds than do the corresponding **1E**. Not only do these observations support the molecular-orbital interpretation, but it is claimed (*1ade*) that they cannot be rationalized by the electrostatic explanation. Moreover, various calculations (*15b,16a,17*) have provided support for the molecular-orbital explanation.

Experimental evidence for an electrostatic origin of the anomeric effect is its decrease as the solvent polarity increases (*18*). If charge-separated resonance form **1A'** were responsible for the anomeric effect, it ought to become more important in more polar solvents. Although the solvent dependence is reproduced by molecular-orbital calculations (*19*), these necessarily include electrostatic interactions, and there exist calculations that support the electrostatic interpretation (*20*).

Proposal. To distinguish which of the two interactions is primarily responsible for the anomeric effect, a comparison system is needed in which one of them is definitely stronger and the other definitely weaker. Such a comparison is of an N-C-O system with an analogous O-C-O system. Since the dipole moment of CH_3NH_2 is 1.31 D, smaller than the 1.70 D of CH_3OH (*21a*), dipole-dipole repulsions should decrease on substituting N for O. Since the ionization potential of CH_3NH_2 is 8.97 eV, substantially lower than the 10.85 eV of CH_3OH (*21b*), substitution of N for O should raise the energy of the filled n orbital and strengthen n-σ^* interactions. Alternatively, resonance form **1A'** contributes more for Y = N than for Y = O. Thus if electrostatic interactions are more important, substitution of N for O would reduce the anomeric effect, whereas if n-σ^* interaction is more important, substitution of N for O would strengthen the anomeric effect.

Data on N-C-O systems are extremely sparse (*22*), owing to their high reactivity, and there are only two that bear on the origin of the anomeric effect. The axial conformer of aqueous 5-amino-5-deoxy-D-glucopyranose (nojirimycin) (*23*) is stabilized by 2.9 kJ mol^{-1} relative to glucopyranose. The increased anomeric effect is consistent with n-σ^* interactions as origin. Yet the conformer of 4,8,9,10-tetramethyl-1,5-dioxa-4,8-diazadecalin (*24*) with both oxygens axial is favored over the one with both equatorial by only 3.8 kJ mol^{-1}, and part of this is due to the fact that an ether group is less bulky than a tertiary amine. Here the anomeric effect must be < 1.9 kJ mol^{-1} per N-C-O. So small an anomeric effect is consistent with an electrostatic origin. Therefore we have prepared 2-methoxy-1,3-dimethylhexahydropyrimidine (**6**) and determined its conformational equilibrium. The N-methyls are essential to prevent elimination to an amidine. There are several conformers (**6Eee, 6Eae, 6Aae, 6Aee**), permitting equilibration of axial and equatorial methoxy groups by successive ring inversions and nitrogen inversions. The comparison is 2-methoxy-1,3-dioxane (**7**), studied by Nader and Eliel (*25*). The n-σ^* interactions are stronger in **6** than in **7**, whereas dipole-dipole interactions are weaker. Therefore a larger proportion of axial conformer for **6** than for **7** would suggest that orbital interactions are the primary origin of the anomeric effect, whereas a smaller proportion would suggest that electrostatic interactions are dominant.

A patented method (*26*) was adapted for the preparation of 2-methoxy-1,3-dimethylhexahydropyrimidine (**6**). The material is quite unstable. In most solvents, especially polar ones, it reacts or decomposes instantly. This is a severe limitation on the studies that can be performed.

Results. As the temperature is lowered, each of the ^{13}C NMR signals of **6** broadens and decoalesces, with a $\Delta G^{\ddagger}$ of 36±1 kJ mol^{-1}. Figure 2 shows the spectrum at -98 °C. There are pairs of signals in unequal proportion, with the minor one downfield, except for the OCH$_3$ signals. Since compression by an axial substituent usually shifts all ^{13}C signals upfield, the major conformer is assigned as **6A**.

Assignment of the two conformers may also be made on the basis of $^1J_{CH}$ for C2, which should be larger for **6A** than for **6E**, owing to weakening of the axial C-H bond by n-σ* delocalization (27). The $^1J_{C2Hs}$ of the major, upfield signal and of the minor, downfield signal are 164 Hz and 159 Hz, respectively. This evidence is the strongest for assigning the major conformer as **6A**.

The room-temperature ^{1}H NMR spectrum of **6** is so simple that gross assignments, without stereochemistry, could be made on the basis of the chemical shift, intensity, and multiplicity of each of the signals, as well as a ^{13}C-^{1}H 2D correlation spectrum.

The stereochemistry is more problematic. Each CH$_2$ has inequivalent protons, since one is cis to OCH$_3$ and the other is trans. Sequential selective decoupling of each of these four signals simplified the remaining three and provided all the coupling constants, especially a large 9.8-Hz coupling between the downfield signals of each pair. This anti coupling is the basis for assigning the downfield H4,6 and H5 as axial. Then from the above assignment of methoxy as predominantly axial, it follows that the axial H4 is cis, etc.

As the temperature is lowered, each signal broadens and begins to decoalesce into signals of unequal proportion. The $\Delta G^{\ddagger}$ of ca. 38 kJ mol^{-1} agrees with the value from ^{13}C NMR. Correlation of exchanging sites could be confirmed by saturation transfer. The assignment of the H2 protons is consistent with the general observation that axial protons are upfield of equatorial and also identifies the major conformer as **6A**.

The percentage of the minor conformer **6E**, as evaluated from each of the reporter nuclei, is listed in Table VII. Also included are the corresponding $\Delta G^{\circ}_{A \to E}$. There is good agreement among all reporter nuclei, and values averaged over all reporter nuclei are included in Table VII.

Discussion. Comparison of **6** with **7** provides an assessment of the relative importance of the factors responsible for the anomeric effect. We find that the axial conformer of **6** is favored by 2.1 kJ mol^{-1} in toluene and by only 1.5 kJ mol^{-1} in

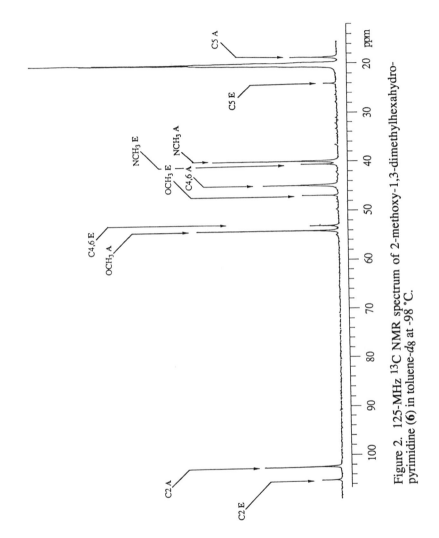

Figure 2. 125-MHz ^{13}C NMR spectrum of 2-methoxy-1,3-dimethylhexahydro-pyrimidine (6) in toluene-d_8 at -98 °C.

Table VII. Conformational equilibrium[a] and $\Delta G°_{A\to E}$ (kJ mol^{-1})[a] for **6**

signal integrated	%E,C_7D_8	$\Delta G°$	%E,CD_2Cl_2	$\Delta G°$
A-^{1}H2/E-OC^{1}H$_3$	20(1)	2.01(5)	24.4[b]	1.7
N^{1}CH$_3$+^{1}H46$_{trans}$	-	-	25(1)	1.6(1)
^{1}H5$_{trans}$	-	-	27.8(5)	1.40(4)
^{13}C2	18(1)	2.2(1)	26(1)	1.5(1)
O^{13}CH$_3$	19(3.5)	2.1(2)	25(2)	1.6(2)
N^{13}CH$_3$	20(2)	2.0(1)	24	1.7(2)
^{13}C4,6	18(2)	2.2(1)	27(2)	1.4(1)
^{13}C5	18(4)	2.2(2.5)	27(3)	1.4(2)
average	19(1)	2.1(1)	26(1)	1.5(1)

[a]Standard deviation of last digit in parentheses. [b]H2+OCH$_3$.

methylene chloride. In comparison the axial conformer of **7** is favored by 1.7 kJ mol^{-1} (*25*). Thus substitution of N for O has *not* resulted in a substantial increase in the proportion of axial conformer.

In view of the popularity of the molecular-orbital interpretation, this result is surprising. The stronger n-σ^* interactions with nitrogen lone pairs ought to have led to more axial conformer for **6**. This suggests that n-σ^* interactions are not dominant for the anomeric effect. This conclusion relies on the assignment of axial and equatorial methoxyls, based on the low-temperature chemical shifts of C2 and H2 and on the $^1J_{CHS}$. However, if it is reversed, the dominant conformer of **6** would have OCH$_3$ equatorial, so that the conclusion would become even stronger.

According to the data in Table VII, $\Delta G°_{A\to E}$ is nearly the same for **6** as the 1.7 kJ mol^{-1} for **7**. However, comparison of anomeric effects requires correction for steric effects according to eq 1 and 2. From A_{Me}^{cHx} (*13b*), $A_{OMe}^{cHx} = 3.1\pm0.2$ kJ mol^{-1} (*28*), and $A_{Me}^{Diox} = 16.6\pm0.4$ kJ mol^{-1} (*25*), A_{OMe}^{Diox} can be estimated as 7.0 ± 0.6 kJ mol^{-1}. For **6** we take A_{Me}^{HPym} as 4.3 kJ mol^{-1} from the equilibrium in 1,2,3-trimethyl-hexahydropyrimidine (*29*). Then eq 2 leads to $A_{OMe}^{HPym} = 1.83$ kJ mol^{-1}. Alternatively, AM1 calculations suggest that interatomic distances in a hexahydro-pyrimidine are quite similar to those in a cyclohexane and considerably longer than those in a 1,3-dioxane. Then we can ignore eq 2 and simply model the steric effects in a hexahydropyrimidine by those in a cyclohexane, or $A_{OMe}^{HPym} = 3.1$ kJ mol^{-1}. To be conservative we take the average, 2.5 ± 0.6 kJ mol^{-1}.

Then E_{An} in **7** is 8.7 ± 0.6 kJ mol^{-1}, whereas for **6** it is 4.3 ± 0.7 kJ mol^{-1}, averaged over the two solvents. Thus the anomeric effect is halved by substituting N for O. It did not increase, as required if it were due to n-σ^* interactions. Therefore we conclude that the anomeric effect in these systems instead arises primarily from electrostatic interactions.

The data in Table VII show a small but significant variation of the proportion of conformer **6E** and $\Delta G°_{A\to E}$ with solvent. The proportion of axial conformer decreases in the more polar methylene chloride, as is consistent with the electrostatic origin of the anomeric effect. If orbital overlap had become more important, we would have observed an increase in the anomeric effect. It must be recognized that these studies are limited to nonpolar solvents and that in polar solvents, where electrostatic effects are reduced, the anomeric effect due to orbital overlap can be larger for the nitrogen case than for the oxygen. This seems to be the case for nojirimycin (23), where the anomeric effect is large. Therefore the conclusion that the anomeric effect in 6 and 7 is due primarily to electrostatics cannot be universal.

Geometric Changes. It is still necessary to account for the structural changes that have been considered the best evidence for the molecular-orbital interpretation and that the electrostatic explanation is claimed (1ade) to be incapable of rationalizing. As a model system we take 1 (Y = O), with two dipoles, one associated with the C-X bond and the other associated with the oxygen lone pairs and the C-O bonds. The electrostatic energy is given by eq 3, where μ_{CX} and μ_O are these dipole moments,

$$E_{dd} = \frac{\mu_{CX}\mu_O[\cos\phi_{12}+3\cos\theta_1\cos\theta_2]}{r_{CO}^3} \qquad (3)$$

assumed situated at carbon and at oxygen, ϕ_{12} is the angle between them, and θ_1 or θ_2 is the angle between a dipole moment and the vector to the other atom, located a distance r_{CO} away. For an axial conformer **1A** with exact tetrahedral bond angles and exactly staggered bonds the angular factor is zero. In contrast, the angular factor makes this dipole-dipole energy positive for the equatorial conformer **1E**, accounting for its destabilization. Eq 4 gives he derivative of this energy with respect to the C1-O bond

$$\frac{\partial E_{dd}}{\partial r_{CO}} = -3\frac{\mu_{CX}\mu_O[\cos\phi_{12}+3\cos\theta_1\cos\theta_2]}{r_{CO}^4} \qquad (4)$$

length. This derivative is negative for **1E**, so that its energy is lowered by lengthening the C1-O bond. Eq 5 gives the derivative of the dipole-dipole energy with respect to the

$$\frac{\partial E_{dd}}{\partial r_{CX}} = \frac{\mu_O[\cos\phi_{12}+3\cos\theta_1\cos\theta_2]}{r_{CO}^3}\frac{\partial\mu_{CX}}{\partial r_{CX}} \qquad (5)$$

C1-X bond length. In CH_3F $\partial\mu/\partial r_{CF}$ is positive, both experimentally and theoretically (30), and this should be general for all C-X. Then the derivative in eq 5 is positive for **1E**, so that its energy is raised by lengthening the C1-X bond.

Thus according to simple electrostatics, the C1-O bond of the equatorial conformer **1E** must be lengthened and its C1-X bond shortened. This means that relative to **1E**, the C1-O bond of the axial conformer **1A** is shorter and its C1-X bond longer. These are exactly the observed structural changes attributed to n-σ* interactions, but they are equally well accounted for by dipole-dipole interactions.

Stereoelectronic Control

Introduction. We turn next to the kinetic counterpart of the anomeric effect, namely, the question of whether antiperiplanar lone pairs facilitate chemical reaction. Deslongchamps' hypothesis of stereoelectronic control (31), sometimes known as the "antiperiplanar lone-pair hypothesis" (32), states that preferential cleavage of a tetrahedral intermediate occurs when there are two lone pairs antiperiplanar to a leaving group. The hypothesis has been widely accepted, with limited but spirited opposition (1b) and occasional counterexamples (33). The above conclusion that n-σ* interactions are less important for the anomeric effect may not be applicable to a transition state, where the bond to the leaving group is being broken, thereby lowering the energy of the σ* orbital and making the interaction stronger.

A prime piece of evidence for stereoelectronic control was the observation (*34*) that a cyclic hemiorthoester cleaves the endocyclic bond to give exclusively the hydroxy ester rather than the lactone. Nevertheless this interpretation has been challenged (*35*). In five-membered rings conformational changes known to be rapid (*36*) ought to have led to the lactone, which is not observed. Therefore an alternative explanation based on the well-known (*37*) destabilization of lactones was proposed (*35*).

A more reliable test of stereoelectronic control is the hydrolysis of a cyclic amidine to a lactam or an aminoamide. There is no destabilization to create a bias against either product. The mechanism is shown in Scheme 1. If the hypothesis of stereoelectronic control is correct, then it is a corollary of the principle of microscopic reversibility that OH^- should attack the amidinium ion antiperiplanar to two nitrogen lone pairs, to produce **A** as initial intermediate. After (rapid) rotation about the exocyclic C-N bond, this intermediate has two lone pairs antiperiplanar to the endocyclic C-N bond but only one lone pair antiperiplanar to the exocyclic C-N bond. According to the hypothesis, this geometry favors cleavage of the endocyclic bond and formation of the ring-opened product, the aminoamide (**B**). Cleavage of the exocyclic bond and formation of the lactam (**C**) could utilize the antiperiplanar lone pair on the oxygen but would require the syn lone pair on the ring nitrogen. In contrast to hemiorthoesters, ring inversion, leading to conformer **D**, does not create a second lone pair antiperiplanar to the exocyclic C-N bond, so this too cannot cleave to lactam. The further requirement is nitrogen inversion, leading to conformer **E**. Even though this conformer could cleave to the lactam, it is inaccessible during the lifetime of the intermediate because nitrogen inversion is slow compared to the rate of cleavage. Thus if stereoelectronic control is operative, the aminoamide **B** is predicted to be the kinetic product.

Scheme 1. Stereoelectronic Control in Hydrolysis of *n*-Membered-Ring Amidines. Two lone pairs antiperiplanar to a leaving group are shown as open lobes. Reactions where there is only one antiperiplanar lone pair, shown as a filled lobe, are designated with a dashed arrow.

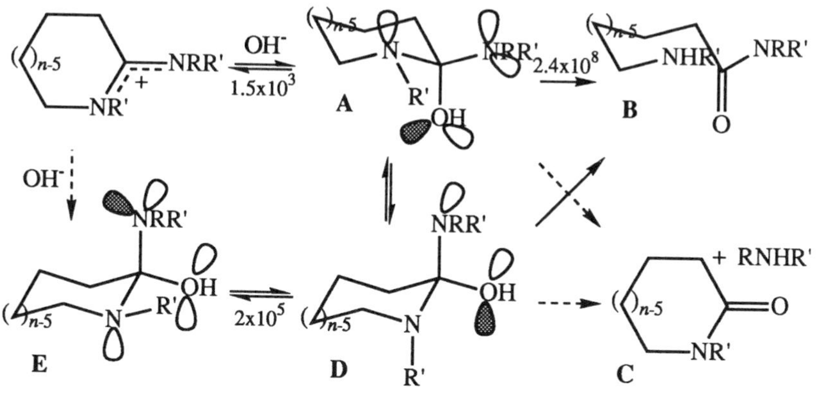

Amidine hydrolysis has a further advantage. In reactions that create or destroy acetals the faster reaction syn to a lone pair on oxygen has been rationalized by assuming reaction via a boat conformer (*38*). Then one of the oxygen's two lone pairs becomes antiperiplanar to the attacking nucleophile or the leaving group. However, in the hemiorthoamide intermediate in amidine hydrolysis the ring nitrogen has only one lone pair, which does not become antiperiplanar even in the boat form. Therefore this rationale is not available.

Initial experiments with R = H = R' showed that the aminoamide is indeed the sole product in the hydrolysis of amidines (*35*). Unfortunately, this result too is ambiguous, owing to a mismatch of leaving abilities (*39*). This can be overcome by methylation at the exocyclic nitrogen (Scheme 1, R' = H, R = CH₃). Again, if stereoelectronic control is operative, the product is expected to be the aminoamide **B**. Yet experimentally (*40*) substantial amounts (>50%) of five- and seven-membered-ring lactams **C** (*n* = 5,7) are observed, implying that stereoelectronic control is not operative in these systems. It is operative but weak in six-membered rings, where 93% of aminoamide **B** (*n* = 6) is produced. Deslongchamps has ignored this evidence counter to his hypothesis.

The evidence depends crucially on the assertion that nitrogen inversion is slow. The rate of nitrogen inversion was taken from tertiary amines, rather than secondary amines as in intermediate **D** (R' = H, R = CH₃). There are additional proton-transfer mechanisms for nitrogen inversion in secondary amines, both stepwise and concerted. The stepwise mechanisms are shown in Scheme 2, but according to pKₐs these are too

Scheme 2. Mechanisms for Nitrogen Inversion via Stepwise Proton Exchange.

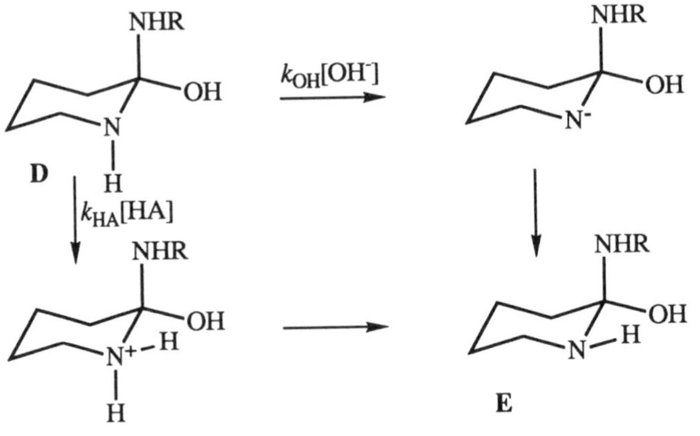

slow. The concerted mechanisms involve simultaneous protonation and deprotonation, catalyzed by OH⁻ (A in Scheme 3) or by *n* water molecules (B in Scheme 3, with *n* arbitrarily equal to 2). Indeed, Deslongchamps (*41*) has assumed (by mistaken analogy to hydroxyl proton exchange) that an NH is equivalent to a nitrogen lone pair, and he

Scheme 3. Mechanisms for Nitrogen Inversion via Concerted Proton Exchange.

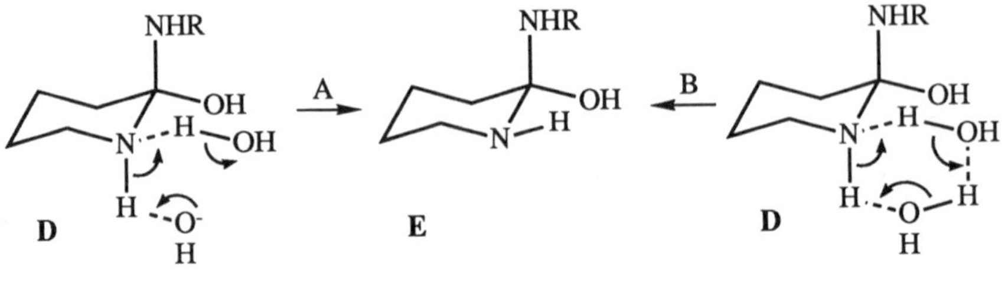

has also postulated the OH⁻-catalyzed mechanism. Then one possible explanation for lactam formation is that nitrogen inversion is catalyzed by proton exchange at the endocyclic nitrogen, leading to conformer **E** of Scheme 1 ($R' = H$), which can cleave to lactam. Did previous results that seem to contradict the hypothesis of stereoelectronic control overlook these mechanisms for nitrogen inversion?

Proposal. Proton-transfer mechanisms for nitrogen inversion can be blocked by substituting a methyl group for the hydrogen of the endocyclic nitrogen. To maintain balanced leaving abilities, it is also necessary to add a second methyl to the exocyclic nitrogen. Therefore we have determined the product distribution from the base-catalyzed hydrolyses of 1-methyl-2-(dimethylamino)pyrrolinium (**8a**), 1-methyl-2-(dimethylamino)-3,4,5,6-tetrahydropyridinium (**8b**), and 1-methyl-7-(dimethylamino)-3,4,5,6-tetrahydro-2H-azepinium (**8c**) iodides in D_2O. In the hemiorthoamide

8a, *n*=5
8b, *n*=6
8c, *n*=7

9

intermediates derived from these amidinium ions uncatalyzed nitrogen inversion is slow, nitrogen inversion by proton transfer is blocked, leaving abilities are matched, and there is no enthalpic preference for amide over lactam. Thus predominant formation of aminoamide would support the hypothesis of stereoelectronic control, whereas formation of any substantial amount of lactam would represent a counterexample to this hypothesis.

As a further control to eliminate the possible influence of steric repulsions, we have studied the product distribution from hydrolysis of 1-methyl-2-(methylimino)-pyrrolidine (**9**), even though this does not have balanced leaving abilities. In this case both stereoelectronic control and leaving abilities ought to favor aminoamide formation, so lactam formation would be even more devastating to the hypothesis.

Results. The products from hydrolysis of the cyclic amidinium ions are the aminoamide **10** and the lactam **11** plus dimethylamine **12** ($R = CH_3$) or methylamine **12** ($R = H$). Product ratios were determined from integration of three characteristic peaks and are tabulated in Table VIII.

8, R=CH₃
9·H⁺, R=H

10

11   **12**

It is clear that substantial amounts of lactam are produced. This agrees with hydrolyses of four-membered-ring amidinium ions (*42*), despite the relief of ring strain on forming the aminoamide. Lactam formation from the hydrolysis of **8b** ($n = 6$) is significantly less than from **8a** ($n = 5$) or **8c** ($n = 7$). This agrees with earlier observations (*40*) on hydrolyses of cyclic amidines.

Table VIII. % Lactam from Hydrolyses of Cyclic Amidinium Ions

integrate	8a[a]	8b[a]	8b[b]	8c[a]	8c[b]	9[a]	9[b]
$CONCH_3$	78	44	72	81	92	26	35
NCH_2	77	38	70	-	-	26	34
β-(CH_2)	81	38	69	83	-	26	34

[a]In excess NaOD (pD>13). [b]In phosphate buffer (pD<12.4).

Discussion. One test of any theory of reactivity is its ability to predict product distributions. In the hydrolyses of these cyclic amidinium ions there is substantial lactam formation (25-90%). Yet a strong prediction of the hypothesis of stereo-electronic control is that lactam ought not to be produced. The following mechanism has been established (43) for the hydrolysis of amidinium ions under strongly basic conditions: initial attack by OH⁻, deprotonation from oxygen, and finally C-N cleavage. Scheme 1 shows the stereochemical aspects of this mechanism for cyclic amidinium ions, but with proton-transfer steps omitted, such that the intermediate that cleaves is actually the conjugate base of what is shown. Moreover, since a nitrogen anion is a terrible leaving group, the nitrogen must be protonated. Therefore basicity is important to leaving ability, and the more basic amine is the one that is cleaved more rapidly (44). In hydrolysis of **8a-c** both leaving groups are secondary amines presumably of the same basicity. Consequently neither product is favored through an imbalance of leaving abilities.

The key to the prediction of stereoelectronic control is that product arises only from conformers **A** and **D**, and that conformer **E** is not accessible during the lifetime of the intermediate. Conformers **A** and **D** permit cleavage only to aminoamide **B**, whereas only **E** has two lone pairs antiperiplanar to the exocyclic nitrogen, to permit cleavage to lactam **C**. There are two possibilities for forming **E**: (1) directly from the original amidinium ion and (2) from **A** by ring inversion plus nitrogen inversion.

Direct formation of **E** from amidinium ion would require OH⁻ attack syn to the lone pair on the ring nitrogen. If instead initial attack must occur antiperiplanar to the lone pairs on both nitrogens, only **A** can be formed as initial intermediate. The alternative possibility for formation of conformer **E** is that the intermediate undergoes both ring inversion and nitrogen inversion (in either order, although Scheme 1 shows only one). Regardless of the rate of ring inversion, the rate of nitrogen inversion in aqueous solution is especially low, about 3×10^5 s⁻¹ (45). Since this is considerably slower than cleavage, which proceeds at $>10^8$ s⁻¹ (40,42), conformer **E** was considered to be inaccessible during the lifetime of the intermediate. Yet there are other mechanisms that might permit faster nitrogen inversion and render **E** accessible.

The purpose of this study was to investigate whether a proton-transfer mechanism for nitrogen inversion might account for products from previous amidine hydrolyses. In the present experiments proton transfer has been blocked by methylation at nitrogen and yet the same products are obtained. Rapid nitrogen inversion (along with ring inversion) in the intermediate therefore cannot be the explanation for lactam formation.

With tertiary amides there is a potential complication from steric effects. Might lactam be formed from amidinium ions **8** simply because exocyclic C-N cleavage relieves steric congestion? Although cleavage of either C-N bond produces a tertiary amide, endocyclic cleavage leads to an N,N-dimethyl amide where the E methyl suffers repulsion. Therefore the product stabilities of amide and lactam are not necessarily balanced.

To eliminate this steric effect, the cyclic N,N'-dimethylamidine **9** was also investigated. The results of its hydrolysis in base are included in Table VIII. The reactive species is the amidinium ion, which takes the EZ configuration (46). The initial conformer resulting from OH⁻ attack is **13**. Rapid rotation about the exocyclic C-N bond then produces **14**, which now has two lone pairs antiperiplanar to the endocyclic

C-N bond. Cleavage of this bond produces the more stable Z amide, depicted in **10** (R = H). This preferential cleavage of the amine whose alkyl group is initially Z has been detected previously in hydrolysis of acyclic amidines (*39*). Yet the competition is between a primary and a secondary amine, which is a better leaving group. Therefore cleavage of the endocyclic C-N bond is favored by both stereoelectronic control and by the imbalance of leaving abilities. Indeed, there is more aminoamide from **9** than from **8a**. Nevertheless there is substantial cleavage of the exocyclic C-N bond, leading to 26-35% lactam, which is not appreciably different from the 3- to 4-fold preference expected (*39,47*) from leaving abilities. Therefore this lactam formation is an even stronger counterexample to stereoelectronic control.

Stereoelectronic control is not responsible for directing the C-N cleavage in either this amidine or the fully *N*-methylated amidinium ions. The formation of lactam demonstrates that the syn lone pair on the ring nitrogen can facilitate cleavage. Although anti elimination in six-membered rings is strongly preferred, syn eliminations in five- and seven-membered rings are quite competitive with anti (*48*). They are not at all unusual. Stereoelectronic control should not have been expected to be so universal as to operate even in five- and seven-membered ring hemiorthoamides.

Even in the six-membered-ring amidines stereoelectronic control is not very strong, since substantial amounts of lactam are produced. This represents a failure of the antiperiplanar lone-pair hypothesis. In reactions that create or destroy acetals such a failure can be avoided by assuming reaction via a boat conformer (*38*) which places one of the oxygen lone pairs antiperiplanar to the attacking nucleophile or to the leaving group. However, a nitrogen has only one lone pair. That lone pair in these hemiorthoamides is trans to the hydroxyl and cis to the other nitrogen, and conformational changes do not change these relationships. Only nitrogen inversion can do so, and this is too slow. Therefore the nitrogen lone pair does not become antiperiplanar to the leaving group even in the boat form, and the failure of the antiperiplanar lone-pair hypothesis cannot be avoided.

It is not certain whether stereoelectronic control is lost in the formation of the tetrahedral intermediate or in its breakdown. These results are equally consistent with OH⁻ addition syn to the two nitrogen lone pairs. However, we suggest that the less selective step is the cleavage of the deprotonated hemiorthoamide, where the push of the -O⁻ may overwhelm any stereoelectronic effect from the nitrogen lone pair. This is equivalent to assuming that this second transition state is early, resembling the intermediate and with little interaction between the bond that is cleaved and the lone pair, whose stereochemistry then does not matter.

These results have implications for the stereoelectronic effect that has been proposed (*49*) (and criticized (*50*)) to account for the enhanced reactivity of five-membered-ring phosphates. The lone pair on the ring oxygen that is antiperiplanar to the leaving group in the pentacovalent intermediate **15** is thought to accelerate cleavage. However, the high-energy *p* lone pair on that oxygen is orthogonal to the bond that is cleaved, and the hybridized lone pair that is antiperiplanar is so buried in energy that it is quite unlikely to be effective in competition with the antiperiplanar *p* lone pairs on the

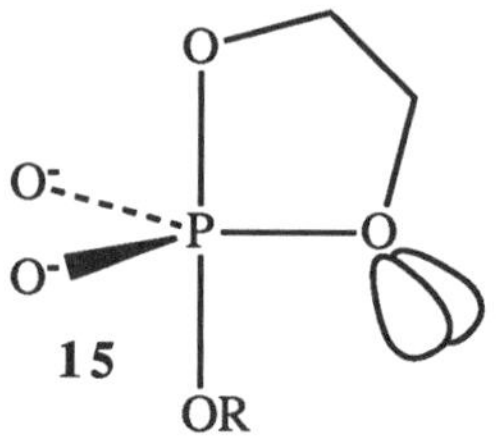

15

oxyanions. Besides, a stereoelectronic effect on the cleavage rate can explain the enhanced reactivity only if cleavage is the rate-limiting step, and it is not.

Methoxy Exchange. Another system where stereoelectronic control might be stronger is the diastereomeric 2-methoxy-cis-4,6-dimethyl-1,3-dioxanes (**16,17**), where the axial methoxy group ought to exchange faster. Indeed, Grignard reagents react only with **17** and not at all with **16** (*51*). To evaluate relative reactivities quantitatively, we have studied the acid-catalyzed exchange of **16** and **17** with methanol. Since the reactive species is the conjugate acid, the reverse anomeric effect, which is discredited above, might seem applicable. However, the only question is of the relative stabilities of the two possible transition states, with methanol either antiperiplanar or syn to lone pairs on the ring oxygens. To avoid complications from the greater stability of **17**, it is advisable to compare relative rates of capture of the intermediate **18**.

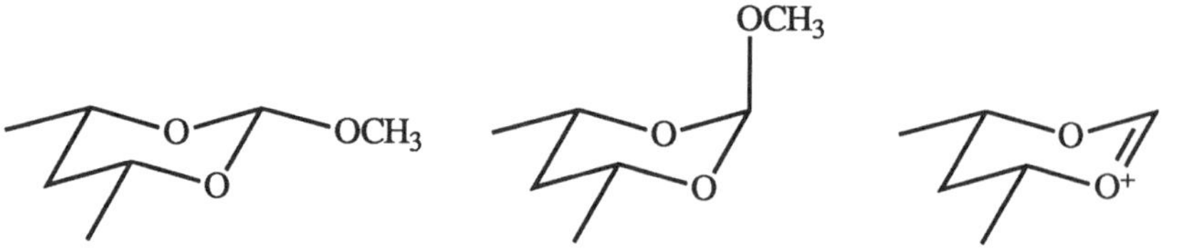

16 **17** **18**

To avoid systematic error due to consumption of catalyzing acid (*52*), the rates should be measured simultaneously in the same sample. A non-selective EXSY method (*53*) is most suitable. The pulse sequence is $90°-t_1-90°-t_m-90°-t$(acquire), as in 2D EXSY (*54*), except with a set of mixing times t_m and for N sites only N evolution times t_1, chosen so that the second $90°$ pulse will invert successive peaks. In the experiment with the jth t_1, the time course of m_{ij}, the deviation of the ith-site magnetization from its equilibrium value, is given by matrix eq 6, where M_0 is a matrix of initial values and R

$$M = e^{-t_m R} M_0 \qquad (6)$$

is a square matrix with $-R_{ij} = k_{ji}$, the rate constant for exchange from site j to site i (Methanol, Equatorial OMe, Axial OMe). The solution to this differential equation is given in eq 7, where X is the matrix that diagonalizes MM_0^{-1} to Λ. A plot of

$$\ln(MM_0^{-1}) = X(\ln\Lambda)X^{-1} = -t_m R \qquad (7)$$

$-(X(\ln\Lambda)X^{-1})_{ij}$ vs t_m is a straight line with slope k_{ji}. The weighted linear least-squares method (*55*) then compensates for the error increasing with t_m. Such a plot is shown in Figure 3.

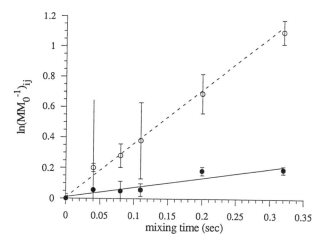

Figure 3. Plot of methoxyl exchange rates for orthoesters **16** and **17** in benzene-d_6 vs mixing time t_m, with the weighted least-squares lines, of slopes k_{ME} (—) and k_{MA} (- -).

Results and Discussion. Figure 4 shows an example of the time course of methoxyl signals following inversion of the methanol magnetization. Preliminary site-to-site rate constants k_{ji} in benzene and in a mixture of methanol and DCCl$_3$ are listed in Table IX. Direct exchange between the axial and equatorial methoxy peaks, which

Table IX. Site-to-site rate constants for exchange of **16** (E OCH$_3$) + **17** (A OCH$_3$) with CH$_3$OH (M)

j	i	k_{ji}, MeOH/CDCl$_3$	k_{ji}, benzene-d_6
M	E	0.8 ± 0.1	0.19 ± 0.05
M	A	4.3 ± 0.4	1.7 ± 0.2
E	M	5.4 ± 0.8	0.13 ± 0.06
A	M	8.8 ± 0.9	0.8 ± 0.1

could arise by a ring-opening mechanism, was not detectable. The fact that $k_{ME} \neq k_{MA}$ means that the reaction is not S$_N$2. Although the method provides both forward and reverse rate constants, the relative magnitudes of k_{ME} and k_{MA} reflect the stereoelectronic effect on the capture of the intermediate. There is indeed a preference for methanol to add antiperiplanar to lone pairs on the ring oxygens, to form **17**. The data indicate that in benzene this preference is (9±3)-fold but in the more polar medium it is only (5±1)-fold. The detectable slope for R_{EM} in Figure 3 shows that equatorial attack does occur at an appreciable rate. Even in this six-membered ring, where the antiperiplanar oxygen lone pairs might have activated strongly the axial position, the stereoelectronic effect is not large.

Conclusions

The proportions of axial anomers of several *N*-alkylglucopyranosylamines and some of their tetra-*O*-acetyl and 4,6-benzylidene derivatives were determined by ^{1}H NMR in a variety of solvents, including acidic media. These proportions are all quite small, so the assignments were confirmed by coupling constants, saturation transfer, reequilibration, and decoupling difference spectroscopy. The values for the neutral amines can be accounted for primarily on the basis of the steric bulk of NH$_2$ and NHR substituents, corrected for a small conformational entropy effect. What is more significant is that the reduction in the proportion of axial anomers that occurs upon *N*-protonation is also small. It can be accounted for largely, but not entirely, on the basis of the slightly greater steric bulk of solvated NH$_3^+$ and NH$_2$R$^+$ substituents. However, there is also a small enhancement of the normal anomeric effect, owing to the greater electronegativity of these cationic substituents. There is definitely no increased tendency, beyond that due to a greater steric bulk, for a positively charged substituents to prefer the equatorial position. Therefore we conclude that the so-called reverse anomeric effect does not exist.

Experimentally the anomeric effect in **6** is only half that in **7**. Since the stronger *n*-σ* interactions that accompany substitution of nitrogen for oxygen do not result in an increase in the anomeric effect, these cannot be the dominant interaction. We therefore conclude that the anomeric effect in nonpolar solvents arises primarily from electrostatic interactions. Moreover, the geometric changes that have long been considered as strong evidence for *n*-σ* interactions can be accounted for on the basis of dipole-dipole interactions. These results ought to stimulate a rethinking of the origins of the anomeric effect, and more attention should be paid to electrostatics.

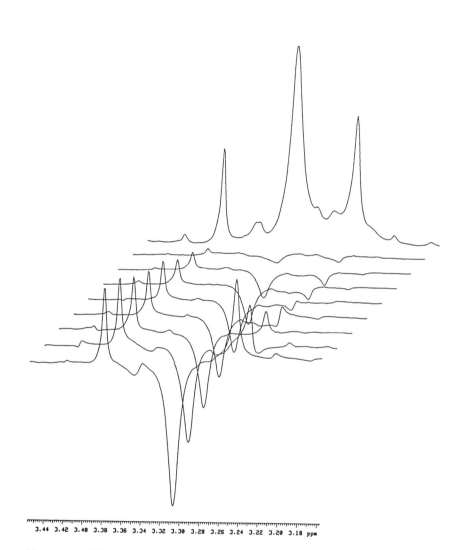

3.44 3.42 3.40 3.38 3.36 3.34 3.32 3.30 3.28 3.26 3.24 3.22 3.20 3.18 ppm

Figure 4. 500-MHz ^{1}H NMR spectra of **16** + **17** following semiselective inversion of methanol in benzene-d_6. Signals from left to right are equatorial methoxyl of **16**, methanol, and axial methoxyl of **17**.

Hydrolysis of 5-, 6-, and 7-membered cyclic N-methylated amidinium iodides in base involves substantial (25-90%) exocyclic C-N cleavage, leading to lactam. This is not the product predicted by the hypothesis of stereoelectronic control. Lactam cannot arise through a conformer with two lone pairs antiperiplanar to the exocyclic nitrogen, since that would require rapid nitrogen inversion. Nor can it arise by reaction of a boat conformer, since the lone pair of the ring nitrogen does not become antiperiplanar even in the boat form. We therefore affirm that stereoelectronic control is not operative in the cleavage of the hemiorthoamide intermediate. Moreover, there is only a 5-9-fold preference for methanol capture antiperiplanar to oxygen lone pairs in the acid-catalyzed methanolysis of 4,6-dimethyl-2-methoxy-1,3-dioxanes. This is not a large stereoelectronic effect. These results raise further question about the operation of stereoelectronic control in other hydrolyses, including phosphate esters.

Further details and discussion of all these topics will be provided in full manuscripts to be published.

Acknowledgments. I am grateful to the excellent graduate students, Kathleen B. Armstrong, John D. Thoburn, Richard E. Engler, and Miles A. Fabian, who carried out these studies. This research was supported by National Science Foundation Grants CHE87-14451 and CHE90-25113.

Literature Cited
1. (a) Kirby, A. J. *The Anomeric Effect and Related Stereoelectronic Effects at Oxygen*; Springer-Verlag: Berlin, 1983. (b) Sinnott, M. L. *Adv. Phys. Org. Chem.* **1988**, *24*, 113. (c) *The Anomeric Effect: Origin and Consequences*; Szarek, W. A.; Horton, D., Eds.; ACS Symposium Series No. 87; Washington, DC., 1979. (d) Tvaroška, I.; Bleha, T. *Adv. Carbohy. Chem.* **1989**, *47*, 45. (e) Juaristi, E.; Cuevas, G. *Tetrahedron* **1992**, *48*, 5019.
2. Sallam, M. A. E.; Whistler, R. L.; Markley, J. L. *Carbohydr. Res.* **1980**, *87*, 87. Taira, K.; Fanni, T.; Gorenstein, D. G. *J. Org. Chem.* **1984**, *49*, 4531. Bennet, A. J.; Sinnott, M. L. *J. Am. Chem. Soc.* **1986**, *108*, 7287.
3. Lemieux, R. U.; Morgan, A. R. *Can. J. Chem.* **1965**, *43*, 2205. Lemieux, R. U. *Pure Appl. Chem.* **1971**, *25*, 527.
4. (a) Finch, P.; Nagpurkar, A. G. *Carbohydr. Res.* **1976**, *49*, 275. (b) Hosie, L.; Marshall, P. J.; Sinnott, M. L. *J. Chem. Soc., Perkin Trans. II* **1984**, 1121. (c) Dauben, W. G.; Köhler, P. *Carbohydr. Res.* **1990**, *203*, 47.
5. Paulsen, H.; Györgydeák, Z.; Friedmann, M. *Chem. Ber.* **1974**, *107*, 1590.
6. Lehn, J.-M.; Wipff, G. *J. Am. Chem. Soc.* **1980**, *102*, 1347. Anders, E.; Markus, F.; Meske, H.; Tropsch, J.; Maas, G. *Chem. Ber.* **1987**, *120*, 735. Wu, Y.-D.; Houk, K. N. *J. Am. Chem. Soc.* **1991**, *113*, 2353. Andrews, C. W.; Fraser-Reid, B.; Bowen, J. P. *J. Am. Chem. Soc.* **1991**, *113*, 8293. Ratcliffe, A. J.; Fraser-Reid, B. *J. Chem. Soc., Perkin Trans. I* **1990**, 747. Mikolajczyk, M.; Graczyk, P.; Wieczorek, M. W.; Bujacz, G. *Angew. Chem., Int. Ed. Engl.* **1991**, *30*, 578. Noe, C. R.; Knollmüller, M.; Steinbauer, G.; Jangg, E.; Völlenkle, H. *Chem. Ber.* **1988**, *121*, 1231.
7. (a) Batchelor, J. G. *J. Chem. Soc., Perkin Trans. II* **1976**, 1585. (b) Booth, H.; Jozefowicz, M. L. *J. Chem. Soc., Perkin Trans. II* **1976**, 895.
8. DeBruyn, A.; Anteunis, M.; Kováč, P. *Coll. Czech. Chem. Comm.* **1977**, *42*, 3057.
9. De Bruyn, A.; Anteunis, M.; Claeyssens, M.; Saman, E. *Bull. Soc. Chim. Belg.* **1976**, *85*, 605.
10. Sanders, J. K. M.; Hunter, B. K. *Modern NMR Spectroscopy*; Oxford University Press: New York, 1987; (a) p 54. (b) p 100.
11. Horton, D.; Walaszek, A. *Carbohydr. Res.* **1982**, *105*, 145.
12. Franck, R. W. *Tetrahedron* **1983**, *39*, 3251.

13. (a) Booth, H.; Everett, J. R. *J. Chem. Soc., Chem. Commun.* **1976**, 278. (b) Eliel, E. L.; Hargrave, K. D.; Pietrusiewicz, K. M.; Manoharan, M. *J. Am. Chem. Soc.* **1982**, *104*, 3635.
14. Edward, J. T. *Chem. & Ind.* **1955**, 1102.
15. (a) Lucken, E. A. C. *J. Chem. Soc.* **1959**, 2954. (b) Romers, C.; Altona, C.; Buys, H. R.; Havinga E. *Top. Stereochem.* **1969**, *4*, 39.
16. (a) Jeffrey, G. A.; Pople, J. A.; Binkley, J. S.; Vishveshwara, S. *J. Am. Chem. Soc.* **1978**, *100*, 373. (b) Fuchs, B.; Schleifer, L.; Tartakovsky, E. *Nouv. J. Chim.* **1984**, *8*, 275. (c) Briggs, A. J.; Glenn, R.; Jones, P. G.; Kirby, A. J.; Ramaswamy, P. *J. Am. Chem. Soc.* **1984**, *106*, 6200. (d) Cossé-Barbi, A.; Dubois, J. E. *J. Am. Chem. Soc.* **1987**, *109*, 1503. Cossé-Barbi, A.; Watson, D. G.; Dubois, J. E. *Tetrahedron Lett.* **1989**, *30*, 163.
17. Wolfe, S.; Rauk, A.; Tel, L. M.; Csizmadia, I. G. *J. Chem. Soc. B* **1971**, 136. Jeffrey, G. A.; Pople, J. A.; Radom, L. *Carbohydrate Res.* **1972**, *25*, 117. David, S.; Eisenstein, O.; Hehre, W. J.; Salem, L.; Hoffmann, R. *J. Am. Chem. Soc.* **1973**, *95*, 3806. Zhdanov, Yu. A.; Minyaev, R. M.; Minkin, V. I. *J. Mol. Struct.* **1973**, *16*, 357. Wolfe, S.; Whangbo, M.; Mitchell, D. J. *Carbohydrate Res.* **1979**, *69*, 1. Radom, L.; Hehre, W. J.; Pople, J. A. *J. Am. Chem. Soc.* **1972**, *94*, 2371. Zefirov, N. S. *Tetrahedron* **1977**, *33*, 3193. Jeffrey, G. A.; Yates, J. H. *J. Am. Chem. Soc.* **1979**, *101*, 820. Ewig, C. S.; Van Wazer, J. R. *J. Am. Chem. Soc.* **1986**, *108*, 4774.
18. Eliel, E. L.; Giza, C. A. *J. Org. Chem.* **1968**, *33*, 3754. Praly, J.-P.; Lemieux, R. U. *Can. J. Chem.* **1987**, *65*, 213.
19. Tvaroška, I.; Bleha, T. *Coll. Czech. Chem. Comm.* **1980**, *45*, 1883. Montagnani, R.; Tomasi, J. *Int. J. Quantum Chem.* **1991**, *39*, 851. Kysel, O.; Mach, P. *J. Mol. Struct.* **1991**, *227* (THEOCHEM 73) 285.
20. Anderson, C. B.; Sepp, D. T. *Tetrahedron* **1967**, *24*, 1707. Tvaroška, I.; Kožár, T. *J. Am. Chem. Soc.* **1980**, *102*, 6929. Pichon-Pesme, V.; Hansen, N. K. *J. Mol. Struct.* **1989**, *183* (THEOCHEM 52) 151.
21. Gordon, A. J.; Ford, R. A. *The Chemist's Companion*; Wiley-Interscience: New York, 1972; (a) p 237. (b) p 4.
22. Senderowitz, H.; Aped, P.; Fuchs, B. *Helv. Chim. Acta* **1990**, *73*, 2113.
23. Pinto, B. M.; Wolfe, S. *Tetrahedron Lett.* **1982**, *23*, 3687.
24. Alcaide, B.; Jimenez-Barbero, J.; Plumet, J.; Rodriguez-Campos, I. M. *Tetrahedron* **1992**, *48*, 2715.
25. Nader, F. W.; Eliel, E. L. *J. Am. Chem. Soc.* **1970**, *92*, 3050.
26. Winberg, H. E., E. I. du Pont de Nemours & Co. 1966, US Patent #3239518; *Chem. Abs.* *64*: P15898c.
27. Perlin, A. S.; Casu, B. *Tetrahedron Lett.* **1969**, 2921. Jennings, W. B.; Boyd, D. R.; Watson, C. G.; Becker, E. D.; Bradley, R. B.; Jerina, D. M. *J. Am. Chem. Soc.* **1972**, *94*, 8501. Bock, K.; Pedersen, C. *J. Chem. Soc., Perkin Trans. II* **1974**, 293. Fraser, R. R.; Bresse, M. *Can. J. Chem.* **1983**, *61*, 576.
28. Schneider, H.-J.; Hoppen, V. *Tetrahedron Lett.* **1974**, 579.
29. Katritzky, A. R.; Baker, V. I.; Ferguson, I. J.; Patel, R. C. *J. Chem. Soc., Perkin Trans. II* **1979**, 143.
30. Newton, J. H.; Levine, R. A.; Person, W. B. *J. Chem. Phys.* **1977**, *67*, 3282. Kim, K.; Park, C. W. *J. Mol. Struct.* **1987**, *161*, 297.
31. Deslongchamps, P. *Stereoelectronic Effects in Organic Chemistry*; Pergamon: Oxford, 1983.
32. Kirby, A. J. *Acc. Chem. Res.* **1984**, *17*, 305.
33. Ratcliffe, A. J.; Mootoo, D. R.; Andrews, C. W.; Fraser-Reid, B. *J. Am. Chem. Soc.* **1989**, *111*, 7661. Somayaji, V.; Brown, R. S. *J. Org. Chem.* **1986**, *51*, 2676. Caserio, M. C.; Souma, Y.; Kim, J. K. *J. Am. Chem. Soc.* **1981**, *103*, 6712. Caserio, M. C.; Shih, P; Fisher, C. L. *J. Org. Chem.* **1991**, *56*, 5517.

34. Deslongchamps, P.; Atlani, P.; Fréhel, D.; Malaval, A.; Moreau, C. *Can. J. Chem.* **1974,** *52,* 3651. Deslongchamps, P.; Chênevert, R.; Taillefer, R. J.; Moreau, C.; Saunders, J. K. *Can. J. Chem.* **1975,** *53,* 1601.
35. Perrin, C. L.; Arrhenius, G. M. L. *J. Am. Chem. Soc.* **1982,** *104,* 2839.
36. Perrin, C. L.; Nuñez, O. *J. Chem. Soc., Chem. Commun.* **1984,** 333.
37. Huisgen, R.; Ott, H. *Tetrahedron* **1959,** *6,* 253.
38. Pothier, N.; Goldstein, S.; Deslongchamps, P. *Helv. Chim. Acta* **1992,** *75,* 604.
39. Perrin, C. L.; Nuñez, O. *J. Am. Chem. Soc.* **1987,** *109,* 522.
40. Perrin, C. L.; Nuñez, O. *J. Am. Chem. Soc.* **1986,** *108,* 5997.
41. Ref. 31, p. 109, p. 149. Deslongchamps, P.; Barlet, R.; Taillefer, R. J. *Can. J. Chem.* **1980,** *58,* 2167.
42. Page, M. I.; Webster, P.; Ghosez, L. *J. Chem. Soc., Perkin Trans. II* **1990,** 805, 813.
43. Robinson, D. R. *J. Am. Chem. Soc.* **1970,** *92,* 3138.
44. Benkovic, S. J.; Barrows, T. H.; Farina, P. R. *J. Am. Chem. Soc.* **1973,** *95,* 8414. Burdick, B. A.; Benkovic, P. A.; Benkovic, S. J. *J. Am. Chem. Soc.* **1977,** *99,* 5716. Slebocka-Tilk, H.; Bennet, A. J.; Keillor, J. W.; Brown, R. S.; Guthrie, J. P.; Jodhan, A. *J. Am. Chem. Soc.* **1990,** *112,* 8507. Ono, M.; Todoriki, R.; Araya, I.; Tamura, S. *Chem. Pharm. Bull.* **1990,** *38,* 1158.
45. Saunders, M.; Yamada, F. *J. Am. Chem. Soc.* **1963,** *85,* 1882. Lambert, J. B.; Keske, R. G.; Carhart, R. E.; Jovanovich, A. P. *J. Am. Chem. Soc.* **1967,** *89,* 3761. Lambert, J. B.; Oliver, W. L., Jr.; Packard, B. S. *J. Am. Chem. Soc.* **1971,** *93,* 933.
46. Perrin, C. L. In *The Chemistry of Amidines and Imidates*; Patai, S.; Rappoport, Z., Eds.; Wiley: Chichester, 1991, Vol. 2; 147.
47. Löfås, S.; Ahlberg, P. *J. Heterocycl. Chem.* **1984,** *21,* 583.
48. Hückel, W.; Tappe, W.; Legutke, G. *Liebigs Ann. Chem.* **1940,** *543,* 191. Cristol, S. J. *J. Am. Chem. Soc.* **1947,** *69,* 338. Cristol, S. J.; Hause, N. L.; Meek, J. S. *J. Am. Chem. Soc.* **1951,** *73,* 674. Hughes, E. D.; Ingold, C. K.; Rose, J. B. *J. Chem. Soc.* **1953,** 3839. DePuy, C. H.; Thurn, R. D.; Morris, G. F. *J. Am. Chem. Soc.* **1962,** *84,* 1314. Baciocchi, E.; Ruzziconi, R.; Sebastiani, G. V. *J. Am. Chem. Soc.* **1983,** *105,* 6114.
49. Gorenstein, D. G. *Chem. Rev.* **1987,** *87,* 1047.
50. Thatcher, G. R. J.; Kluger, R. *Adv. Phys. Org. Chem.* **1989,** *25,* 99.
51. Eliel, E. L.; Nader, F. W. *J. Am. Chem. Soc.* **1970,** *92,* 584.
52. Desvard, O. E.; Kirby, A. J. *Tetrahedron Lett.* **1982,** *23,* 4163. A. J. Kirby, personal communication.
53. Engler, R. E.; Johnston, E. R.; Wade, C. G. *J. Magn. Reson.* **1988,** *77,* 377.
54. Perrin, C. L.; Dwyer, T. J. *Chem. Rev.* **1990,** *90,* 935.
55. Perrin, C. L.; Engler, R. E. *J. Magn. Reson.* **1990,** *90,* 363.

RECEIVED June 28, 1993

Chapter 6

No Kinetic Anomeric Effect in Reactions of Acetal Derivatives

Michael L. Sinnott

Department of Chemistry (M/C 111), University of Illinois at Chicago, 801 West Taylor Street, Chicago, IL 60607–7061

The extension of the n-σ^* model of the static anomeric effect to the reactions of acetals fails to account either for the relative rates of reaction of epimeric pairs of compounds, or for their relative rates of formation from oxocarbonium ion precursors. Attempts to reconcile the kinetic and product data with stereoelectronically correct transition state geometry in the case of pyranose-type acetals commonly invoke reaction through skew and boat conformations. Transition state conformation, though, can be determined from geometry-dependent kinetic isotope effects. β-Deuterium and ring ^{18}O effects in the reaction of glycosyl derivatives provide no evidence for stereoelectronically correct transition states in the hydrolysis of glucosides or glycosides of N-acetyl neuraminic acid, either spontaneously or with the appropriate glycosidase.

Axial-Equatorial Rate Ratios in the Hydrolysis of Tetrahydropyranyl Derivatives

The "theory of stereoelectronic control" asserts that the departure of a leaving group from a carbon atom α to a lone-pair-bearing heteroatom is assisted in an additive way by the presence of lone pairs antiperiplanar to the leaving group (1). It represents a simple extension of the n-σ^* overlap model of the static anomeric effect (2) to transition states as well as ground states, and successfully rationalizes a significant body of data concerned with reactions at sp^3 hybridized carbon at the acyl level of oxidation. However, application of the theory to the

0097–6156/93/0539–0097$06.00/0
© 1993 American Chemical Society

reactions of epimeric pairs of tetrahydropyranyl and pyranosyl derivatives results in predictions apparently in conflict with experiment. The ring conformation in the transition state for reaction of the axial epimer is predicted to be simply derived from the ground-state chair conformation, but the ring conformation in the transition state for reaction of the equatorial epimer is predicted to be derived from a normally disfavored conformation (be it the other chair or skew) in which there is a lone pair antiperiplanar to the leaving group. It is reasonable to suppose that such a conformational change would require energy. Therefore any "kinetic anomeric effect" should result in axial leaving groups being lost faster than equatorial leaving groups

They do not. It has been known for many years that β (equatorial) glycosyl halides are much more labile than their α–anomers, and that the effect is especially pronounced in the *manno* series (therefore neighbouring group participation is not the origin of observed differences) (3). The effect was quantitated when O'Connor and Barker (4) studied the hydrolyses of a series of anomeric pairs of glycopyranosyl phosphates and found that the equatorial anomer lost $H_2PO_4^-$ faster than the axial by factors of 2-4. Confirmation that this result did not in some way reflect inadequate conformational control by sugar hydroxy- and hydroxymethyl groups came from work in Kirby's laboratory (5) on the spontaneous hydrolyses of epimeric pairs of 2-(p-nitrophenoxy)-*trans*-1-oxadecalins, in which it was found that the leaving-group-equatorial epimers reacted ~3-fold faster than the leaving-group-axial epimers.

Reactions in which oxocarbonium ions are generated by unassisted departure of the leaving group are generally characterized by values of $\beta_{lg} \leq -1$ (6-9), suggesting that the transition state is very late, with the leaving group largely departed. If this is so, then the *faster* hydrolysis of the equatorial epimers is simply explained by their higher ground state energy, a consequence of the operation of the ground state anomeric effect. Both epimers go to a common intermediate through transition states which are similar in energy. The reaction profile can be summarised qualitatively in Figure 1.

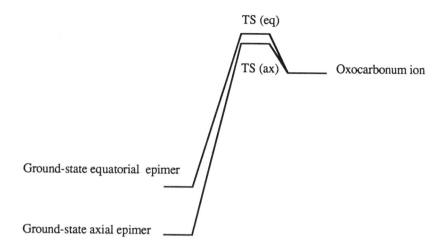

Figure 1
Reaction profile for epimeric pairs of tetrahydropyranyl derivatives

Three predictions can be made on the basis of such an explanation of the apparent "anti-stereoelectronic" behaviour of epimeric pairs of tetrahydropyranyl derivatives:

1. Increasing the strength of the ground state anomeric effect will increase the equatorial/axial rate ratio.

2. There will be little or no stereochemical discrimination in the reactions of tetrahydropyranyl cations with nucleophiles.

3. Reversal of the sense of the ground state anomeric effect of the leaving group will result in axial leaving groups departing faster than equatorial leaving groups.

These predictions are fulfilled. We have recently shown that the equatorial/axial rate ratio for loss of fluoride from glucopyranosyl fluorides (I) in water is around 40 (the precise number depending on temperature and buffer), in accord with the greater electronegativity of fluorine compared to oxygen (*11*).

In an incisive set of recent experiments, Deslongchamps *et al.* (*12*) showed that acid-catalyzed cyclization of compound (II) yielded (III) and (IV) in a 3:2 mixture, but that acid-catalyzed equilibration of the two products gave (III) exclusively. This is exactly what would be expected for a reaction going from right to left on the reaction profile of Fig.1.

The situation in respect of leaving groups subject to a reverse anomeric effect is complicated by the tendency of such groups bearing a full positive charge to force the tetrahydropyranyl ring to which they are attached to a ground state conformation with the leaving group equatorial, even against usually effective conformational constraints, such as the all-equatorial substituent pattern of glucopyranose. Thus, α-D-xylopyranosyl pyridinium ions adopt the $^1C_4$ conformation, and α-D-glucopyranosyl pyridinium ions appear to adopt the $^1S_3$ conformation (9) in water, and the specific-acid-catalyzed hydrolysis of methyl α-glucopyranoside proceeds through a transition state conformation derived from the $^1S_3$ skew form (13). However, empirical conformational energy calculations suggested that the factor by which α-glycopyranosyl pyridinium ions hydrolyze faster than their β anomers is too small for the ring conformations at the transition state to have the leaving group axial (9).

1,8-Dioxadecalin-derived Systems.   A number of patterns of reactivity of axial and equatorial tetrahydropyranyl derivatives are known which are not readily accounted for by the qualitative free energy profile of Figure 1, but they all involve the generation or decomposition of the 1,8 -dioxadecalin systems (14-16). Thus, acid-catalyzed cyclization of (V) gives (VI) as the exclusive kinetic product which slowly changes to the equilibrium 5:6 mixture of (VI) and (VII) (14). A system which in retrospect is highly informative is (VIII), in which different solvent isotope effects on the pH-independent reaction suggested that the rate-determining step in the equatorial case is C-O fission, but in the axial case it is diffusion apart of the tethered tetrahydropyranyl cation-p-nitrophenolate pair (15). Quite clearly the three-carbon tether holding the nucleophile to the tetrahydropyranyl ring in the 1,8-dioxadecalin system is affecting differential reactivity very substantially, and in 1988 I wrote "Examination of Dreiding models reveals that the solid angle in which the nucleophilic oxygen can approach the electron-deficient carbon without introducing strain into the chain is much smaller for *trans* attack than for *cis*. It seems probable that this, rather than any antiperiplanar lone pair effect, is responsible for the results with 1,8-oxadecalin systems" (*17*). It is pleasing that these ideas have recently been dignified by molecular modeling (*18*).

It seems therefore that the 1,8-oxadecalin system is a special case, biased towards stereoelectronically correct patterns of reactivity by the three-carbon tether, that does not affect the general applicability of the free energy profile of Figure 1.

Bimolecularity of Reactions of Acetals and Glycosides: Concerns about the general applicability of the free energy profile of Figure 1 that center around the question of whether the reactions of glycosides and acetals do indeed involve oxocarbonium ions as true intermediates are less readily allayed. Early work on the lifetimes of oxocarbonium ions (*19*) led to an estimate of the lifetime of the methoxymethyl cation in water of 10^{-15}s, less than the period of a bond vibration. The methoxymethyl cation was therefore too unstable to exist, so that all nucleophilic displacement reactions at the methoxymethyl center in water were predicted to be necessarily bimolecular. This was subsequently confirmed by the characterization of bimolecular reactions involving departure of 2,4-dinitrophenolate (*20*) and *N, N*-dimethyl anilines (*21*). At the same time, displacements of tetrahydropyranyl derivatives were shown to be unimolecular (*6*). Further extrapolation to the glucopyranosyl system, on the assumption that the relative importance of resonance and inductive effects was the same on both the generation of oxocarbonium ions and their reaction with water, suggested that reactions of glucopyranosyl derivatives were also likely to bimolecular. The glucopyranosyl cation is undoubtedly too unstable to exist in mixtures of ethanol and trifluorethanol, in the presence of both neutral and anionic leaving groups (22). The situation in water, though, is more complex. The spontaneous hydrolyses of 2,4-dinitrophenyl β-D-glucopyranoside and of α- and β-xylopyranosyl pyridinium ions are affected by 1.0M concentrations of various salts, but the effects parallel basicity rather than nucleophilicity, and the neutral nucleophile thiourea has no effect (13). Measurements of the lifetimes of oxocarbonium ions of various structural types suggested that the assumption of equal relative importance of inductive and resonance effects on the forward and reverse of reactions generating oxocarbonium ions was oversimplified (23); shorter extrapolations, in which the two effects were separated, yielded the result that the glucopyranosyl cation has a short but significant lifetime in water. At the same time, clearly bimolecular reactions in acetal systems where the oxocarbonium ion had a real existence were observed (24).

The discovery of clearly bimolecular reactions on α–glucopyranosyl fluoride (25), but only with anionic nucleophiles present at high concentrations, was not therefore totally unsurprising. If the substrate reacts through a conformation simply derived from the $^4C_1$ ground-state, the approach of the incoming nucleophile is comparatively ready. However, it has long been known from classical solvolysis chemistry that displacement of equatorial leaving groups from conformationally rigid six-membered rings is exceedingly difficult, to the extent that 2-adamantyl p-toluenesulfonate is used as a standard for "limiting" solvolyses (26). Therefore one would expect nucleophilic reactions in the sugar series to be less readily observed with equatorial leaving groups, so that there is no experimental conflict between the results with α-glucopyranosyl fluoride (25) and 2,4-dinitrophenyl β-D-glucopyranoside (13). By the same token the increase in equatorial/axial rate ratio on changing the leaving group to fluoride can still be taken as evidence for the reaction profile of Figure 1, since for the foregoing reasons nucleophilic participation would be expected to decrease this ratio.

Nonetheless, the conclusion, that the glucopyranosyl cation is too unstable to exist in contact with an anion (25) complicates analysis of transition state geometry, by virtue of the consequent requirement for a preassociated nucleophile in displacements of anionic leaving groups. A probe of the nucleophilic character of a displacement reaction is the carbon kinetic isotope effect at the reaction center. The interpretation of non-hydrogen kinetic isotope effects is complicated by the greater relative effect of isotopic substitution on the molecular moment of inertia compared to zero-point energy, and also by the fact that the contribution of the isotopically substituted atom to the reduced mass of the system is smaller, and more dependent on the rest of the molecule, than for hydrogen. Nonetheless, it appears that for reactions involving only first-row atoms, the larger the carbon isotope effect, the greater the associative character of the reaction (*27-31*).

In Table 1 are set out kinetic isotope effects, obtained by the isotopic quasi-racemate method (*13*, and references therein) for the reactions of D-glucopyranosyl fluoride in water (measured by Y. Zhang, except for some secondary hydrogen effects measured by J. Bommuswamy) and for acid catalyzed hydrolyses of methyl α- and β glucopyranoside (*13*).

Table 1
Kinetic Isotope Effects in the Reactions of Glucopyranosyl Derivatives

Site & Isotope→ ↓ Reaction	αD (1-^{2}H)	βD (2-^{2}H)	γD (5-^{2}H)	5-^{18}O	1-^{13}C
αGlcpOMe, 2.0M HClO$_4$ 80°C (*13*)	1.137	1.073	0.987	0.996	1.007
βGlcpOMe, 2.0M HClO$_4$ 80°C (*13*)	1.08_9	1.04_5	0.97_1	0.991	1.011
αGlcpF, 0.2M succinate, $I = 1$ (NaClO$_4$), 80°C	1.14_2	1.06_7	0.97_9	0.98_6	1.03_2
αGlcpF, 0.2M succinate, 2.0 M NaN$_3$, 50°C.	1.16_9				1.08_5
βGlcpF, 0.3M succinate, $I = 1$ (NaClO$_4$), 50°C	1.10_5	1.05_9	0.98_1	0.98_8	1.06_4
βGlcpF, 0.3M bis-tris, $I = 1$ (NaClO$_4$), 50°C	1.08_6	1.03_2			1.01_7

All fluoride reactions at pH 6.0

The effect of the cosolutes in the fluoride reactions deserve comment. High concentrations of buffer are necessary since high concentrations of substrate, the hydrolysis of which liberates HF, are required for the isotopic quasi-racemate technique. The sterically hindered cationic buffer bis-tris has no kinetic effect, and the anionic succinate buffer has negligible effect on the hydrolysis of the α–fluoride. These two sets of data therefore refer to the water reaction. The reaction with 2.0M sodium azide is known to be a bimolecular reaction of the glycosyl fluoride with azide ion (25). The reaction of the β-fluoride in 0.3M sodium succinate buffer, pH 6.0, is, formally largely that of the fluoride with succinate monoanion.

The ^{13}C effects on the reactions of α–glucopyranosyl fluoride are very much in line with the conclusion (23, 25) that the glucopyranosyl cation is too unstable to exist in contact with an anion, but has a real existence in contact with neutral nucleophiles. The water reaction of the fluoride is characterized by a much higher effect than that for the acid-catalyzed hydrolysis of methyl α-glucopyranoside, in which the the incipient glucopyranosyl cation is in contact only with neutral water and methanol molecules. The azide reaction is characterized by a very high, but not unprecedented effect. Ando *et al.* (29) reported reaction center ^{14}C effects (k_{12}/k_{14}) of up to 1.162 for reactions of *N, N*-dimethylanilines with benzyl benzenesulfonates, coresponding to 7.8-8.7 % effects for ^{13}C (27). It appears that particularly high reaction-center carbon isotope effects may be associated with undoubtedly bimolecular reactions in which there is extensive resonance stabilization of positive charge at the reaction center. It is also noteworthy that the α-deuterium kinetic isotope effect increases as the incoming nucleophile changes from water to the more highly polarizable azide ion, in accord with previous observations (21).

The results of the reactions of the β fluoride are less readily interpreted. The ^{13}C effect for the water reaction is little higher than for the acid-catalyzed hydrolysis of the methyl glycoside, so that the possibility appears to exist that, whilst the glucopyranosyl cation is too unstable to exist in contact with a fluoride ion on its α face, the contact ion pair on the β face has a fleeting existence, because of steric hindrance involved in bringing a nucleophilic water molecule up to reaction center in the axial sense. The identification of the reaction of the β fluoride with succinate monoanion as a nucleophilic displacement, unlikely in any event since the normally more nucleophilic dianion is less effective, is ruled out since the optical rotation change on complete reaction is the same as for the water reaction, so the product is glucose, yet glucose acylals are known to accumulate even in the reactions of the more stable α fluoride (25). Its identification as a close analogue of the water reaction, in which the succinate monoanion donates a proton to the leaving fluoride, is made unlikely by the radically different ^{13}C and β–deuterium isotope effects for the two processes. The transition state that best fits the facts now available is a kinetically equivalent one (IX), in which neutral succinic acid acts on the substrate monanion, to yield 1,2 -anhydroglucose as first product, which rapidly hydrolyzes. Transition state (IX) is very similar to that for the well-known base-catalyzed hydrolysis of 1,2-*trans* glycosides (*51*), which is also known for the 1,2-*trans* fluorides (*52*). This *trans* relationship of the intramolecular nucleophile and leaving group of the β fluoride accounts for the comparative unimportance of the succinate-catalyzed process for the α fluoride. It has been found that the hydrolysis of α–glucopyranosyl fluoride is subject to general acid catalysis (*32*), but only by anionic acids: the kinetically equivalent reaction of the neutral acid with the substrate anion might be taking place here also.

IX

Determination of Transition State Geometry from Geometry-Dependent Kinetic Isotope Effects. The general applicability of the free energy profile of Figure 1 for reactions of epimeric tetrahydropyranyl derivates therefore cannot realistically be disputed, although complexities introduced by tethered nucleophiles and bimolecularity of apparently unimolecular reactions must be recognized. The "theory of stereoelectronic control" can however be rescued if it is argued that because TS (eq) is very late (in the left to right direction), or very early (in the right to left direction), it can have stereoelectronically correct geometry without energetic consequences being manifest (*12*). A related idea was advanced by Kirby (*5*), who considered stereoelectronically correct boat conformations to be accessible in the case of ground state equatorial derivatives because of the operation of the ground state anomeric effect.

Happily, transition state conformations are experimentally accessible from geometry-dependent kinetic isotope effects. We have concentrated on β-deuterium and ring ^{18}O effects. The ring ^{18}O effects arise from a tightening of the O5-C1 bond as it acquires double bond character. There is ample evidence that effective orbital overlap is obtained only with the p-type lone pair on oxygen, so that planar oxocarbonium ions result (33, 34). The effect is inverse, and if to a first approximation it is a zero-point energy effect arising simply from the acquisition of double bond character, we can write

$$\ln (k_{16}/k_{18}) = \cos^2\omega \ln (k_{16}/k_{18})_{max} \qquad (1)$$

where $(k_{16}/k_{18})_{max}$ is the maximal isotope effect arising from a fully developed π-bond between the oxygen and the carbon, and ω is the dihedral angle between the p-type lone pair and the electron-deficient p orbital on C1. The problem is that there is one measurement (the effect) and two unknowns (ω and $(k_{16}/k_{18})_{max}$). In work on the hydrolysis of methyl glucosides (13), we resolved this problem by using upper and lower limits of $(k_{16}/k_{18})_{max}$ to calculate upper and lower limits for ω. The upper limit was a figure of 0.988 obtained for the acid-catalyzed hydrolysis of 2-propyl 1-[^{18}O]-α–arabinofuranoside, which hydrolyzes by endocyclic C-O fission (35); the lower limit was a value of 0.976-0.978 calculated for the isotope effect on the equilibrium between the appropriate rotamers of dihydroxymethane and a proton, and protonated formaldehyde and water.

β-Deuterium kinetic isotope effects have their origin in two phenomena, hyperconjugation and the inductive effect of deuterium, and can be considered in terms of equation (2):

$$\ln (k_H/k_D) = \cos^2 \theta . \ln(k_H/k_D)_{max} + \ln(k_H/k_D)_i \qquad (2)$$

The first term represents hyperconjugation of the C-L σ orbital with an electron-deficient p orbital on an adjacent carbon atom: θ is the dihedral angle between the C-L bond and this orbital. $\ln (k_H/k_D)_{max}$ is the maximal hyperconjugative effect obtained when the C-L bond and the p orbital are exactly eclipsed, and increases as the positive charge on the adjacent carbon atom increases, with the associated weakening of the C-L bond. The second term in equation(2) represents a small, geometry-independent inductive deuterium effect. A single measurement of such an effect encounters the problem that there is one experimental parameter (the effect) and three unknowns (θ, $(k_H/k_D)_{max}$, and $(k_H/k_D)_i$). In work on the glucopyranosyl system, one is perforce confined to one effect, and we estimated bounds for θ by using experimental values of the equilibrium constant for an otherwise degenerate hydride shift between CH_3-C^+ and CD_3-C-H sites (35) for the lower bound, and the kinetic effect for the hydrolysis of $CL_3CH(OEt)_2$ for the upper bound (36). Where however one has two diastereotopic β-hydrons, as with the glycosides of N-acetyl neuraminic acid and related compounds, no assumptions need be made about $(k_H/k_D)_{max}$, since the effects for each of the pair of diastereotopic hydrogen atoms at C-3 give us two simultaneous equations (3 and 4) and two unknowns for the hyperconjugative portion of the effect:

$$\ln (k_H/k_D)_R = \cos^2 \theta.\ln(k_H/k_D)_{max} \qquad (3)$$

$$\ln (k_H/k_D)_S = \cos^2 (\theta + 120^0).\ln(k_H/k_D)_{max} \qquad (4)$$

Experimental values of θ and ω obtained in this way, will define more closely the ring conformations of tetrahydropyranyl derivatives at the transiton state.

Non-Enzymic Reactions of Glucopyranosyl Derivatives. The two extreme values of both θ and ω obtained for both anomers of methyl glucoside enabled eight possible transition state structures to be determined. These angles were used as constraints in a modified MM2 calculation of the rest of the molecule (*13*). The results for the α-glucoside were unambiguous, and very largely determined by the small ring ^{18}O effect: all four derived transition state structures were simply derived from the $^1S_3$ skew conformation, and are shown diagrammatically in (X). This result was interpreted as a consequence of the operation of the reverse anomeric effect on the leaving group, which bore a full positive charge. The derived transition state structures were profoundly stereoelectronically incorrect. The derived transition state structures for the methyl β-glucoside were more ambiguous, in that whilst the modified MM2 program minimized to two flattened $^4C_1$ chairs if the theoretical value of $(k_{16}/k_{18})_{max}$ was used in the calculation of ω, if the experimental estimate was used it did not minimize to a unique structure. Nonetheless, we considered the balance of evidence to lay very heavily towards the stereoelectronically incorrect flattened $^4C_1$ chair (XI).

Because of the necessity for a nucleophilic water molecule in the transition state, analogous calculations are not possible for the water reaction of α-glucosyl fluoride. The fact that the ring ^{18}O effect, even though the reaction is nucleophilic, is much higher than for α methyl glucoside implies that ω is much smaller (using the same extreme estimates of $(k_{16}/k_{18})_{max}$, between 0^0 and 37^0). Since the β–deuterium isotope effect is within experimental error the same for the two α leaving groups, the range of θ is therefore similar (43–31^0). Such angles, qualitatively, indicate a flattened $^4C_1$ chair, shown diagrammatically in (XII). The finding that the transition state conformation alters radically when the reverse anomeric effect no longer operates on the leaving group supports our explanation of the origin of transition state structure X for the methyl glucoside.

As indicated above, the reaction of succinate monoanion with the β-fluoride is not an A_ND_N (S_N2) reaction and appears not to be an analogue of the water reaction. Detailed reasons for drawing transition state (IX) are:

(i) Higher ^{13}C effect than for the water reaction, indicating a reaction with a local S_N2 -like transition state, even if one involving direct displacement by succinate monoanion has to be rejected.

(ii) Substantial ring ^{18}O effect, indicating extensive delocalization of charge onto the ring oxygen atom.

(iii) Higher β–deuterium kinetic isotope effect than for the water reaction. This is is a key piece of evidence for transition state (IX), since the pre-equilibrium forming the oxyanion at position 2 would be subject to an isotope effect because of the inductive effect of deuterium. If the equilibrium effect on the ionization of the 2-OH of β–glucopyranosyl fluoride is the same as that on the ionization of formic acid (38), then this could account for the whole of the observed kinetic effect.

The water reaction of β-glucopyranosyl fluoride requires the ring ^{18}O effect for full transition state characterization, but similarity of the effects so far measured suggests a similar transition state to the one deduced for methyl β–glucopyranoside.

Non-Enzymic Reactions of Glycosides of N-Acetyl Neuraminic Acid

We have recently found that the liberation of aglycon from glycosides of N-acetyl neuraminic acid (XIII) takes place by four processes: the proton-catalyzed reaction of the neutral molecule, the proton-catalyzed reaction of the anion, the spontaneous reaction of the anion, and a process which may involve attack by the ionized 9-hydroxyl group on C6 with displacement of the hemiacetal anion (39).

We measured those kinetic isotope effects on the first three processes, that arose from deuterium substitution at C4 and separately at the *proR* and *proS* positions. The purpose of the measurement of the secondary effect at C4 was to provide an estimate of $(k_H/k_D)_{i.}$, from the assumption, apparently also valid for isotopes (40), that inductive effects decrease by a factor of 2.5-3 per carbon-carbon bond interposed between substituent and reaction center. In the event no such γ effect was detectable in the reactions of the anion, so $\ln (k_H/k_D)_i$ was set equal to zero. The separate *proR* and *proS* effects then enabled a value of θ of 30° and a value of $(k_H/k_D)_{max}$ of 1.1 to be obtained for both the acid-catalyzed and spontaneous reactions of the anion. For the acid-catalyzed reaction of the neutral molecule, a value of θ of between 51° and 57°, and a value of $(k_H/k_D)_{max}$ of between 1.19 and 1.5 were obtained, depending on whether the (large) inductive correction was applied or not.

XIII

Since $(k_H/k_D)_{max}$ measures the electron-deficiency of an orbital, not the reaction center as a whole, the lower value of this parameter when the C1 carboxylic acid group is ionized requires some form of nucleophilic participation by the ionized carboxylate group, despite the strained α-lactone ring that results. In our view this provides quite striking confirmation of the conclusions of Banait and Jencks, that glucosyl cations are too unstable to exist in contact with an anionic nucleophile.

The values of θ, obtained without any assumptions, are not really compatible with stereoelectronically correct transition states for any of the three reactions for which they were obtained. The only transition state conformation which is stereoelectronically correct and compatible with larger kinetic isotope effects for the *proR* deuterium at C3 than for the *proS* is derived from the $^{4,0}B$ classical boat, which is likely to be prohibitively high in energy since the trihydroxypropane side chain and the leaving group are in a 1,3 - di-pseudoaxial relationship.

Enzymic Reactions. Attempts to draw stereoelectronically correct geometries for the transition states of reactions catalyzed by glycosyl transferring enzymes have been made over the years, but they have for the most part encountered difficulties, either of conflict with experiment, or of neglect of subsequent steps in the enzyme turnover sequence.

The case of those glycoside hydrolases which yield the sugar product in the same anomeric configuration as the substrate are particularly instructive. They operate via glycosyl enzyme intermediates which are reached and left by oxocarbonium-ion-like transition states (*41*). In the case of the enzymes cleaving simple sugars, the glycosyl enzyme intermediates are acylals derived from side-chain glutamate or aspartate residues, of the opposite anomeric configuration to the substrate (for a recent detailed study see 42). The requirement for a plethora of implausible changes in substrate ring conformation, if the transition states for these reactions are to be stereoelectronically correct, has been pointed out previously (43). More recently, the paradigmal case of lysozyme has been addressed by X-ray crystallographic examination of an enzyme-saccharide complex in which the D subsite, containing the saccharide residue at which reaction takes place, is occupied by an N-acetyl muramic acid residue (44). Although distortion of the sugar ring is apparent, apart from interaction of the protein with the C5 hydroxymethyl group pushing it into a pseudoaxial position, the distortions are not large, and more towards the geometry of the oxocarbonium ion than to stereoelectronic correctness.

More direct measures of transition state geometry have been confined to measurements of β-deuterium kinetic isotope effects, since ring ^{18}O effects are accessible only with great difficulty. Yeast α-glucosidase (a retaining glycosidase) was uncooperative, in that the rate-determining step in the hydrolysis of aryl α-glucosides was a noncovalent event (45), though bond-breaking was kinetically accessible for α-glucopyranosyl pyridinium salts, which are however in the $^1S_3$ conformation in free solution. Very high effects (1.13 for the 4-bromoisoquinolinum ion and 1.08_5 for the pyridinium ion) were obtained, indicating a very low value of θ. By themselves, these effects were compatible with reaction through the $^4C_1$ conformation, but the puzzle of the non-covalent event preceding catalysis, which was slow for the aryl glucosides and fast for the glucosyl pyridinium salts, even though k_{cat} values for the two sets of substrates overlapped, led us to favor the $^{2,5}B$ conformation (XIV) as the reactive ground-state conformation (a meaningful concept in enzymic reactions, since ES complexes with the substrate in a normally disfavored conformation are not systems to which the Curtin-Hammett principle applies). The evidence for stereoelectronic incorrectness with this enzyme remained, however, indirect.

More direct evidence was obtained with a study of the second chemical step in the ebg β-galactosidase of $Escherichia\ coli$. With three experimental evolvants, ebg^a, ebg^b, and ebg^{ab}, this was kinetically accessible. The hydrolysis of the galactosyl-ebg enzymes was characterized by an α-deuterium kinetic isotope effect of 1.10, but a β–deuterium kinetic isotope effect of 1.00 (46, 47). This is very difficult to reconcile with stereoelectronically correct reaction through the $^4C_1$ conformation, since in such a conformation the β-deuterium kinetic isotope effect should be maximal, and we tentatively suggested reaction through the $B_{2,5}$ conformation (XV), which would lead directly to a local minimum for the oxocarbonium ion.

It is with $Vibrio\ cholerae$ neuraminidase, however, that the evidence (obtained by X. Guo) for a stereoelectronically incorrect transition state is most compelling, and surprising in the light of the interpretation of the electron density map of the influenza neuraminidase - sialic acid complex in terms of the sialic acid being in a $B_{2,5}$ conformation (with sialic acid numbering: see XIII) (*48*). The viral and the bacterial sialidases (with the exception of the *Salmonella typhimurium* sialidase) are however very different (*49*).

V. *cholerae* neuraminidase catalyzes the hydrolysis of compounds of structure (XIII) with retention of configuration, so that in principle either the formation or hydrolysis of the intermediate could govern k_{cat}. However, there is a detectable ^{18}O leaving group kinetic isotope effect on V_{max} for the hydrolysis of p-nitrophenyl N-acetyl neuraminic acid (1.040 ± 0.017), indicating that glycon-aglycon cleavage is still rate-determining. The secondary deuterium kinetic isotope effects for the two diastereotopic hydrons on C3 show directly that, whatever is seen in the X-ray structure of the influenza enzyme, the reactive conformation in the V. *cholerae* enzyme is not $B_{2,5}$, since the effect for the *proR* deuterium (1.037 ± 0.014) is bigger than that for the *proS* deuterium (1.018 ± 0.015).

We have therefore have uncovered evidence for the stereoelectronically incorrect reactive ground state conformations (XIV) for an α-glucoside bound to yeast α-glucosidase, (XV) for galactosyl-*ebg*, and (XIII - i.e. the preferred conformation in free solution) for an α-N-acetyl neuraminide bound to V. *cholerae* neuraminidase.

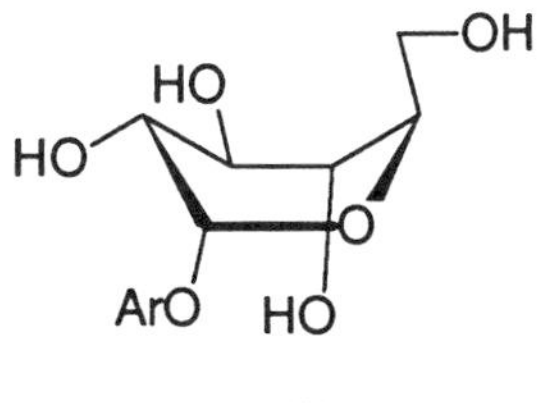

XIV

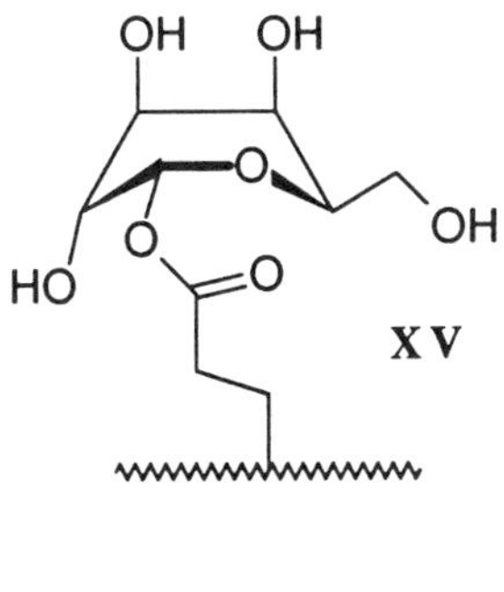

XV

There is thus no evidence for any energetic or stereochemical consequences of a kinetic anomeric effect in free solution, nor any evidence that natural selection has taken advantage of a stereoelectronic effect so subtle as to be manifested only in enzyme chemistry (*cf* the stereoelectronic rationale of the A- and B-side coenzyme stereochemistry of alcohol dehydrogenases: *50*).

Acknowledgement
Financial support from NIH grant GM 42469 is gratefully acknowledged.

Literature Cited

1. Deslongchamps, P. *Stereoelectronic Effects in Organic Chemistry*, Pergamon, Oxford 1983

2. Romers, C.; Altona, C.; Buys, H. R.; Havinga, E. *Top. Stereochem.* **1969**, *4*, 73.

3. Haynes, L. J.; Newth, F. H. *Adv. Carbohydr. Chem.* **1955**, *10*, 207.

4. O'Connor, J. V.; Barker, R. *Carbohydr. Res.* **1979**, *73*, 227.

5. Chandrasekhar, S.; Kirby, A. J.; Martin, R. J. *J. Chem. Soc. Perkin Trans. 2* **1983**, 1619.

6. Craze, G.-A.; Kirby, A. J. *J. Chem. Soc. Perkin Trans. 2* **1978**, 354

7. Lönnberg, H.; Pohjola, V. *Acta Chem. Scand.* **1976**, *30A*, 669.

8. Jones, C. C.; Sinnott, M. L.; Souchard, I. J. L. *J. Chem. Soc. Perkin Trans. 2* **1977**, 1191.

9. Hosie, L.; Marshall, P.J.; Sinnott, M. L. *J. Chem. Soc. Perkin Trans. 2* **1984**, 1121.

10. Käppi, R.; Kazimierczuk, Z.; Järvinen, P.; Seela, F.; Lönnberg, H. *J. Chem. Soc. Perkin Trans 2* **1991**, 595.

11. Konstantinidis, A. K.; Sinnott, M. L. *Biochem. J.* **1991**, *279*, 587.

12. Pothier, N.; Goldstein, S.; Deslongchamps, P. *Helv. Chim. Acta* **1992**, *75*, 604.

13. Bennet, A.J.; Sinnott, M. L. *J. Am. Chem. Soc.* **1986**, *108*, 7287.

14. Beaulieu, N.; Dickinson, R. A.; Deslongchamps, P. *Can. J.Chem.* **1980**, *58*, 2531.

15. Kirby, A. J.; Martin, R. J. *J. Chem. Soc. Perkin Trans. 2* **1983**, 1627.

16. Kirby, A. J.; Martin, R. J. *J. Chem. Soc. Perkin Trans. 2* **1983**, 1633.

17. Sinnott, M. L. *Adv. Phys. Org. Chem.* **1988**, *24*, 113.

18. Deslongchamps, P. *This volume.*

19. Young, P. R.; Jencks, W. P. *J. Am. Chem. Soc.* **1977**, *99*, 8238.

20. Craze, G.-A.; Kirby, A.J.; Osborne, R. *J. J. Chem. Soc. Perkin Trans 2* **1978**, 357.

21. Knier, B. L.; Jencks, W. P. *J. Am. Chem. Soc.* **1980**, *102*, 6789.

22. Sinnott, M. L.; Jencks, W. P. *J. Am. Chem. Soc.* **1980**, *102*, 2026.

23. Amyes, T. L.; Jencks, W.P. *J. Am. Chem. Soc.* **1989**, *111*, 7888.

24. Amyes, T. L.; Jencks, W. P. *J. Am. Chem. Soc.* **1989**, *111*, 7900.

25. Banait, N. S.; Jencks, W. P. *J. Am. Chem. Soc.* **1991**, *113*, 7951.

26. Schleyer, P. von R.; Fry, J. L.; Lam, L. K. M.; Lancelot, C. J. *J. Am. Chem. Soc.* **1970**, *92*, 2542.

27. Axelsson, B. S.; Matsson, O.; Långström, B. *J. Am. Chem. Soc.* **1990**, *112*, 6661-6668

28. Yamataka, H.; Tamura, S.; Hanafusa, T.; Ando, T. *J. Am. Chem. Soc.* **1985**, *107*, 5429.

29. Ando, T.; Tanabe, H.; Yamataka, H. *J. Am. Chem. Soc.* **1984**, *106*, 2084.

30. Ando, T. Yamataka, H,; Tamuram S.; Hanafusa, T. *J. Am. Chem. Soc.* **1982**, *104*, 5493

31. Wong, S.-L.; Schowen, R. L. *J. Am. Chem. Soc.* **1983**, *105*, 1951.

32. Banait, N. S.; Jencks, W. P. *J. Am. Chem. Soc.* **1991**, *113*, 7958.

33. Briggs, A. J.; Evans, C. M.; Glenn, R.; Kirby, A. J. *J. Chem. Soc. Perkin Trans. 2* **1983**, 1637.

34. Cremer, D.; Gauss, J.; Childs, R. F.; Blackburn, C. *J. Am. Chem. Soc.* **1985**, *107*, 2435.

35. Bennet, A. J.; Sinnott, M. L.; Wijesundera, W. S. S. *J. Chem. Soc. Perkin Trans. 2* **1985**, 1233.

36. Siehl, H. U.; Walter, H. *J. Am. Chem. Soc.* **1984**, *106*, 5355.

37. Kresge, A. J.; Weeks, D. P. *J. Am. Chem. Soc.* **1984**, *106*, 7140.

38. Streitweiser, A.; Klein, H. S. *J. Am. Chem. Soc.* **1963**, *85*, 2759.

39. Ashwell, M.; Guo, X.; Sinnott, M. L. *J. Am. Chem. Soc.* **1992**, *114*, 10158.

40. Melander, L.; Saunders, W. H. *Reaction Rates of Isotopic Molecules*, John Wiley, New York, Chichester, Brisbane and Toronto, 1980, p.199.

41. Sinnott, M. L. *Chem. Rev.* **1990**, *90*, 1171.

42. Street, I. P.; Kempton, J. B.; Withers, S. G. *Biochemistry* **1992**, *31*, 9970.

43. Sinnott, M. L. *Biochem. J.* **1984**, *224*, 817.

44. Strynadka, N. C. J.; James, M. N. G. *J. Mol. Biol.* **1991**, *220*, 401.

45. Hosie, L.; Sinnott, M. L. *Biochem. J.* **1985**, *226*, 437.

46. Li, B. F. L., Holdup, D.; Morton, C. A. J.; Sinnott, M. L. *Biochem. J.* **1989**, *260*, 109.

47. Elliott, A. C.; K, S.; Sinnott, M. L.; Smith, P. J.; Bommuswamy, J.; Guo, Z.; Hall, G.; Zhang, Y. *Biochem. J.* **1992**, 155.

48. Varghese, J. N.; McKimm-Breschkin, J. L.; Caldwell, J. B.; Kortt, A. A.; Colman, P. M. *Proteins* **1992**, *14*, 327.

49. Henrissat, B. *Biochem. J.* **1991**, *361*, 697.

50. Nambiar, K. P.; Stauffer, D. M.; Kolodziej, P. A.; Benner, S. A. *J. Am. Chem. Soc.* **1983**, *105*, 5886.

51. Capon, B. *Chem. Rev.* **1969**, *69*, 407.

52. Micheel, F.; Klemer, A. *Adv. Carbohydr. Chem.* **1961**, *16*, 89.

RECEIVED May 12, 1993

Chapter 7

Involvement of nσ* Interactions in Glycoside Cleavage

C. Webster Andrews[1], Bert Fraser-Reid[2], and J. Phillip Bowen[3]

[1]Division of Organic Chemistry, Burroughs Wellcome, 3030 Cornwallis Road, Research Triangle Park, NC 27709
[2]Department of Chemistry, Duke University, P.O. Box 90346, Durham, NC 27708-0346
[3]Computational Center for Molecular Structure and Design, Department of Chemistry, University of Georgia, Athens, GA 30602

Proton-induced cleavage of acetals and glycosides has been investigated by *ab initio* methods. Bond lengths and bond angles are described as a function of conformation for protonated and unprotonated versions of dimethoxymethane. The path of the cleavage reaction is described in terms of conformation as well as hybridization of the donor oxygen. Lone pair orbitals on the donor oxygen are considered with respect to cleavage. The results are then projected onto pyranosides using our $^4H_3$ (axial anomer) and ^4E (equatorial anomer) transition state models for glycoside hydrolysis.

Evidence continues to accumulate about the crucial roles that oligosaccharides play in various biochemical processes relating, frequently, to molecular recognition or energy storage. Formation and degradation of oligosaccharides involve cleavage of glycosidic bonds at the anomeric center. The mechanism of this cleavage reaction has been the subject of our investigations.

From the standpoint of cleavage, an important mechanistic contribution was the theory of stereoelectronic control enunciated by Deslongchamps[1] (Scheme 1). This was an extension of the nσ* delocalization rationalization of the anomeric effect by Altona.[2] In the context of intersaccharide (i.e., glycosidic) cleavage the theory has been renamed[3] the antiperiplanar lone pair hypothesis (ALPH), since it proposes that an axial glycoside, that contains an antiperiplanar lone pair, **1**, is poised for cleavage to give the cyclic oxocarbenium ion **2**, whereas the equatorial counterpart, **4**, must first proceed to a boat conformation, **3**, wherein the ALPH requirement is met (Scheme 1).

An apparent anomaly of ALPH noted by Benner is that an antiperiplanar lone pair can be involved both in (a) stabilization of an

a   R´ = –CH$_2$CH$_2$– (β = 20h; α = 40h)

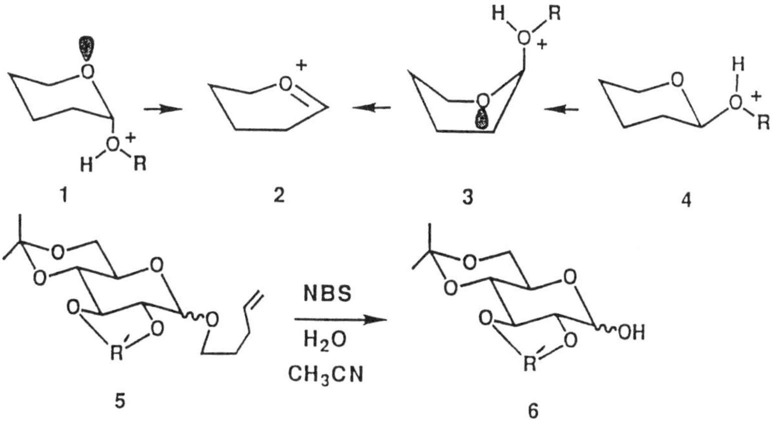

b   R´ = –CMe$_2$– (α = β = 65h)

Scheme 1

axial glycoside as well as in (b) facilitation of its proton-induced cleavage, e.g. **1--->2**.[4]

Hydrolysis of Conformationally Restrained Pyranosides[5]

Our interest in this question emanated from the discovery in our laboratory that n-pentenyl glycosides could be hydrolyzed under neutral conditions by the use of electrophilic reagents.[6] Accordingly cyclic acetal protecting groups, such as are present in **5**, would not be affected during hydrolysis to **6** (Scheme 1) and could therefore be utilized to restrain the pyranoside ring thereby making boat conformations energetically unfavorable.[5] It therefore became possible to specifically address the occurrence of boat forms during the hydrolysis of β-glucosides.

Examination of the chair to boat flexibility of **5a** and **5b** with MM3 showed that the effect of the constraining groups was to destabilize the boat conformations by 5-8 kcal relative to the unconstrained monosaccharide (Figure 1). Nonetheless, the β-anomers of the 6·6·6 system **5a** and the 6·6·5 analogue **5b** were found to react at appreciable rates. Furthermore, the cleavage rates for the β-anomers were not dramatically slower than those for the analogous α-anomers, which according to theory,[1] should have reacted from chair forms without affect by the cyclic restraining groups.

These results raised doubts that β-anomers react *via* boat conformations and prompted us to undertake *ab initio* studies of the cleavage mechanism for α-and β-glycosides.

Ab Initio Studies Of Glycoside Cleavage Involving Protonated Dimethoxymethane[7]

Glycoside cleavage is subject to acid catalysis (Scheme 2). Consider the fragment CO_n-COH+ where O_n is an oxygen bearing an n-type lone pair, OH+ is the protonated oxygen, and the σ* orbital is localized on the C-OH+ bond. There are two cases to consider depending on whether the CO-COH+ dihedral angle is gauche or anti as in **8** and **11** respectively. The results summarized below pertain to *in vacuo* equilibrium structures obtained by geometry optimization using the 6-31G* basis set.

If the angle is gauche, e.g. **8**, the O_n-C bond length is near 1.31 Å while the C-OH+ bond length is near 1.56 Å. (Scheme 2). In the unprotonated acetal, **7**, these bond lengths would be 1.37 and 1.39 Å respectively. *It is clear that the effect of the proton is to enhance the nσ* interaction that pre-existed in 1, and to drive the system toward cleavage of the C-OH⁺ bond.* Other indicators of progress toward cleavage are (a) that the C-O_n-C angle expands towards 120° (118.3°) consistent with rehybridization from sp³ to sp², and that the CH_3O--COH+ dihedral angle moves towards the 90° mark. Structure **8** shows

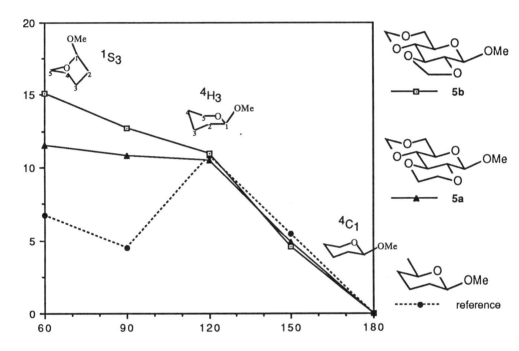

C5-O5--C1-O1 Dihedral Angle (degrees)

Figure 1: MM3 Energies for Chair-->Twist Boat Conformational Changes of Pyranosides

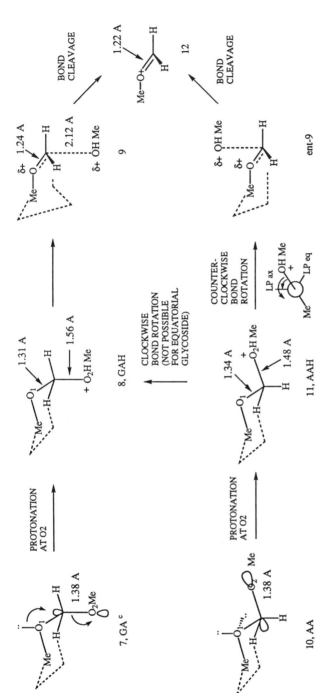

Scheme 2[a,b]

a Broken lines are used to show the relationship of the pyranose ring to the dimethoxymethane conformer in question.
b Tabulation of bond angles and dihedral angles that indicate hybridization changes for rotamer GA:

	7	8	9
C-O1-C bond angle	114.1^0	118.3^0	122.5^0
C-O1-C-O2 dihedral angle	65.8^0	86.4^0	90.0^0

c Although GG is the lowest energy rotamer, we have used GA to avoid the influence of the exo anomeric effect, and in order to focus on a single no* interaction.

the critical bond-lengths of protonated dimethoxymethane in GA conformation (hereafter referred to as GAH, Figure 2) manifesting a $n\sigma^*$ interaction.[7]

For conformations in which CO_n-COH^+ is anti, as in **11**, the bond length changes upon protonation are significantly diminished. In this case, the O_n-C bond length is near 1.34 Å while the C-OH^+ bond length is near 1.48 Å. Since the n-type orbital is not antiperiplanar to the σ^* orbital, there is no possibility of an $n\sigma^*$ interaction and the small changes bond lengths reflect the effect of protonation alone. The shortening of the O_n-C bond is probably a coulombic effect as described by Wolfe and coworkers.[8] In addition, rehybridization of the C-O_n-C bond angle does not occur (115.6°).

There are two cleavage options available to the protonated conformation **11** which lacks an $n\sigma^*$ interaction: it can cleave in its current conformation to yield a high energy out-of-plane carbenium ion (i.e. no oxostabilization), or it can undergo bond rotation to give a conformation with an $n\sigma^*$ interaction, and then cleave with oxostabilization.

Figures 3 and 4 illustrate bond length and energy variations respectively for the conversion of a protonated conformation without a $n\sigma^*$ into a conformation with a $n\sigma^*$ interaction i.e., reading from right to left AAH--->GAH. Figure 3 shows that as the $n\sigma^*$ interaction develops, there are synchronized changes of length in the bonds undergoing reorganization. Figure 4 shows that energy stabilization is concomitant with development of $n\sigma^*$ overlap. *It is also seen that protonation lowers the energy barrier for the conversion (AAH to GAH, solid curve), relative to the unprotonated conversion (AA to GA, dashed curve).*

Advancement To the Transition State for Cleavage

In structure **8**, Scheme 2 it is seen that protonation has increased the C-OH^+ bond length to 1.56 Å only. A study by Burgi and Dubler-Steudle predicts that the bond length in the transition state should be near 2.0 Å.[9] A species meeting this requirement was found by building an idealized transition state structure **9** with a 90° C-O_n--C-OH dihedral angle, and a 180° O_n-C--OH-C dihedral angle. The 90° value for the CO_n--COH^+ dihedral angle places the protonated leaving group orthogonal to the planar oxocarbenium ion. Geometry optimization with the 6-31G* basis set and the constraints produced structure **9**, which lies only 2.8 kcal above the equilibrium structure **8**. Prominent features of this model are an optimized O_n-C bond length of 1.24 Å and a C-OH^+ bond length of 2.12 Å. Also prominent is the planar oxocarbenium ion fragment consisting of C-O_n=C with an angle of

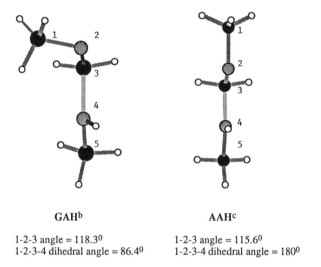

GAH[b]   AAH[c]

1-2-3 angle = 118.3^0 1-2-3 angle = 115.6^0
1-2-3-4 dihedral angle = 86.4^0 1-2-3-4 dihedral angle = 180^0

a Figure adapted from reference 7.
b GAH is the same as GAH2 in reference 7.
c 1-2-3-4 constrained to 180^0; changes to GAH with full optimization.

**Figure 2: 6-31G* Geometry Optimized Rotamers
of Protonated Dimethoxymethane[a]**

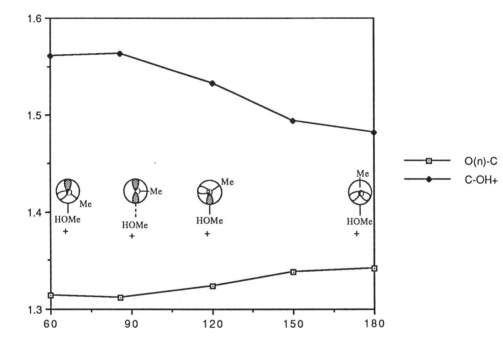

Me-O--C-OH+ Dihedral Angle (degrees)

Figure 3:   Bond Length Variation (6-31G* opt)
For AAH-->GAH Conversion

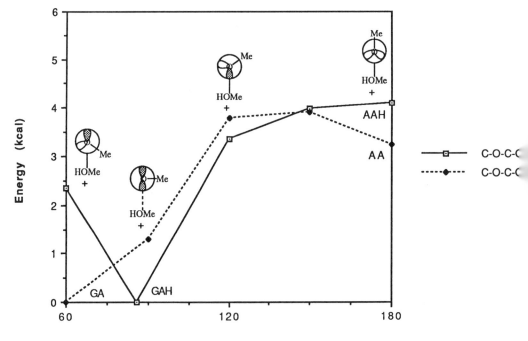

Figure 4:   Energy Variation (6-31G* opt) For AA-->GA
Conversion Before and After Protonation

122.5°, indicating that O_n, the donor oxygen, is sp^2 hybridized in the transition state.

Extending the Dimethoxymethane Study to Pyranosides

The relevance of dimethoxymethane to the study of glycoside cleavage is indicated by means of the broken lines in **7** and **10** (Scheme 2). All pyranoside structures were geometry optimized using the 6-31G* basis set. Structure **7** mimics the gauche C5-O5--C1-O1 fragment of an axial glycoside. The C-O bond lengths are near 1.38 Å. Protonation of the *exo*cyclic oxygen leads directly to activated structures **8**, and **9**, described above, and cleavage product **12**. The cyclic counterpart of **9** is a $^4H_3$ half-chair.

However the picture for the equatorial pyranoside, **10**, was more complicated. Protonation of the exocyclic oxygen gives **11** which shows some C-O bond lengthening and shortening relative to **10**, but not to the same extent as occurs in the axial counterpart, **8**. The difference is that there is no ring oxygen lone pair that is antiperiplanar to the σ* orbital of the protonated C-O bond. Therefore, there is no nσ* interaction present in **11** that can drive the system to cleavage.

To generate an nσ* interaction in structure **11**, bond rotation must occur. Clockwise rotation would convert **11** into **8**. This is possible for dimethoxymethane but not for the pyranoside because of restrictions imposed by the ring.

On the other hand, *counter*clockwise rotation in structure **11** offers two possibilities for nσ* interactions. Counterclockwise rotation by 120° generates a conformation with an antiperiplanar nσ* interaction involving the "equatorial" lone pair orbital. The conformation thereby obtained is effectively a boat (c.f. **3**). **However, counterclockwise rotation by only 60° generates a conformation in which the protonated oxygen is *syn*periplanar to the "axial" lone pair.** This structure is the enantiomer of **9**, ie. **ent-9** and hence it leads to oxocarbenium ion **12** *via* activated cleavage. The cyclic counterpart of **ent-9** is a ^4E sofa.

The energy profile for the counterclockwise rotation **11--->ent-9** is shown in Figure 4. Structure **11** corresponds to AAH with a CO_n--COH^+ dihedral angle of 180°. We interpret Figure 4 as follows. The 180° structure on the right has no nσ* interaction and cannot cleave. It therefore begins rotation toward the 60° conformation. At the 120° point a synperiplanar nσ* interaction develops. However, the nσ* interaction does not mature until the 90° mark is reached, and with activation, it converts to transition state **ent-9** which now undergoes cleavage to **12**.

Stereoelectronic Requirements for Cleavage

The ALPH would dictate that the cleavage must occur at the 60° point where the antiperiplanar nσ* alignment is reached. However Figure 4 shows that were the 60° point to be reached, the system would have to return to the 90° angle for cleavage to occur. Figure 4 further indicates that cleavage is equally facile from the 60° and 120° conformations.

One can postulate from this a general rule for stereoelectronic control, that a **periplanar** lone pair is needed for cleavage, whether it be anti or syn.   For an equatorial glycoside, the antiperiplanar relationship corresponds to the $^{1,4}B$ (or $^{1}S_3$) boat, **3**, invoked by Deslongchamps.   The synperiplanar relationship occurs in a ^4E sofa conformation.[10]

The conclusion that a synperiplanar lone pair orbital can be a precursor to cleavage could not have been reached without studying nσ* systems with a protonated leaving group. Indeed, unprotonated acetal fragments always show optimum nσ* interactions at the 60° (gauche) conformation involving an antiperiplanar lone pair orbital. Figure 4 shows the rotational energy curve for the unprotonated C-O-C-O analog (dashed curve), showing that the synperiplanar alignment is not energetically favorable.

Effect of sp² vs sp³ Hybridization of Oxygen

Thus far, the n type lone pairs have been represented as sp^3 hybrid orbitals **13**, Figure 5, since this was a feature of Deslongchamp's view of stereoelectronic control. However, a more accurate picture of the lone pair orbitals on oxygen has emerged in which the oxygen lone pair orbitals are considered to be energetically nonequivalent, **14**, with orbitals of σ and π symmetry[8,11] The π orbital is higher in energy by 20-45 kcal, and is the frontier orbital.

An important structural aspect of the transition state (e.g. **9**, Scheme 2) is that the donor oxygen is sp^2 hybridized based on the fact that the $C-O_n-C$ angle is 122°. This suggests that the n-type donor orbital should be pictured as a π or p orbital at the transition state **15**.

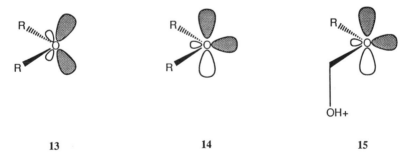

| 13 | 14 | 15 |

Figure 5: n-Type Orbital Representations for the Donor Oxygen

Another important structural aspect is that the CO_n--$COH+$ dihedral angle is $90°$. Based on this, the leaving group ($OH+$) and the donor π orbital are coplanar, meaning that the n and σ^* orbitals are periplanar by the Klyne and Prelog definition[12] (the lobes of the p orbital are simultaneously antiperiplanar and synperiplanar to the σ^* orbital). This is simply another perspective on the more general rule for stereoelectronic control, stated above.

Acknowledgements

This work was supported by grants from the National Science Foundation to B.F.R. (CHE 8920033) and from the North Carolina Supercomputing Center to B.F.R., J.P.B. and C.W.A. We thank the Biological Instrumentation Program (NSF) for use of the convex C220 at the University of North Carolina, Chapel Hill.

References

1. Deslongchamps, P. "Stereoelectronic Effects in Organic Chemistry", Pergamon Press, New York, 1983, pp 30-35.
2. Altona C., Havinga, E. *Tetrahedron*, **1966**, *22*, 2275
3. Sinnott, M.L. *Adv. Phy. Org. Chem.*, **1988**, *24*, 113.
4. Benner, S.A. *Mol. Struct. Energ.*, **1988**, *9*, 27.
5. Ratcliffe, A.J.; Mootoo, D.R.; Andrews, C.W.; Fraser-Reid, B. *J. Am. Chem. Soc.*, **1989**, *111*, 7661.
6. Mootoo, D.R.; Date, V.; Fraser-Reid, B. *J. Am. Chem. Soc.*, **1988**, *110*, 2662.
7. Andrews, C.W.; Bowen, J.P.; Fraser-Reid, B. *J. Chem.Soc., Chem. Commun.*, **1989**, 1913.
8. Wolfe, S.; Rauk, A.; Tel L.M.; Csizmadia, I.G. *J. Chem. Soc., B*, **1971**, 136; Wolfe, S.; Whangbo, M.-H.; Mitchell, D.J. *J. Carbohyd. Res.*, **1979**, *69, 1.*
9. Burgi, H.-H.; Dubler-Steudle, K.C. *J. Am. Chem. Soc.*, **1988**, *110*, 7291.
10. Andrews, C.W.; Fraser-Reid, B.; Bowen, J.P. *J. Am. Chem. Soc.*, **1991**, *113*, 8293.
11. David, S.; Eisenstein, O.; Heher, W.J.; Salena, L.; Hoffman, R. *J. Am. Chem. Soc.*, **95**, 3806, 1973; A Pictorial Approach To Molecular Structure and Reactivity, Hout, R.F. Jr.; Pietro, W. J.; Hehre, W.J. John Wiley & Sons, New York, 1984, p. 55 and pp. 73-79).
12. W. Klyne and V. Prelog, Experientia, 16, 521 (1960)

RECEIVED June 24, 1993

Chapter 8

X–C–Z Anomeric Effect and Y–C–C–Z Gauche Effect (X,Y = O,S; Z = O,N)
Evaluation of the Steric, Electrostatic, and Orbital Interaction Components

B. Mario Pinto and Ronald Y. N. Leung

Department of Chemistry, Simon Fraser University, Burnaby, British Columbia V5A 1S6, Canada

A combined experimental and theoretical protocol has yielded the steric, electrostatic, and orbital interaction components of the X-C-Z anomeric and Y-C-C-Z gauche interactions between O/O, S/O, O/N, and S/N heteroatoms. The anomeric effect and gauche effect have been studied in 2-substituted heterocyclohexanes and 5-substituted 1,3-diheterocyclohexanes, respectively. The composite conformational effects operating in 2- or 3-substituted diheterocyclohexanes have then been evaluated in light of the former individual effects. Low temperature ^{13}C NMR spectroscopy has yielded the conformational free energies and molecular mechanics (MM2) calculations have yielded the electrostatic and steric components. The orbital interaction component is obtained by difference. The orbital interaction terms of the X-C-Z anomeric effect and the Y-C-C-Z gauche effect are found to be attractive for all combinations studied. The combination of an endo and exo anomeric interaction is always found to be more favorable than the exo interaction alone. The concept of a "reverse anomeric effect" observed with NHR substituents is discounted. The conformational preferences in O-C-N and S-C-N fragments are attributed mainly to a large steric component that outweighs the attractive orbital interaction component. Additivity of the orbital interaction component of the anomeric and gauche effects in the 1,4-diheterocyclohexanes is observed for the systems with oxygen substituents but generally not for those with nitrogen substituents. An additional hydrogen bonding component is invoked to explain the discrepancy in the latter systems. A proposal for the additivity of orbital interaction effects is advanced.

A great deal of effort by both experimental and theoretical chemists has focused on the investigation of the nature and "origins" of seemingly anomalous conformational

0097–6156/93/0539–0126$08.50/0

behavior. Such activity has resulted in the definition of special conformational effects whose understanding continues to be of fundamental significance. Two such effects, the anomeric effect (*1-3*) and gauche effect (*4*) have have gained particular notoriety in the scientific community (*5-8*). The effects have emerged as phenomenological effects (*9*) whose "origins" remain somewhat elusive. A large part of the problem rests with the decomposition of a total effect into its various electronic and steric components. Thus, whereas it is relatively easy to establish the result of a perturbation such as the replacement of one heteroatom by a cognate atom, it is a more difficult task to deduce the cause of the effect. It is even more of a challenge to make predictions regarding an unknown system and one has to resort usually to an experimental or theoretical study of the latter prior to a rationalization.

It was our intention to gain some insight into the "origins" of the X-C-Z anomeric and Y-C-C-Z gauche interactions between the heteroatom pairs O/O, O/N, S/O, and S/N in terms of their steric, electrostatic, and orbital interaction components. Furthermore, we intended to probe the interplay of the anomeric and gauche effect when present concurrently in a given system. Thus, the systems that were of ultimate interest to us were the 2- or 3-substituted-1,4-diheterocyclohexanes (Scheme 1) which could exhibit both anomeric and gauche effects. We have chosen the 2-substituted heterocyclohexanes and 5-substituted 1,3-diheterocyclohexanes (Scheme 1) as models with which to probe the anomeric and gauche effects, respectively.

Attractive and Repulsive Gauche Effects

The gauche effect was originally defined by Wolfe (*4*) as the tendency for a molecule to adopt that structure which has the maximum number of gauche interactions between adjacent electron pairs and/or polar bonds. In the present work, a preference for the gauche or anti form in a Y-C-C-Z fragment in excess of that predicted on the basis of steric and polar factors is termed an attractive or repulsive gauche effect, respectively (*10*).

The gauche conformations of certain 1,2-diheterosubstituted ethanes (Scheme 2) are favored in spite of the unfavorable steric and electrostatic interactions (*10*). The gauche conformation of ethylene glycol itself is favored but in this case there is stabilization due to hydrogen bonding (*11*). Dimethoxyethane also shows a strong gauche preference in solution although a smaller gauche preference is indicated in the gas phase (*11*). Ab initio calculations at a high level of computation, including electron correlation, show a very small gauche/anti energy difference (*11*).

A preference for the axial orientation of the substituents in the 5-substituted 1,3-dioxanes (Scheme 2), corresponding to an attractive gauche effect, has been observed (*12*). In contrast, Eliel and Juaristi (*13*) have provided evidence for repulsive S/O and S/S gauche effects from a study of the equilibria of 5-methoxy- and 5-methylthio-1,3-dithianes (Scheme 2).

The conformational equilibria of 1,2-disubstituted cyclohexanes have been compared with those calculated on the basis of the known conformational energies (ΔG^0_X, ΔG^0_Y) in the monosubstituted cyclohexanes (Scheme 3) (*14,15*). The interaction between substituents ($\Delta G^0_{X/Y}$), in the absence of additional conformational effects, should be approximated by the sum of the steric (E_v) and polar (ΔE_D) interactions of X and Y, and a plot of $\Delta G^0_{X/Y} - E_v$ vs ΔE_D should yield

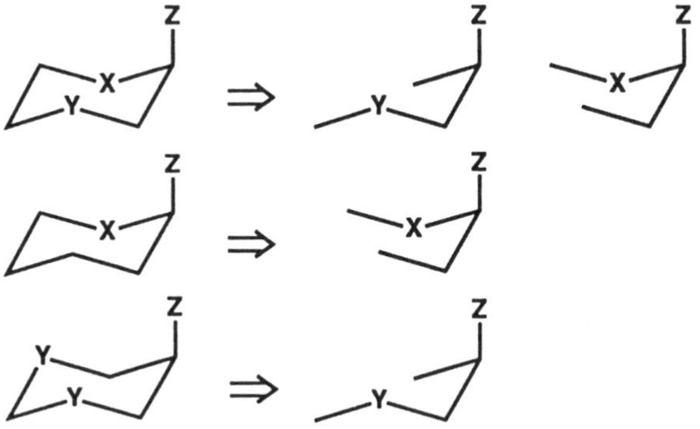

SCHEME 1

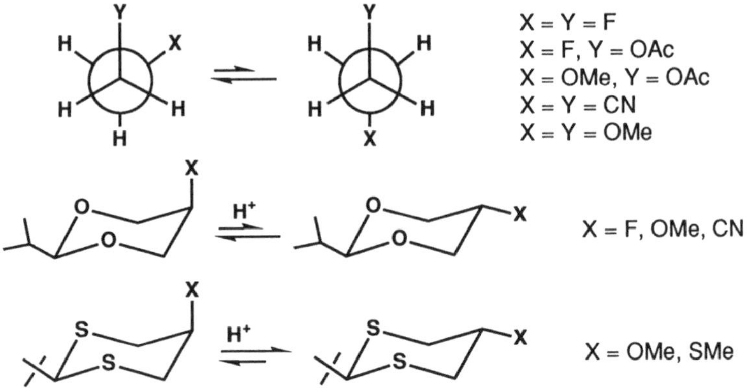

$X = Y = F$
$X = F, Y = OAc$
$X = OMe, Y = OAc$
$X = Y = CN$
$X = Y = OMe$

$X = F, OMe, CN$

$X = OMe, SMe$

SCHEME 2

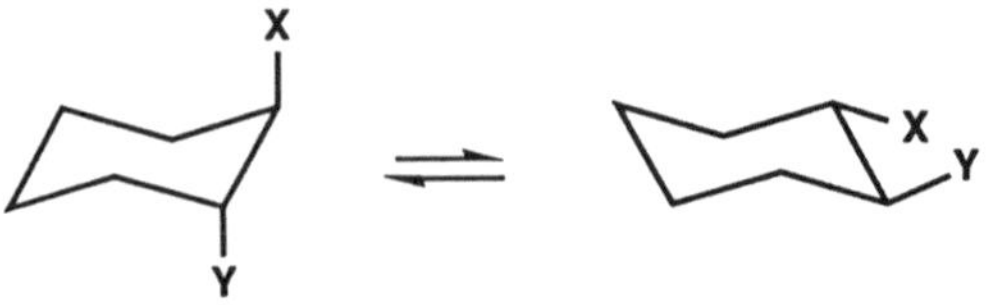

$$\Delta G°_{exp} = \Delta G°_{X/Y} + \Delta G°_X + \Delta G°_Y$$
$$\Delta G°_{X/Y} = E_V + \Delta E_D$$
$$\Delta G°_{X/Y} - E_V = \Delta E_D$$

SCHEME 3

a straight line of unit slope. If additional effects were operative, points in the field above the line would indicate additional repulsion while those in the field below the line would indicate additional attraction. Use of this approach led Zefirov *et al.* (*14,15*) to conclude that the O/Cl, O/Br, O/I, and Cl/I interactions could be interpreted adequately in terms of steric and polar interactions whereas the O/O, F/I, F/Br, F/Cl, and F/O systems exhibited additional gauche attraction, and the S/S, S/Cl, S/Br, and S/O systems exhibited additional gauche repulsion. Extension of this work by Carreno *et al* (*16*) to other 1,2-disubstituted cyclohexanes (X= SMe; Y= OMe, OH, F, Cl, Br, I) has suggested that, with the exception of the S/F interaction, all others are repulsive.

The 2-substituted-1,4-diheterocyclohexanes (Scheme 4) have also been examined by Zefirov (*17*) and here too, additional attractive and repulsive gauche effects were proposed. The repulsive effects were attributed to the through-space interaction between lone pair orbitals (Figure 1). The net interaction is destabilizing (*18*). The decrease in the proportion of the axial conformer in the 1,4-oxathiacyclohexanes relative to the corresponding dioxacyclohexanes was taken as evidence for stronger repulsive S/S than O/S interactions. However, in a subsequent photoelectron spectroscopic study of the analogous lone pair interactions in 3,7,9-triheterobicyclo [3.3.1] nonanes (Scheme 5), Gleiter *et al.* (*19*) showed that the interaction between lone pairs has a significant through-bond component, as advocated by Hoffmann (*20*), and cautioned against a simple interpretation in terms of through-space interactions.

According to Hoffmann's explanation, the non-bonding orbitals on the heteroatoms interact through-space to give in phase and out-of-phase combinations (Figure 1) but these combinations interact in turn with σ and σ^* orbitals of the σ framework to give bonding and antibonding combinations, as shown in Figure 2. One of these interactions is a two electron, two orbital interaction and is stabilizing whereas the other is a four electron, net destabilizing interaction (*18*). Since the two interactions oppose one another, their relative contributions can result in a lone pair interaction that is net attractive or net repulsive. Through-bond and through-space interactions have been suggested as the "origin" of additional gauche attraction (*21*).

An alternative explanation for the gauche effect has been advanced by Dionne and St. Jacques (*22*) on the basis of a conformational study of the 3-X-substituted-1,5-benzodioxepins (X= I, Br, Cl, F, OMe) (Scheme 6). Orbital interactions of the $\sigma \rightarrow \sigma^*$ type were shown to be more important than other factors such as electrostatic effects. The increase in the T_B and C_a conformers from I to Cl was attributed to greater $\sigma_{C-H} \rightarrow \sigma^*_{C-X}$ interactions, and the corresponding decrease in the C_e conformer to less important $\sigma_{C-X} \rightarrow \sigma^*_{C-O}$ interactions. Similar $\sigma \rightarrow \sigma^*$ interactions were also shown to be important in dictating the conformational preferences of the 3-X-substituted-1-benzoxepins (where one of the oxygen atoms is replaced by carbon) (*23*). The explanation was also later forwarded by Alcudia *et al.* (*24*) to account for the gauche preference of the methylthio group in 3-methylthio derivatives of 2-methoxyoxacyclohexane and also by Juaristi and Antunez (*25*) to account for the preferences in 5-O-substituted-1,3-dioxanes.

A further explanation has been suggested recently by Wiberg *et al.* (*26*) to account for the gauche preference in 1,2-difluoroethane. The gauche preference is attributed to destabilization of the anti form which results from more severely bent bonds and less bond overlap in the latter conformer.

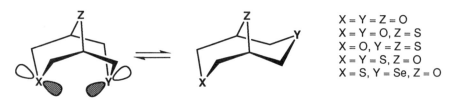

X = OBu^n, Y = O; X = OMe, Y = S
X = SBu^n, Y = O; X = OEt, Y = S
X = OBu^n, Y = S; X = OBu^t, Y = O
X = SBu^n, Y = S; X = SPr^n, Y = O
X = OMe, Y = O; X = SPr^n, Y = S
X = OEt, Y = O

SCHEME 4

X = Y = Z = O
X = Y = O, Z = S
X = O, Y = Z = S
X = Y = S, Z = O
X = S, Y = Se, Z = O

SCHEME 5

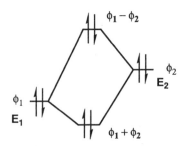

Destabilization α S^2 (E$_1$ + E$_2$)

Figure 1. Energy level diagram for a two orbital, four electron destabilizing orbital interaction.

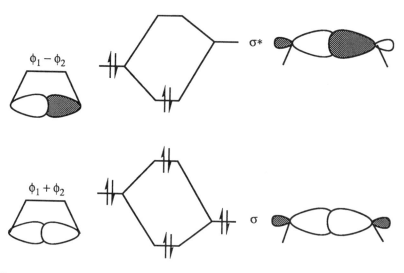

Figure 2. Energy level diagram for through-space and through-bond orbital interactions.

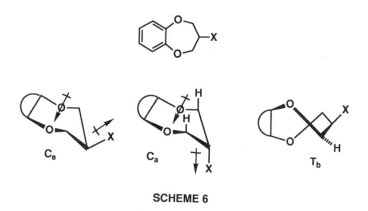

SCHEME 6

There have been several additional studies of the gauche effect. For example, the conformational analysis of 2-alkylthio derivatives of 1,1-dimethoxyethane (Scheme 7) indicated that the sterically most hindered, gauche conformation was preferred. Analysis of the rotamer populations led the authors to conclude that it was the $\sigma_{C-H} \rightarrow \sigma^*_{C-O}$ and $\sigma_{C-S} \rightarrow \sigma^*_{C-H}$ interactions that were of importance in controlling the preferences (27). The O-C-C-S gauche effect has also been investigated in 3-substituted thiacyclohexanes, and the effects of sulfoxide and sulfone functions have also been noted (28,29). The S-C-C-N gauche effect has been studied in 1,2-disubstituted phenylethanes and also in the 3-substituted thiacyclohexanes (30,31). The existence of an attractive gauche effect between acetoxy or acetamido functions and C-N or C-O bonds has also been demonstrated by Bernet et al. (32).

Endo and Exo Anomeric Effects

The concept of endo and exo anomeric interactions (33-37) appears to have been widely accepted although there is some question as to the magnitude and influence of the exo anomeric effect (38).

The systems containing exocyclic nitrogen substituents are of particular interest since they exhibit an equatorial preference (39,40). It has been suggested (41) that the observed preference results from the dominance of $n_N \rightarrow \sigma^*_{C-O}$ orbital interactions in the equatorial conformer, and can be attributed to a "reverse anomeric effect". Booth et al. (42) also concluded that the equatorial preference of the methylamino group in 2-methylaminooxacyclohexane results from a compromise between the endo and exo anomeric effects (the N-exo anomeric effect in the equatorial conformer is not opposed by the O-endo anomeric effect, as in the axial conformer).

The widely accepted explanation for endo and exo anomeric interactions is one in terms of stabilizing charge transfer interactions between the donor lone pair orbitals and the adjacent σ^* acceptor orbitals. The interaction is illustrated in Figure 3 for the case of RY-CH$_2$-X, within the framework of Perturbational Molecular Orbital (PMO) theory (41). The stabilizing interaction derives from the p-type orbital on the heteroatom Y with the σ^*_{C-X} fragment orbital in the gauche conformation; the stabilization is proportional to the square of the overlap between interacting orbitals and is inversely proportional to their energy difference (18). When X is a rotor, similar interactions between the p-type orbital on X and σ^*_{C-Y} will lead to a preference for the gauche, gauche conformation. These orbital interactions are now associated with endo and exo anomeric interactions (33-37).

The existence of anomeric interactions involving sulfur atoms is of relevance to the present work and has been confirmed both by experimental (43) and theoretical (44) studies.

Methodology ((45); Leung, R. Y. N.; Pinto, B. M.; manuscripts submitted)

Whereas the studies described in the foregoing sections have provided valuable information on the nature of the conformational effects, the exact evaluation of the individual steric, electrostatic, and orbital interaction components has been difficult.

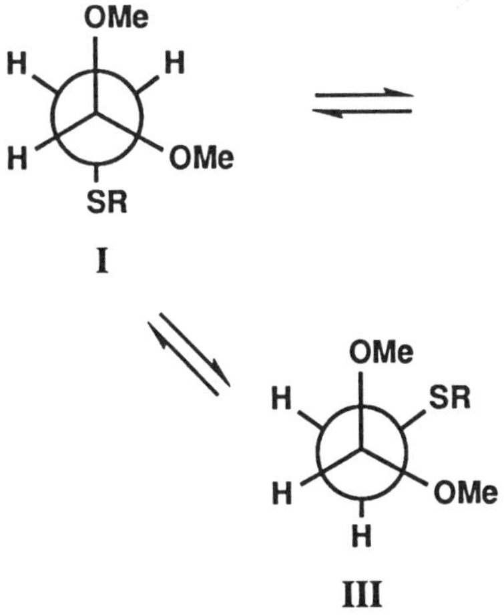

SCHEME 7

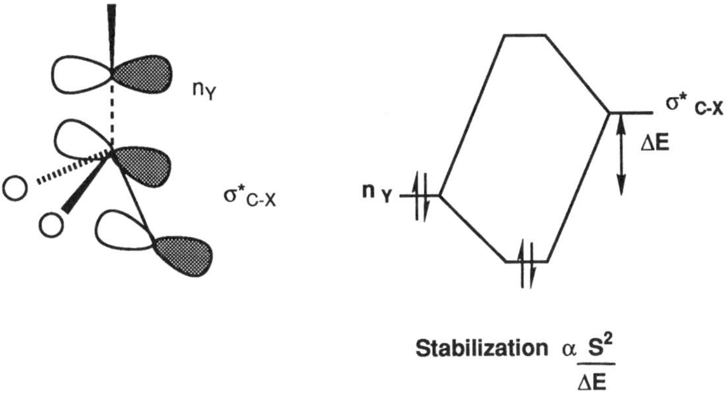

Stabilization $\alpha \dfrac{S^2}{\Delta E}$

Figure 3. Combinations of through-space and through-bond orbital interactions in 5-methylamino-1,3-dioxacyclohexane **4**.

In order to provide an evaluation of such components for the O/O, O/N, S/O, and S/N gauche effect and anomeric effect, the systems shown in Scheme 8 were synthesized. Of particular interest is an understanding of the individual components associated with the interplay of the anomeric and gauche effects.

Assuming additivity, the experimental conformational free energy differences can be represented by equation 1, where ΔG^0_{steric} and $\Delta G^0_{electrostatic}$ are the steric and electrostatic components, respectively, and $\Delta G^0_{orbital}$ is attributed here to the additional free energy difference when the summation of ΔG^0_{steric} and $\Delta G^0_{electrostatic}$

$$\Delta G^0_{exptl} = \Delta G^0_{steric} + \Delta G^0_{electrostatic} + \Delta G^0_{orbital} \tag{1}$$

cannot account for the experimentally determined free energy difference. The experimental conformational free energies were evaluated by means of low temperature ^{13}C NMR spectroscopy. In certain cases, the original equilibria were too biased to permit an accurate determination of the equilibrium constant and counterpoise groups were necessary.

The steric interactions were evaluated in the following manner. Initially, the steric effect of the methyl group was evaluated computationally (MM2) (46) in the heterocycle of interest and in cyclohexane ($-\Delta G^0_{Me}$ or the "A_{Me} value"). The process gave a coefficient, α, that estimated the steric effects relative to cyclohexane (47). Then, for the substituent, Z of interest, the conformational free energy in the substituted cyclohexane ($-\Delta G^0_Z$ or the "A_Z value") was multiplied by the coefficient, α, for a given heterocycle. It must be emphasized that this "A value" was derived computationally for particular rotamers about the exocyclic linkage (see Scheme 9). The method is illustrated here for the case of 2-methylamino-1,4-dioxacyclohexane **18**. The value of α (2.06/1.74 = 1.18) indicates that the steric interaction of a methyl group in 1,4-dioxacyclohexane is 1.18 times more severe than in cyclohexane. Therefore, the estimated steric interaction of the methylamino substituent in **18** will be 1.83 × 1.18 = 2.16 kcal mol^{-1}, where the value of 1.83 is calculated for methylaminocyclohexane in which the rotamer about the exocyclic linkage is set to mimic a hypothetical exo anomeric interaction (42).

$\Delta G^0_{electrostatic}$ was estimated by use of a point charge model with Abraham's formula (48), as implemented in the PCMODEL program (49); a dielectric constant of 1.5 was used. The remaining orbital interaction component in each system was then evaluated by subtracting the steric and electrostatic components from the experimental conformational free energy.

Initially, the individual two-centered conformational effects were evaluated: the gauche effect in the 5-substituted-1,3-diheterocyclohexanes **3-15** and the anomeric effect in 2-substituted-oxacyclohexanes **21, 28** and -thiacyclohexanes **17, 23**. The additivity of these two effects was then assessed by comparison with the composite effect observed in the 2- or 3-substituted-1,4-diheterocyclohexanes. For the purposes of the present work, only the methylamino- and methoxy- substituent effects will be considered, and only the data obtained from direct observation of the conformational equilibria at the slow exchange limit will be discussed.

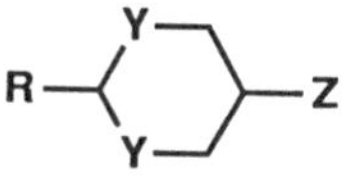

1.	Y = O	R = H	Z = H
2.	Y = O	R = i-Pr	Z = H
3.	Y = O	R = H	Z = NH$_2$
4.	Y = O	R = H	Z = NHMe
5.	Y = O	R = H	Z = NMe$_2$
6,9.	Y = O	R = i-Pr	Z = NH$_2$
7,10.	Y = O	R = i-Pr	Z = NHMe
8,11.	Y = O	R = i-Pr	Z = NMe$_2$
12.	Y = O	R = H	Z = OMe
13.	Y = S	R = H	Z = NHMe
14.	Y = S	R = H	Z = OMe
15.	Y = S	R = Me	Z = OMe

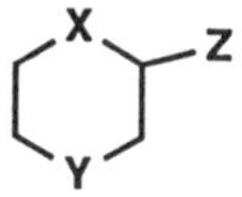

16.	X = S	Y = S	Z = NHMe
17.	X = S	Y = CH2	Z = NHMe
18.	X = O	Y = O	Z = NHMe
19.	X = S	Y = O	Z = NHMe
20.	X = O	Y = S	Z = NHMe
21.	X = O	Y = CHMe	Z = NHMe
22.	X = S	Y = S	Z = NHMe
23.	X = S	Y = CH2	Z = OMe
24.	X = S	Y = CHMe	Z = OMe
25.	X = O	Y = S	Z = OMe
26.	X = S	Y = O	Z = OMe
27.	X = O	Y = O	Z = OMe
28.	X = O	Y = CH2	Z = OMe

SCHEME 8

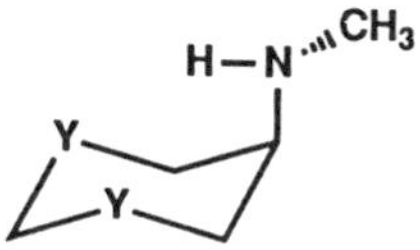

29 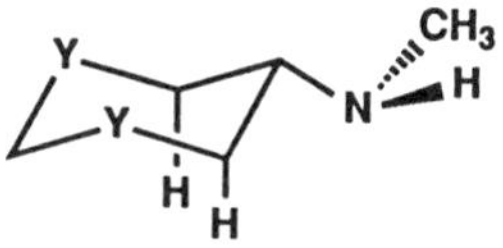   30 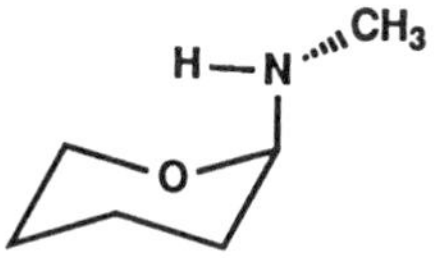   31

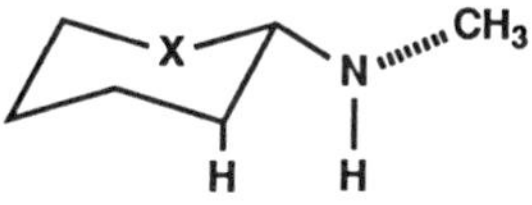

32 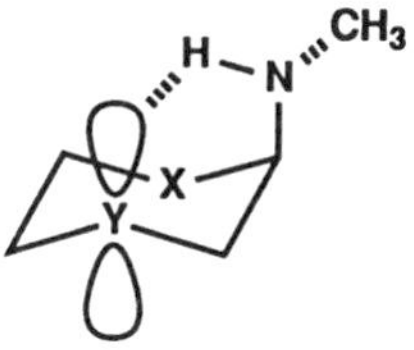   33 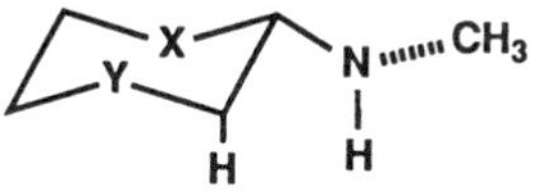   34

SCHEME 9

Gauche Effects

Table I gives the equilibrium data for the 5-methoxy- and 5-methylamino-1,3-diheterocyclohexanes. There are several noteworthy trends. For the 5-methoxy compounds **12**, **14**, the equatorial conformers are more stable whereas the axial conformers are more stable for the 5-methylamino compounds **4**, **13**. This is certainly not due to a steric effect since the A values of NHMe and OMe are 1.70 (*51*) and 0.6 kcal mol^{-1} (*52*), respectively. The greater equatorial preference of 5-methoxy-1,3-dithiacyclohexane **14** vs the corresponding dioxacyclohexane **12** is also inconsistent with expectations based on a consideration of electrostatic effects. Clearly, the preferences cannot be rationalized readily on steric or electrostatic grounds.

The component analysis is indicated in Table II. Since the ^{1}H NMR coupling constant data and the solvent dependence of the equilibrium in 5-methylamino-1,3-dithiacyclohexane **13** indicated a preponderance of the rotamer in which the N-H was pointing inside the ring (*45*), rotamer **29** (Scheme 9) was chosen for the calculation of steric and electrostatic effects in **13**, and also in **4**. For the equatorial conformer, rotamer **30** minimized the non bonded interactions (Scheme 9). For the 5-methoxy-1,3-dioxa and dithiacyclohexanes **12**, **14**, the steric interactions only make a minor contribution to the different conformational behavior. The corresponding 5-methylamino compounds **4**, **13** show an increased steric effect yet the the axial conformers are favored experimentally. In terms of electrostatic effects, there is a strong destabilization of the axial conformers in the 5-methoxy compounds **12**, **14**, as expected from a consideration of dipolar interactions (Scheme 10). However, there is an attractive electrostatic interaction in the axial conformers of the methylamino compounds **4**, **13**. This is likely due to a hydrogen bonding interaction (*45*) between the N-H hydrogen and the lone pairs on sulfur or oxygen.

Analysis of the Orbital Interaction Component. The orbital interaction component indicates that the gauche interactions between O/O, O/N, S/O, and S/N are all attractive. In contrast, the studies on 1,2-disubstituted cyclohexanes (*14, 15*) and 5-methoxy-1,3-diheterocyclohexanes (*13*) have suggested that although O/O has an attractive gauche effect, the S/O interaction is repulsive. It should be noted, however, that the O/S gauche interaction in 5-methylthio-1,3-dioxacyclohexane was judged to be attractive (*13*). The discrepancy in the S/O gauche effect may be due to the limitation in assessing the charges on atoms. Although the same methodology is adopted in the present study and the literature studies (*13-15*), lone pairs are included specifically in the PCMODEL calculations (*49*). In addition, previous workers have estimated the "theoretical" values for electrostatic and steric interactions by "hand calculations" of non-optimized geometries (Dreiding models) (*13*). As indicated by Eliel and Juaristi (*13*), the calculated energy difference of the axial and equatorial methylthiocyclohexane was overestimated more than two-fold relative to the experimental value; however, good agreement between calculated and experimental conformational free energies was obtained by use of the MM2 force field (*53*).

Through-Bond and Through-Space Effects. The interpretation of gauche interactions according to Hoffmann's theory (*20*), considers the effect of both through-bond and through-space orbital interactions. The interactions are illustrated

Table I. Equilibrium Data[a] for 5-Substituted-1,3-Dihetero-cyclohexanes (Scheme 8)

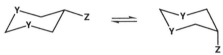

Compd.	Z	Y	Ke⇌a (error)	$\Delta G°$ (error) kcal mol^{-1}
4	NHMe	O	3.21 (0.04)	-0.36 (0.01)
12	OMe	O	0.53 (0.02)	-0.19 (0.01)
13	NHMe	S	4.90 (0.19)	-0.48 (0.03)
15 (14)	OMe	S	4.81[b] (0.21)	-1.44[c] (0.05)

a) In $CD_2Cl_2/CFCl_3$ (15/85) at 153 K.
b) The equilibrium constant is for 5-methoxy-2-methyl-1,3-dithiacyclo-hexane.
c) The $\Delta G°$ reported here is for 5-methoxy-1,3-dithiacyclohexane **14** after the $\Delta G°$ of the 2-methyl group ($\Delta G° = -1.92 \pm 0.02$)*(50)* has been subtracted from the $\Delta G°$ of 5-methoxy-2-methyl-1,3-dithiacyclohexane **15** ($\Delta G° = -0.48 \pm 0.03$).

**Table II. Component Analysis of Ethane-Type Gauche Effect in
5-Substituted-1,3-Diheterocyclohexanes (Scheme 8)**

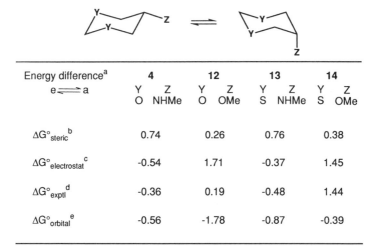

Energy difference[a] e ⇌ a	4 Y Z O NHMe	12 Y Z O OMe	13 Y Z S NHMe	14 Y Z S OMe
ΔG°_{steric}[b]	0.74	0.26	0.76	0.38
$\Delta G^\circ_{electrostat}$[c]	-0.54	1.71	-0.37	1.45
ΔG°_{exptl}[d]	-0.36	0.19	-0.48	1.44
$\Delta G^\circ_{orbital}$[e]	-0.56	-1.78	-0.87	-0.39

a) In kcal mol^{-1}.
b) Differences between the axial and equatorial conformers were calculated
   by multiplying the proportionality constants, obtained from comparing the
   methyl substituted cyclohexane and 5-methyl-1,3-dioxa and dithiacyclohexanes, with the "A-values" of the respective substituents (See text).
c) Differences in electrostatic energies were taken from the charge/charge
   interaction terms for the optimized structures, using the interactive molecular
   modelling program PCMODEL(49). This program uses a variant of the MM2
   force field.
d) Experimental ΔG° values from Table I.
e) Differences in orbital interaction energies were calculated by subtracting
   the ΔG°_{steric} and $\Delta G^\circ_{electrostat}$ from ΔG°_{exptl}.

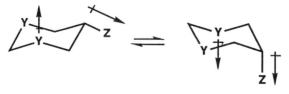

SCHEME 10

for the case of 5-methylamino-1,3-dioxacyclohexane **4** (Figure 4). The $(n_1\text{-}n_2) \rightarrow$ $\sigma^*_{C\text{-}C}$ interaction is stabilizing and the $(n_1\text{+}n_2) \rightarrow \sigma_{C\text{-}C}$ interaction is destabilizing. For simplicity, only one of the ring heteroatoms is considered. In fact, orbitals on the two ring heteroatoms interact through-space to give an n_+ and n_- combination (*54, 55*). The $\Delta G^0_{orbital}$ components in Table II indicate that whereas the O/O interaction is more attractive than the O/N interaction, the S/O interaction is less attractive than the S/N interaction (Scheme 11). To a first approximation, with the assumption that the $\sigma_{C\text{-}C}$ and $\sigma^*_{C\text{-}C}$ orbitals remain approximately constant in energy in the series of compounds, and given the relative energies of the p-type orbitals on O, N, and S (*56*), one predicts the relative ordering of the stabilizing and destabilizing interactions shown in Scheme 11. The prediction derives from the fact that the stabilizing interaction is inversely proportional to the energy difference (ΔE) between interacting orbitals and the destabilizing interaction is proportional to the sum of their orbital energies ($E_1\text{+}E_2$) (*18*). The analysis of the destabilizing orbital interactions requires further clarification. The original $n_N \rightarrow n_{O(S)}$ interaction results in a higher-lying (n_1 + n_2) combination than that resulting from an $n_O \rightarrow n_{O(S)}$ interaction (Figure 4). It follows that the relative magnitudes of the subsequent $(n_1\text{+}n_2) \rightarrow \sigma_{C\text{-}C}$ interaction will be O/O > O/N; S/O > S/N. It would appear that the $\Delta G^0_{orbital}$ component is dominated by the destabilizing interaction in the oxygen series and by the stabilizing interaction in the sulfur series. Comparisons between different pairs of heterocycles with the same substituent and different ring heteroatoms is difficult since, in addition to energy differences between interacting orbitals, overlap factors between orbitals could also be changing significantly. Nonetheless, the results indicate that the O/O interaction is more attractive than the S/O interaction, and the O/N interaction less attractive than the S/N interaction.

Through-Bond $\sigma \rightarrow \sigma^*$ Interactions. According to St. Jacques' interpretation of the gauche effect, the role of σ and σ^* orbitals is considered. The relevant interactions are shown in Scheme 12. The attractive orbital interaction component in the axial conformers suggests that four $\sigma_{C\text{-}H} \rightarrow \sigma^*_{C\text{-}X}$ interactions in the axial conformers are more stabilizing than the two sets of $\sigma_{C\text{-}Y(Z)} \rightarrow \sigma^*_{C\text{-}Z(Y)}$ interactions in the equatorial conformers. A more detailed analysis of pairs of molecules follows, for ease of comparison.

A comparison of the O/O vs O/N interaction is given by the pair **12** and **4** (Scheme 12). The greater O/O vs O/N interaction is consistent with the relative order of orbital energies since a lower-lying $\sigma^*_{C\text{-}O}$ than $\sigma^*_{C\text{-}N}$ fragment orbital is predicted to result in a greater interaction with the donor $\sigma_{C\text{-}H}$ orbital (*18*) in the axial conformer (Scheme 13). The trends in orbital interactions in the equatorial conformers also follow from the given order of orbital energies (Scheme 13). In the case of the O/N vs S/N interaction, given by the pair **4**, **13**, the interaction is greater in the sulfur analogue. This is consistent with a stronger $\sigma_{C\text{-}H} \rightarrow \sigma^*_{C\text{-}S}$ than $\sigma_{C\text{-}H} \rightarrow \sigma^*_{C\text{-}O}$ interaction (*44*) (Scheme 14). The interactions in the equatorial conformers can also be ordered according to the orbital energies (*44*), but both these interpretations must be treated with caution since overlap factors could also be changing significantly.

A comparison of the $\Delta G^0_{orbital}$ values for the 5-methoxy-1,3-dioxa vs - dithiacyclohexane **12** vs **14** (O/O vs S/O) indicates a stronger net stabilizing interaction in the axial conformer of **12** despite the weaker $\sigma_{C\text{-}H} \rightarrow \sigma^*_{C\text{-}O}$ than $\sigma_{C\text{-}H}$

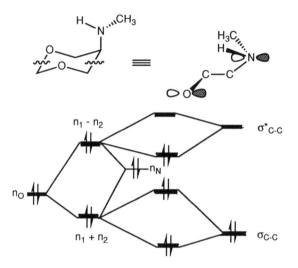

Figure 4. Perturbational Molecular Orbital (PMO) analysis of the stabilizing orbital interactions in the gauche conformation of RY-CH_2-X molecules.

$$-\Delta G^{\circ}_{orbital} \quad O/O > O/N$$
$$-\Delta G^{\circ}_{orbital} \quad S/O < S/N$$

$$n_O < n_N < n_S$$

$$(n_1 - n_2) \longrightarrow \sigma^*_{CC} \qquad (n_1 + n_2) \longrightarrow \sigma_{CC}$$

$$O/O < O/N \qquad\qquad O/O > O/N$$
$$S/O < S/N \qquad\qquad S/O > S/N$$

SCHEME 11

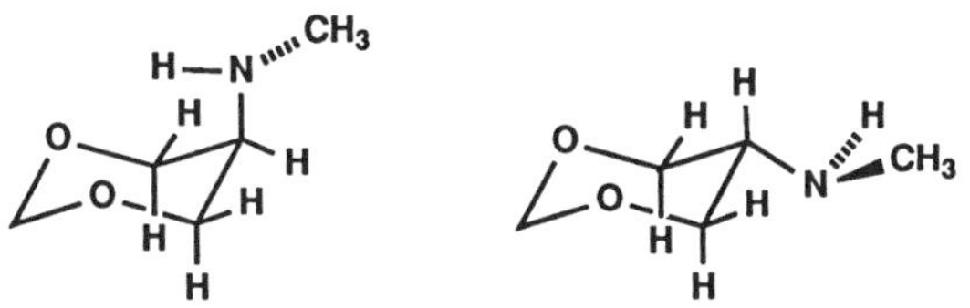

4

$2 \left[\sigma_{C-H} \longrightarrow \sigma^*_{C-O} \right]$
$2 \left[\sigma_{C-H} \longrightarrow \sigma^*_{C-N} \right]$

2 exo $\left[\sigma_{C-N} \longrightarrow \sigma^*_{C-O} \right]$
2 endo $\left[\sigma_{C-O} \longrightarrow \sigma^*_{C-N} \right]$

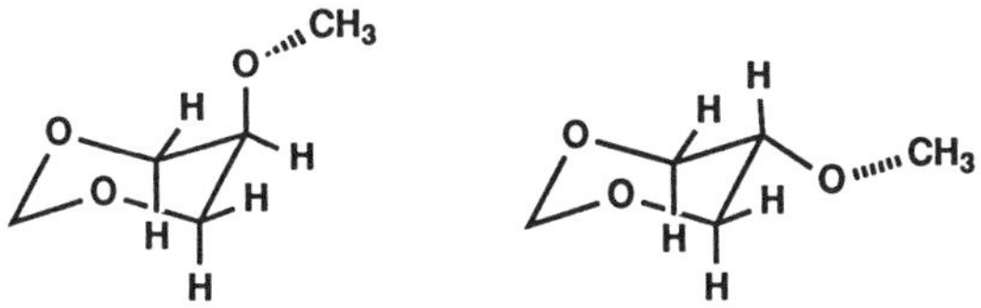

12

$2 \left[\sigma_{C-H} \longrightarrow \sigma^*_{C-O} \right]$
$2 \left[\sigma_{C-H} \longrightarrow \sigma^*_{C-O} \right]$

2 exo $\left[\sigma_{C-O} \longrightarrow \sigma^*_{C-O} \right]$
2 endo $\left[\sigma_{C-O} \longrightarrow \sigma^*_{C-O} \right]$

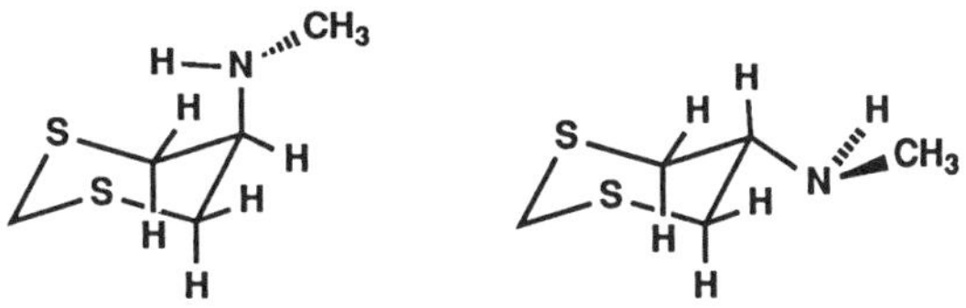

13

$2 \left[\sigma_{C-H} \longrightarrow \sigma^*_{C-S} \right]$
$2 \left[\sigma_{C-H} \longrightarrow \sigma^*_{C-N} \right]$

2 exo $\left[\sigma_{C-N} \longrightarrow \sigma^*_{C-S} \right]$
2 endo $\left[\sigma_{C-S} \longrightarrow \sigma^*_{C-N} \right]$

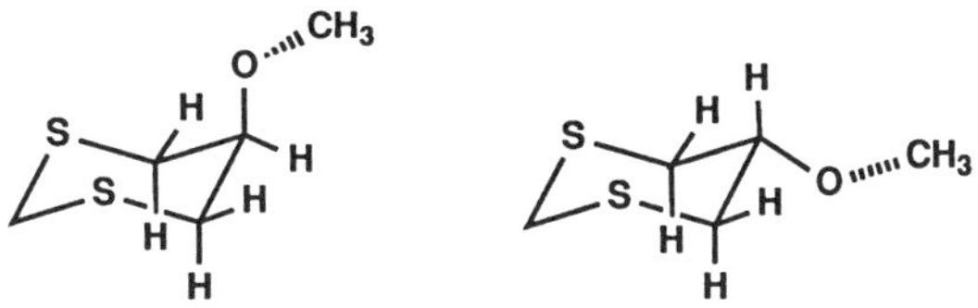

14

$2 \left[\sigma_{C-H} \longrightarrow \sigma^*_{C-S} \right]$
$2 \left[\sigma_{C-H} \longrightarrow \sigma^*_{C-O} \right]$

2 exo $\left[\sigma_{C-O} \longrightarrow \sigma^*_{C-S} \right]$
2 endo $\left[\sigma_{C-S} \longrightarrow \sigma^*_{C-O} \right]$

SCHEME 12

$\rightarrow \sigma^*_{C-S}$ orbital interaction (Scheme 15). Here too, the effects of differential overlap must be treated with caution. Finally, a comparison of S/N vs S/O gauche effects in **13** vs **14**, based on a consideration of orbital energies should be valid. However, the S/O interaction is less stabilizing than the S/N interaction despite the greater $\sigma_{C-H} \rightarrow \sigma^*_{C-O}$ interaction (Scheme 16). The inconsistencies might be due to the relative importance of the $\sigma_{C-Y(Z)} \rightarrow \sigma^*_{C-Z(Y)}$ interactions in the equatorial conformers (Schemes 13-16). An exact evaluation will have to await a detailed quantitative PMO analysis of these orbital interactions.

Analysis of Endo and Exo Anomeric Interactions

Table III summarizes the experimental conformational free energies for the 2-methoxy- and 2-methylamino-oxacyclohexanes and -thiacyclohexanes **17, 21, 23,** and **28,** and Table IV gives the component analysis. Whereas the axial conformers are favored for the methoxy-substituted compounds, the equatorial conformers are favored for the methylamino compounds.

O/O Interactions. The axial conformer of 2-methoxyoxacyclohexane **27** is favored by 0.46 kcal mol^{-1}. Similar values have been reported by previous workers (*35,59*). The electrostatic component favors the axial conformer, as predicted from a consideration of dipolar interactions (Scheme 17) (*2,3*). After subtraction of the steric and electrostatic components, the orbital interaction component has a value of 0.42 kcal mol^{-1} in favor of the axial conformer. *Therefore, a combination of one exo $n_O \rightarrow \sigma^*_{C-O}$ and one endo $n_O \rightarrow \sigma^*_{C-O}$ orbital interaction is more stabilizing than one exo $n_O \rightarrow \sigma^*_{C-O}$ orbital interaction by 0.42 kcal mol^{-1}* (Scheme 18). Thus, although the two hyperconjugative interactions oppose one another, the summation of the endo and exo interaction is more stabilizing than the exo interaction alone. Moreover, the magnitude of the endo interaction should be smaller than the exo interaction in the axial isomer since the p-type lone pair orbital of the endocyclic oxygen is not optimally aligned due to geometric constraints (*60,61*). It also follows that the exo interaction in the equatorial conformer will be greater than that in the axial conformer (*34*).

O/N Interactions. The equatorial conformer of 2-methylaminooxacyclohexane is favored by 1.66 kcal mol^{-1} but the orbital interaction component favors the axial conformer by 1.40 kcal mol^{-1}. According to Booth *et al.* (*35*) and Lemieux (*62*), the equatorial preference in 2-methylaminooxacyclohexane arises because of the competition between endo and exo anomeric interactions, in conjunction with the steric effect. Our data indicate that differences in the electrostatic components are small (0.35 kcal mol^{-1}) but that the steric energy differences are large (2.71 kcal mol^{-1}). The steric component is unusually large because the orientation of the methylamino substituent for an optimal exo anomeric effect in the axial conformer forces the N-H hydrogen into the oxacyclohexane ring (Booth *et al.*(*42*) have shown by NOE measurements and ^{1}H NMR coupling-constant data that the preferred rotamer in solution is as shown in **31** (Scheme 9)). In the equatorial conformer, the $n_N \rightarrow \sigma^*_{C-O}$ orbital interaction is present whereas in the axial conformer, both $n_O \rightarrow \sigma^*_{C-N}$ and $n_N \rightarrow \sigma^*_{C-O}$ interactions are present (Scheme 18).

$-\Delta G°_{orbital}$ O/O > O/N

Axial: $\sigma_{CH} \longrightarrow \sigma^*_{CO}$ > $\sigma_{CH} \longrightarrow \sigma^*_{CN}$

Equatorial: exo $\sigma_{CO} \longrightarrow \sigma^*_{CO}$ > $\sigma_{CO} \longrightarrow \sigma^*_{CN}$
             endo $\sigma_{CO} \longrightarrow \sigma^*_{CO}$ < $\sigma_{CN} \longrightarrow \sigma^*_{CO}$

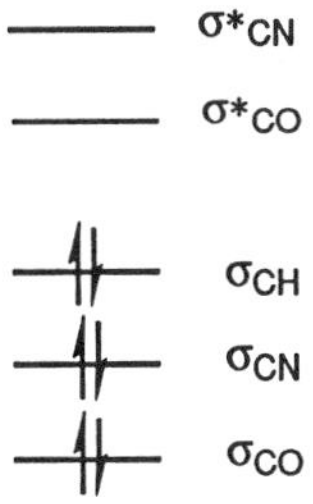

———— σ^*_{CN}

———— σ^*_{CO}

⊣⊦ σ_{CH}

⊣⊦ σ_{CN}

⊣⊦ σ_{CO}

SCHEME 13

$-\Delta G°_{orbital}$ O/N < S/N

Axial: $\sigma_{CH} \longrightarrow \sigma^*_{CS}$ > $\sigma_{CH} \longrightarrow \sigma^*_{CO}$

Equatorial: exo $\sigma_{CN} \longrightarrow \sigma^*_{CS}$ > $\sigma_{CN} \longrightarrow \sigma^*_{CO}$
             endo $\sigma_{CS} \longrightarrow \sigma^*_{CN}$ > $\sigma_{CO} \longrightarrow \sigma^*_{CN}$

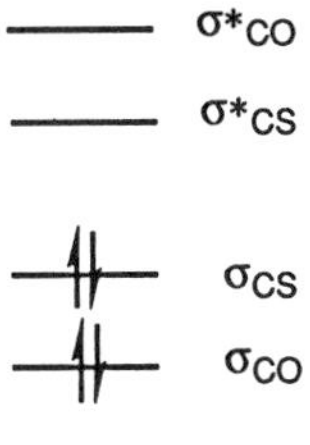

———— σ^*_{CO}

———— σ^*_{CS}

⊣⊦ σ_{CS}

⊣⊦ σ_{CO}

SCHEME 14

$-\Delta G°_{orbital}$ O/O > S/O

Axial: $\sigma_{CH} \longrightarrow \sigma^*_{CS}$ > $\sigma_{CH} \longrightarrow \sigma^*_{CO}$
Equatorial: $\sigma_{CO} \longrightarrow \sigma^*_{CS}$ > $\sigma_{CO} \longrightarrow \sigma^*_{CO}$

SCHEME 15

$$-\Delta G^\circ_{orbital} \ \ S/O < S/N$$

Axial: $\sigma_{CH} \longrightarrow \sigma^*_{CO} > \sigma_{CH} \longrightarrow \sigma^*_{CN}$

Equatorial: $\sigma_{CO} \longrightarrow \sigma^*_{CS} < \sigma_{CN} \longrightarrow \sigma^*_{CS}$

SCHEME 16

Table III. Equilibrium Data[a] for Substituted -1,4-Diheterocyclohexanes and Heterocyclohexanes (Scheme 8)

Compd.	X	Y	Z	Ke ⇌ a (error)	ΔG° (error) kcal mol^{-1}
16	S	S	NHMe	31.90 (1.60)	-1.06 (0.05)
17	S	CH$_2$	NHMe	0.37 (0.01)	0.31 (0.01)
18	O	O	NHMe	0.06 (0.00)	0.89 (0.04)
19	S	O	NHMe	b	<-1.1[c]
20	O	S	NHMe	0.19 (0.01)	0.51 (0.02)
21	O	CHMe	NHMe	2.60[d] (0.15)	1.66[e] (0.07)
22	S	S	OMe	28.60 (2.07)	-1.02 (0.09)
23	S	CH$_2$	OMe	b	-1.53[f]
25	O	S	OMe	0.63 (0.02)	0.14 (0.01)
26	S	O	OMe	b	<-1.1[c]
27	O	O	OMe	6.56 (0.20)	-0.57 (0.03)
28	O	CH$_2$	OMe	4.53 (0.11)	-0.46 (0.02)

a) In CD$_2$Cl$_2$/CFCl$_3$ (15/85) at 153K.
b) Only the axial conformer was observed.
c) See discussion.
d) The equilibrium constant (K) is for 4-methyl-2-methylaminooxacyclohexane **21.**
e) This was calculated by subtracting the ΔG°_{4-Me} of the 4-methyl substituent in oxacyclohexane (ΔG°_{4-Me} = -1.95 ±0.05 kcal mol^{-1}) (57) from the ΔG° of **21** (ΔG°_{21} = -0.29 ±0.02 kcal mol^{-1}).
f) From the equilibration of 2-methoxy-4-methylthiacyclohexane in CCl$_4$ (58).

Table IV. Component Analysis[a] of Conformational Effects in Substituted 1,4-Diheterocyclohexanes and Heterocyclohexanes (Scheme 8)

Compd.	X	Y	Z	$\Delta G°_{steric}$[b]	$\Delta G°_{electrostat}$[c]	$\Delta G°_{exptl\,(error)}$	$\Delta G°_{orbital}$[d]
16	S	S	NHMe	0.95	0.08	-1.06 (0.05)	-2.09
17	S	CH$_2$	NHMe	1.23	0.24	0.31 (0.01)	-1.16
18	O	O	NHMe	2.20	0.13	0.89 (0.04)	-1.44
19	S	O	NHMe	0.60	-0.05	<-1.1[e]	<-1.65
20	O	S	NHMe	2.47	0.24	0.51 (0.02)	-2.20
21	O	CHMe	NHMe	2.71	0.35	1.66 (0.07)	-1.40
22	S	S	OMe	0.27	0.13	-1.03 (0.09)	-1.43
23	S	CH$_2$	OMe	0.35	-0.63	-1.53[f]	-1.23
25	O	S	OMe	0.70	-0.12	0.14 (0.01)	-0.44
26	S	O	OMe	0.17	0.38	<-1.1[e]	<-1.65
27	O	O	OMe	0.62	0.10	-0.58 (0.03)	-1.30
28	O	CH$_2$	OMe	0.77	-0.81	-0.46 (0.02)	-0.42

a) In kcal mol^{-1}.

b) Calculated according to the corresponding "A values" of the substituents (see discussion).

c) Calculated according to Abraham's formula*(48)* using the program PCMODEL*(49)*.

d) $\Delta G°_{orbital} = \Delta G°_{exptl} - (\Delta G°_{steric} + \Delta G°_{electrostat})$.

e) See discussion.

f) From equilibration of 2-methoxy-4-methylthiacyclohexane in CCl$_4$*(58)*.

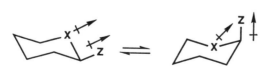

SCHEME 17

We conclude that *the combination of endo and exo anomeric interactions in the axial conformer (Scheme 18) is more stabilizing than the exo anomeric interaction alone in the equatorial conformer by 1.4 kcal mol^{-1}. It is the accentuated steric effect that has the dominant influence on the conformational equilibrium and leads to the equatorial preference (63).* The accentuated steric effect in axial 2-methylaminooxacyclohexane has also been suggested recently by other workers (64,65) as the possible origin of the equatorial preference.

Ab initio calculations of $H_2N-CH_2-OCH_3$ at the 6-31G*//4-21G level (64) indicate a stabilization of the g$^+$g$^-$ relative to the gt form (Scheme 19), as expected since an $n_O \rightarrow \sigma^*_{C-N}$ interaction is only present in the g$^+$g$^-$ conformer. However, a comparison of the tg and tt conformers indicates that the latter is more stable (Scheme 19). The authors suggested that in the tg conformer, the interaction between the nitrogen lone pair orbital and σ^*_{C-O} was counteracted by the $n_O \rightarrow \sigma^*_{C-N}$ interaction. Since nitrogen is a stronger electron donor and the C-N bond a weaker acceptor (66), this results in the prediction of a more stable tt conformer. In our opinion, as in the case of methylaminooxacyclohexane, stabilization obtained by the summation of $n_O \rightarrow \sigma^*_{C-N}$ and $n_N \rightarrow \sigma^*_{C-O}$ orbital interactions (in the tg conformer) is greater than that from one $n_N \rightarrow \sigma^*_{C-O}$ interaction (in the tt conformer). It is likely that the steric interaction between the methoxy group and the two N-H hydrogens leads to the relative destabilization of the tg conformer.

It is also appropriate that we comment on the relative magnitudes of the orbital interaction components in the O/O vs O/N systems. It is clear (Table IV, Scheme 18) that the N/O endo + exo interactions are more stabilizing than the corresponding O/O interactions. The greater anomeric stabilization in the N-C-O fragment than in the O-C-O fragment is also suggested by the greater proportion of the α-isomer of nojirimycin (63%) (67) than of glucose (36%) (68) in aqueous solution. The conclusion is corroborated further by the ΔH^0 values obtained for the ring inversion equilibria in 2-carbomethoxy- oxa- and aza- cyclohexane (69,70).

Reverse Anomeric Effect?   It has been suggested (41,64) that amino and methylamino substituents show a reverse anomeric effect, ie. a preference for the equatorial orientation over and above that described by steric and electrostatic effects. Our data suggest that the conformational preferences can be adequately explained in terms of endo + exo anomeric interactions and steric interactions. It is not a stabilizing electronic effect in the equatorial conformer that results in the overwhelming equatorial preference of the substituent. It is an accentuated steric effect in the axial conformer that results from expression of an exo anomeric effect (42). The electronic or orbital interaction component favors the axial conformer! We suggest that the term "reverse anomeric effect" be reserved for those systems for which it was originally defined by Lemieux and Morgan (71,72), ie. systems containing quaternary, nitrogen aromatic substituents such as pyridinium or imidazolidinium (73,74).

S/O Interactions. The axial conformer of 2-methoxythiacyclohexane 23 is preferred mainly because of the orbital interaction component (1.23 kcal mol^{-1}). The experimental value is derived from data of deHoog (58) from equilibration studies since the conformational equilibrium of **23** was too biased. We recognize the

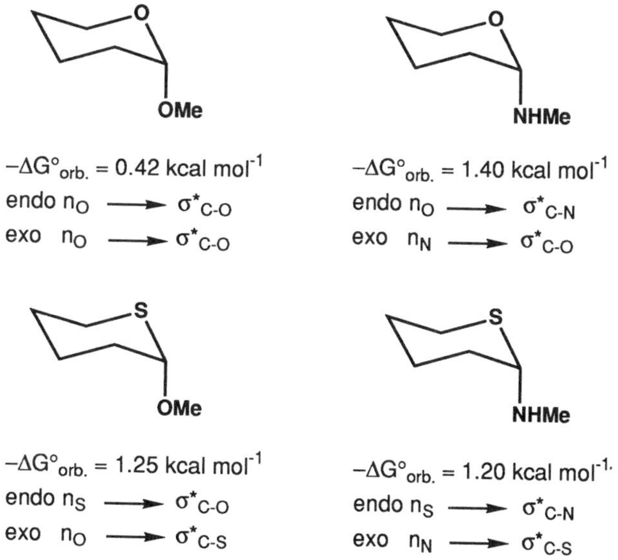

$-\Delta G^{\circ}_{orb.} = 0.42$ kcal mol^{-1}

endo $n_O \longrightarrow \sigma^*_{C-O}$

exo $n_O \longrightarrow \sigma^*_{C-O}$

$-\Delta G^{\circ}_{orb.} = 1.40$ kcal mol^{-1}

endo $n_O \longrightarrow \sigma^*_{C-N}$

exo $n_N \longrightarrow \sigma^*_{C-O}$

$-\Delta G^{\circ}_{orb.} = 1.25$ kcal mol^{-1}

endo $n_S \longrightarrow \sigma^*_{C-O}$

exo $n_O \longrightarrow \sigma^*_{C-S}$

$-\Delta G^{\circ}_{orb.} = 1.20$ kcal mol$^{-1.}$

endo $n_S \longrightarrow \sigma^*_{C-N}$

exo $n_N \longrightarrow \sigma^*_{C-S}$

SCHEME 18

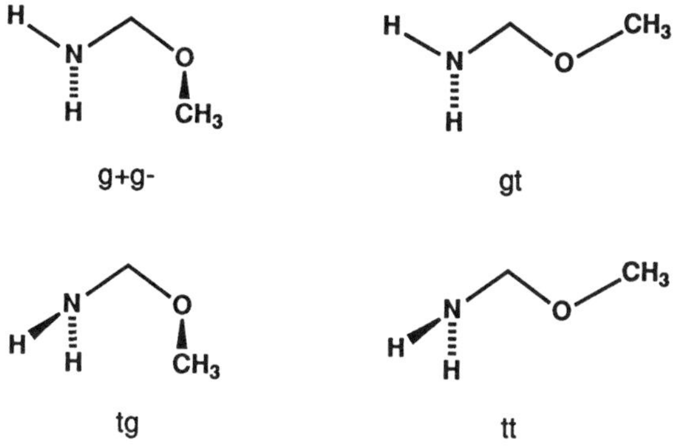

g+g-

gt

tg

tt

SCHEME 19

approximation involved since the two studies have been performed at different temperatures, and entropic effects have not been evaluated. The sum of the endo n_S $\rightarrow \sigma^*_{C-O}$ and exo $n_O \rightarrow \sigma^*_{C-S}$ interaction in the axial conformer is more stabilizing than the exo anomeric $n_O \rightarrow \sigma^*_{C-S}$ interaction in the equatorial conformer (Scheme 18), and there is a greater orbital interaction component than in the oxygen analogue **27**. The result is consistent with greater stabilizing hyperconjugative interactions since a higher lying S lone pair orbital (than O) will result in a greater endo $n_S \rightarrow$ σ^*_{C-O} interaction, and a lower lying σ^*_{C-S} orbital (than σ^*_{C-O}) will give a greater exo $n_O \rightarrow \sigma^*_{C-S}$ interaction; the interpretation must be treated with caution, however, since effects of differential overlap are not readily estimated. Nonetheless, the data permit a comment on the "origin" of the greater axial preference in the thio analogue and in 5-thio sugars in general, relative to their oxygen counterparts. Thus, whereas the electrostatic component favors the axial form of the oxygen congener to a slightly greater extent, the steric component in the thio compound **23** is less than in the oxygen analogue **28**. However, the steric component is less dominant than the orbital interaction component. It follows that the greater preference for the α-isomer in 5-thioglucose (80%) (75) vs glucose (36%) (68) in aqueous solution can be attributed mainly to a greater stabilization from the combination of endo and exo anomeric orbital interactions.

S/N Interactions. The equatorial conformer of 2-methylaminothiacyclohexane **17** is favored, and is due mainly to a large steric energy difference which favors the equatorial conformer. The equatorial preference is less than in the oxacyclohexane analogue.   This difference is accounted for by significantly different steric components.   There is a small electrostatic energy difference and the orbital interaction components are also similar. There are curious differences between the orbital interaction components for a given substituent in the oxygen and sulfur series (Scheme 18) that are not readily interpreted.

The Interplay of the Anomeric and Gauche Effects in Substituted 1,4-Diheterocyclohexanes

The equilibrium data are given in Table III. Since the equilibria of compounds **19** and **26** were highly biased and only the axial conformer peaks could be detected with certainty, the estimates of ΔG^0 were derived from our estimated detection limits. The methoxy-substituted compounds, with the exception of 2-methoxy-1,4-oxathiacyclohexane **25**, show a preference for the axial conformation; compound **25** shows a slight equatorial preference. With respect to the heteroatom pairs at positions 1 and 4, the preference for the axial conformation decreases in the sequence O/O **27** > O/C **28** > O/S **25** and S/C **23** $\geq$ S/O **26** > S/S **22**. The methylamino compounds show interesting behavior. 2-Methylamino-1,4-dithiacyclohexane **16** and 3-methylamino-1,4-oxathiacyclohexane **19**   show an axial preference whereas the other derivatives show equatorial preferences; the equatorial preference decreases in the sequence O/C **21** > O/O **18** > O/S **20** and S/C **17** > S/S **16** > S/O **19**.

Orbital Interaction Component.   The component analysis of the composite conformational effect is given in Table IV.   While it is interesting to compare changes within the series in terms of differences in steric, electrostatic, and orbital interaction components, the present work will focus on a discussion of only the orbital interaction component. The orientation of the methylamino group in the axial **33** and equatorial **34** conformers is shown in Scheme 9; these rotamers have been chosen to maximize exo anomeric interactions, as in the 2-substituted heterocyclohexanes **31, 32**.

The data in Table IV indicate that *in all cases, the combination of anomeric and gauche interactions is attractive with respect to the orbital interaction component, ie., axial conformers are favored.* The composite conformational effect appears to be particularly strong for 2-methylamino-1,4-oxathia- and 1,4-dithiacyclohexane **16, 20**, and quite significant for all molecules, with the possible exception of 2-methoxy-oxacyclohexane **28** and 2-methoxy-1,4-dioxacyclohexane **27**.

A more detailed analysis of the composite orbital interaction component in terms of the individual components is presented in Table V. The anomeric contribution is obtained from the analysis of the 2-substituted heterocyclohexanes, and is shown in column (1). The gauche contribution is taken as half of the orbital interaction component in the 5-substituted 1,3-diheterocyclohexanes (column 2). Justification for the latter approximation comes from the observation that the anomeric effect in 2-phenylseleno-1,3-dithiacyclohexane (*76,77*) is twice that in 2-phenylselenothiacyclohexane (Pinto, B. M.; Johnston, B. D., unpublished data). The composite orbital interaction component in the 1,4-diheterocyclohexanes (column 3) is then compared to the sum of the individual components in column (4) (see Table V). For Z = OMe, good agreement is obtained between the sum of the two individual interactions and the composite interaction. It would appear that the concept of additivity might be valid for these systems. For Z = NHMe, the situation is different. $\Delta G^0_{orbital}$ in the composite systems is greater than the sum of the individual components for all combinations of O and S, although there is good agreement for the 1,4-dioxacyclohexane system. It is conceivable that in the methylamino-substituted compounds, bond length and bond angle variations in the acetal units (*78*) lead to more favorable hydrogen-bonding interactions in the axial conformers than approximated by the 5-substituted-1,3-diheterocyclohexane model compounds (**33** in Scheme 9). The strong $n_N \rightarrow \sigma^*_{C-S(O)}$ exo anomeric interaction in the axial conformer (*67,69*) will lead to a shorter C-N bond and hence, an enhanced hydrogen-bonding interaction. Furthermore, the π-donation from nitrogen will lead to a strong polarization of the N-H bond. The prediction is supported by the Mulliken population analysis of $HO(S)-CH_2-NH_2$ conformers calculated by ab initio MO methods ($6-31G^*$) (Leung, R. Y. N.; Pinto, B. M., unpublished data). The magnitude of the hydrogen-bond interactions between the positively charged N-H hydrogen and the ring heteroatom will thus be increased in the composite systems. It is also possible that the different orientations of the methylamino substituent in the equatorial conformers of the 1,3- and 1,4- diheterocyclohexanes, when assessing steric and electrostatic energies, plays a significant role.

Additivity of Orbital Interactions? Finally, we put forward a proposal regarding the additivity of orbital interactions. The composite systems have been dissected in terms of pairwise, two-center interactions. However, additional three-center interactions may be involved since the interactions of two heteroatoms may be

Table V. Orbital Interaction Component ($-\Delta G°_{orbital}$ e $\rightleftharpoons$ a (kcal mol^{-1}))
in Substituted Heterocyclohexanes and Diheterocyclohexanes

(1)   (2)   (3)

(1)	(2)	Z = OMe	(3)	(4) [a]
1.23	0.39	X = S, Y = S	1.43	1.43
0.42	1.78	X = O, Y = O	1.30	1.31
1.23	1.78	X = S, Y = O	>1.65	2.12
0.42	0.39	X = O, Y = S	0.44	0.62
		Z = NHMe		
1.16	0.23	X = S, Y = S	2.09	1.28
1.40	-0.12	X = O, Y = O	1.44	1.34
1.16	-0.12	X = S, Y = O	>1.65	1.10
1.40	0.23	X = O, Y = S	2.20	1.52

a) (4) = (1) + (2)/2

perturbed by a third. Scheme 20 illustrates the orbital interactions that are major contributors to the exo anomeric and gauche effects in the axial conformers. The two orbital interactions are orthogonal to each other and should have a minimal influence on one another in the composite system. In contrast, the interactions shown in Scheme 21 are operating in the same plane. Here, it is conceivable that these three-center interactions act in concert and will give an effect that is not approximated by the sum of the individual components. Therefore, the dominance of the interactions shown in Scheme 20 will lead to additivity whereas the dominance of orbital interactions shown in Scheme 21 will result in non-additivity of the individual effects. It is also clear from the foregoing section that additional electrostatic stabilization, eg. N-H hydrogen bonding will also lead to non-additivity.

Conclusions

Partitioning of the experimental conformational free energies of 5-substitued-1,3-diheterocyclohexanes, 2-substituted heterocyclohexanes, and 2- or 3-substituted-1,4-diheterocyclohexanes into a sum of physically intuitive terms has yielded the steric, electrostatic, and orbital interaction components. It is recognized that the protocol gives an approximation to reality, particularly since specific conformations have been selected for the calculation of steric and electrostatic interactions, and factors such as the entropy of mixing have not been considered. Otherwise, the interplay of enthalpic and entropic components is reflected in the ΔG^0 values (*34-37,42,43,77*). The treatment offers some insight into the "origin" of the conformational effects.

The orbital interaction terms of the X-C-Z anomeric effect and the Y-C-C-Z gauche effect are found to be attractive for all combinations of X,Y = O,S; Z = O,N. The combination of an endo and exo anomeric interaction is always found to be more favorable than the exo interaction alone. The concept of a "reverse anomeric effect" with NHR substituents is discounted. The conformational preferences observed in O-C-N and S-C-N fragments are attributed instead to the dominance of steric components that outweigh the attractive orbital interaction components.

Analysis of the Y-C-C-Z gauche effect indicates a significant orbital interaction component that can be analyzed in terms of a combination of through-space and through-bond orbital interactions or, alternatively, in terms of through-bond interactions operating through the σ framework. In the case of the axially-substituted, methylamino compounds, the electrostatic interactions contain a significant hydrogen-bonding component.

The orbital interaction component in the substituted 1,4-diheterocyclohexanes is also attractive in the axial conformers in all cases studied. Additivity of the orbital interaction components of the individual anomeric and gauche effects in the 1,4-diheterocyclohexanes is observed for the systems with oxygen substituents but generally not for those with nitrogen substituents. An enhanced hydrogen-bonding component is invoked to explain the discrepancy in the latter systems. A proposal for the additivity of orbital interaction effects in terms of the orthogonal or parallel nature of pairwise orbital interactions is advanced.

The analysis presented for the composite systems has focused on the stabilizing orbital interactions. A more detailed understanding of the role of stabilizing and destabilizing orbital interactions in these systems will have to await a quantitative Perturbation Molecular Orbital analysis (*79*).

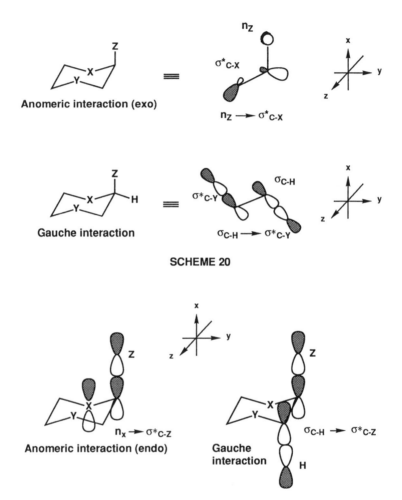

SCHEME 20

SCHEME 21

Orbital interactions play a significant role in the endo and exo anomeric effect and the gauche effect; their understanding for the conformational analysis of heterocycles is essential.

Acknowledgments. We are grateful to the Natural Sciences and Engineering Research Council of Canada for financial support and to S. Mehta for drawing the diagrams.

Literature Cited

1. Jungius, C. L. *Z. Phys. Chem.* **1905**, *52*, 97.
2. Edward, J. T. *Chem. Ind. (London)* **1955**, 1107.
3. Lemieux, R. U. In *Molecular Rearrangements*, de Mayo, P., Ed.; Interscience, New York, 1964.
4. Wolfe, S. *Acc. Chem. Res.* **1972**, *5*, 102.
5. *Anomeric Effect, Origin and Consequences*; Szarek, W. A.; Horton, D., Eds.; ACS Symposium Series 87; American Chemical Society; Washington, DC, 1979.
6. Kirby, A. J. *The Anomeric Efffect and Related Stereoelectronic Effects at Oxygen*; Springer-Verlag: New York, 1983.
7. Deslongchamps, P. *Stereoelectronic Effects in Organic Chemistry*; Pergamon: New York, 1983.
8. Juaristi, E.; Cuevas, G. *Tetrahedron* **1992**, *48*, 5019.
9. Zefirov, N. S. *Tetrahedron* **1977**, *33*, 3193.
10. For a review, see Juaristi, E. *J. Chem. Educ.* **1979**, *56*, 438.
11. For a review of experimental and theoretical data, see Murcko, M. A.; Dipaola, R. A. *J. Am. Chem. Soc.* **1992**, *114*, 10010.
12. Abraham, R. J.; Banks, H. D.; Eliel, E. L.; Hofer, O.; Kaloustian, M. K. *J. Am. Chem. Soc.* **1972**, *94*, 1913.
13. Eliel, E. L.; Juaristi, E. *J. Am. Chem. Soc.* **1978**, *100*, 6114.
14. Zefirov, N, S.; Gurvich, L. G.; Shashkov, A. S.; Krimer, M. Z.; Vorobeva, G. A. *Tetrahedron* **1976**, *32*, 1211.
15. Zefirov, N. S.; Samoshin, V. V.; Subboin, O. A.; Baranenkov, V. I.; Wolfe, S. *Tetrahedron* **1978**, *34*, 2953.
16. Carreno, M. C.; Carretero, J. C.; Garcia Ruano, J. L.; Rodriguez, J. H. *Tetrahedron* **1990**, *46*, 5649.
17. Zefirov, N. S. *J. Org. Chem., USSR* **1970**, *6*, 1768.
18. Albright, T. A.; Burdett, J. K.; Whangbo, M.-H. *Orbital Interactions in Chemistry*; John Wiley: New York, 1985.
19. Gleiter, R.; Kobayashi, M.; Zefirov, N. S.; Palyulin, V. A. *Dokl. Chem.* **1977**, *235*, 396.
20. Hoffmann, R. *Acc. Chem. Res.* **1971**, *4*, 1.
21. Epiotis, N. D.; Sarkanen, S.; Bjorkquist, D; Bjorkquist, L; Yates, R. *J. Am. Chem. Soc.* **1974**, *96*, 4075.
22. Dionne, P.; St. Jacques, M. *J. Am. Chem. Soc.* **1987**, *109*, 2616.
23. Dionne, P.; St. Jacques, M. *Can. J. Chem.* **1989**, *67*, 11.
24. Alcudia, F.; Llera, J. M.; Garcia Ruano, J. L.; Rodriguez, J. H. *J. Chem. Soc. Perkin Trans. 2* **1988**, 1225.

25. Juaristi, E.; Antunez, S. *Tetrahedron* **1992**, *48*, 5941.
26. Wiberg, K. B.; Murcko, M. A.; Laidig, K. E.; MacDougall, P. J. *J. Phys. Chem.* **1990**, *94*, 6956.
27. Alcudia, F.; Campos, A. L.; Llera, J. M.; Zorrilla, F. *Phosphorus and Sulfur* **1988** *36*, 29.
28. Brunet, E.; Eliel, E. L. *J. Org. Chem.* **1986**, *51*, 677.
29. Garcia Ruano, J. L.; Rodriguez, J. H.; Alcudia, F.; Llera, J. M.; Olferowicz, E. M.; Eliel, E. L. *J. Org. Chem.* **1987**, *52*, 4099.
30. Brunet, E.; Carreno, M. C.; Gallego, M. T.; Garcia Ruano, J. L. *Tetrahedron* **1985**, *41*, 1733.
31. Brunet, E.; Azpeitia, P. *Tetrahedron* **1988**, *44*, 1751.
32. Bernet, B.; Piantini, U.; Vasella, A. *Carbohydr. Res.* **1990**, *204*, 11.
33. Lemieux, R. U.; Koto, S. *Tetrahedron* **1974**, *30*, 1933.
34. Praly, J.-P.; Lemieux, R. U. *Can. J. Chem.* **1987**, *65*, 213.
35. Booth, H.; Khedhair, K. A. *J. Chem. Soc., Chem. Commun.* **1985**, 467.
36. Booth, H.; Khedhair, K. A.; Readshaw, S. A. *Tetrahedron* **1987**, *43*, 4699.
37. Booth, H. Dixon, J. M.; Readshaw, S. A. *Tetrahedron* **1992**, *48*, 6151.
38. Wu, T. C.; Goekjian, P. G.; Kishi, Y. *J. Org. Chem* **1987**, *52*, 4819.
39. Ricart, G.; Glacet, C.; Couturier, D. *C. R. Acad. Sci. Ser. C.* **1977**, *284*, 319.
40. Tesse, J.; Glacet, C.; Couturier, D. *C. R. Acad. Sci. Ser. C.* **1975**, *280*, 1525.
41. Wolfe, S.; Whangbo, M.-H.; Mitchell, D. J. *Carbohydr. Res.* **1979**, *69*, 1.
42. Booth, H.; Dixon, J. M.; Khedhair, K. A.; Readshaw, S. A. *Tetrahedron* **1990**, *46*, 1625.
43. For leading references, see: Juaristi, E.; Gonzalez, E. A.; Pinto, B. M.; Johnston, B. D.; Nagelkerke, R. *J. Am. Chem. Soc.* **1989**, *111*, 6745.
44. Wolfe, S.; Pinto, B. M.; Varma, V.; Leung, R. Y. N. *Can. J. Chem.* **1990**, *68*, 1051.
45. Leung, R. Y. N. Ph. D. Thesis, Simon Fraser University, Burnaby, B. C., Canada, 1991.
46. MM2 (85), QCPE Program No. MM2 85. Indiana University, Bloomington, Indiana, USA.
47. Franck, R. W. *Tetrahedron* **1983**, *39*, 3251.
48. Abraham, R. J.; Rossetti, Z. L.; *J. Chem. Soc. Perkin Trans.2* **1973**, 582.
49. PCMODEL, Serena Software, Box 3076, Bloomington, Indiana, USA.
50. Eliel, E. L.; Hutchins, R. O. *J. Am. Chem. Soc.* **1969**, *91*, 2703.
51. Booth, H.; Jozefowicz, M. L. *J. Chem. Soc., Perkin Trans. 2* **1976**, 895.
52. Eliel, E. L.; Gianni, M. H. *Tetrahedron Lett.* **1962**, 97.
53. Allinger, N. L.; Hickey, M. J. *J. Am. Chem. Soc.* **1975**, *97*, 5167.
54. Sweigart, D. A.; Turner, D. W. *J. Am. Chem. Soc.* **1972**, *94*, 5592.
55. Sweigart, D. A.; Turner, D. W. *J. Am. Chem. Soc.* **1972**, *94*, 5599.
56. Kimura, K.; Kesumata, S.; Achiba, Y.; Amazaki, T. Y.; Iwaka, S. *Handbook of the HeI Photoelectron spectra of Fundamental Organic Molecules*; Japan Scientific Societies Press: Tokyo; Halsted Press: New York, 1981.
57. Eliel, E. L.; Hargrave, K. D.; Pietrusiewicz, K. M.; Manoharan, M. *J. Am. Chem. Soc.* **1982**, *104*, 3635.
58. de Hoog, A. J., Ph. D. Thesis, University of Leiden, 1971.

59. Booth, H.; Grindley, T. B.; Khedhair, K. A. *J. Chem. Soc., Chem. Commun.* **1982**, 1047.

60. Dubois, J. E.; Cosse-Barbi, A.; Watson, D. G. *Tetrahedron Lett.* **1989**, *30*, 163.

61. Cosse-Barbi, A.; Watson, D. G.; Dubois, D. G. *Tetrahedron Lett.* **1989**, *30*, 167.

62. Lemieux, R. U. *Proceedings VIIth International Symposium on Medicinal Chemistry, Vol 1*; Swedish Pharmaceutical Press: Stockholm, 1985, p 329.

63. Pinto, B. M. Ph. D. thesis, **1980**, Queen's University, Kingston, Ont. Canada.

64. Krol, M. C.; Huige, J. M. C; Altona, C. *J. Comput. Chem.* **1990**, *11*, 765.

65. Cramer, C. J. *J. Org. Chem.* **1992**, *57*, 7034.

66. Aped, P.; Scheifer, L.; Fuchs, B. *J. Comput. Chem.* **1989**, *10*, 265.

67. Pinto B. M.; Wolfe, S. *Tetrahedron Lett.* **1982**, *23*, 3687.

68. Lemieux, R. U.; Stevens, J. D. *Can J. Chem.* **1966**, *44*, 249.

69. Booth, H.; Dixon, J. M.; Khedhair, K. A. *Tetrahedron* **1992**, *48*, 6161.

70. see also Booth, H.; Lemieux, R. U. *Can. J. Chem.* **1971**, *49*, 777.

71. Lemieux R. U.; Morgan, A. R. *Can. J. Chem.* **1965**, *43*, 2205.

72. Lemieux R. U.; Morgan, A. R. *Pure Appl. Chem.* **1971**, *25*, 527.

73. Paulsen, H.; Gyorgydeak, Z.; Friedmann, M. *Chem. Ber.* **1974**, *107*, 1590.

74. Finch, P.; Nagpurkar, A. G. *Carbohydr. Res.* **1976**, *49*, 275.

75. Lambert, J. B.; Wharry, S. M. *J. Org. Chem.* **1981**, *46*, 3193.

76. Pinto, B. M; Johnston, B. D.; Sandoval-Ramirez, J.; Sharma, R. D. *J. Org. Chem.* **1988**, *53*, 3766.

77. Pinto, B. M; Johnston, B. D.; Nagelkerke, R. *J. Org. Chem.* **1988**, *53*, 5668.

78. Pinto, B. M.; Schlegel, H. B.; Wolfe, S. *Can. J. Chem.* **1987**, *65*, 1658.

79. Whangbo, M.-H.; Schlegel, H. B.; Wolfe, S. *J. Am. Chem. Soc.* **1977**, *99*, 1296.

RECEIVED May 12, 1993

Chapter 9

Origin and Quantitative Modeling of Anomeric Effect

Peter A. Petillo[1] and Laura E. Lerner

Department of Chemistry, University of Wisconsin—Madison,
Madison, WI 53706

The origin of the anomeric effect has been examined by decomposing *ab initio* wavefunctions of dihydroxymethane (**1**) and dimethoxymethane (**2**) with the Natural Bond Orbital (NBO) method. The NBO analysis suggests that the anomeric effect arises from stabilizing 2-electron $n(O) \rightarrow \sigma^*_{CO}$ delocalizations. Geometry optimization of **1** and **2** in the absence of all $n(O) \rightarrow \sigma^*$ interactions lengthens CO bond lengths by 0.06Å and reduces OCO bond angles. Optimization in the absence of all hyperconjugative interactions results in more dramatic changes in geometry ($\Delta r_{CO} = 0.13$Å, $\Delta\angle_{OCO} = 1.5° - 6.0°$). An energetic analysis of the optimized NOSTAR pure Lewis structures indicate that **2-g⁺a** is destabilized relative to **2-g⁺g⁺** by 0.12 kcal/mole. This implies that dipole-dipole destabilization of **2-g⁺a** is an order of magnitude less important than $n(O) \rightarrow \sigma^*_{CO}$ interactions. Geometry optimization of **1** and **2** at MP4SDTQ/6-31G* appear to approach limiting values, suggesting that these wavefunctions are suitable for quantitative modeling of the anomeric effect in **1** and **2**.

Conformations about the glycosidic center in carbohydrates are dictated by stereoelectronic effects generally referred to as the anomeric effect (*1-15*). While experimental evidence overwhelming supports the belief that special bonding interactions exist in acetals and related structures, the exact origin of these effects remains disputed. Two distinct origins of the anomeric effect have been advanced (*11*). The first suggests that the anomeric effect arises from destabilizing electronic interactions (Figure 1), either through dipole-dipole interactions or lone pair-lone pair repulsions (the "rabbit ear" effect - *4,9,12*). By contrast, the second school of thought suggests

---

[1]Current address: Whitehead Institute for Biomedical Research, 9 Cambridge Center, Cambridge, MA 02142

0097–6156/93/0539–0156$06.00/0

that the anomeric effect is due to stabilizing 2e- delocalizations (hyperconjugation) from the oxygen lone pairs into adjacent polarized antiperiplanar σ^*_{CO} bonds, displayed as two single headed arrows (*5,6,11*). Typical orbital energy diagrams for each model are also shown.

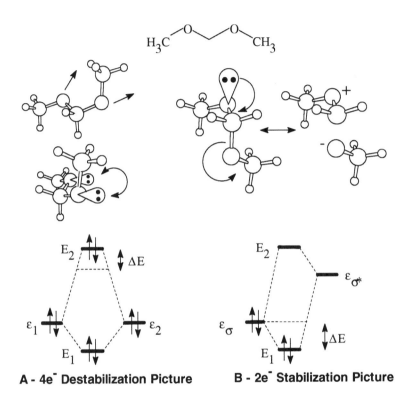

A - 4e⁻ Destabilization Picture   **B - 2e⁻ Stabilization Picture**

Figure 1: Dipole-dipole/rabbit ear (4e- destabilizing) and hyperconjugative (2e-stabilizing) models and MO diagrams justifying the origin of the anomeric effect.

In this discussion, the origin of the anomeric effect in acetals will be identified and guidelines for quantitative modeling of the glycosidic linkage in carbohydrates delineated. To accomplish this, *ab initio* calculations on dihydroxymethane (**1**) and dimethoxymethane (**2**) will be discussed (*16-19*). The relevant conformations about the glycosidic linkage in D-glucose, namely the **g⁺g⁺** conformation (α-D-glucose anomer-like or +*sc*, +*sc* conformer) and the **g⁺a** conformation (β-D-glucose anomer-like or +*sc*, *ap* conformer) will be considered (Figure 2). The **aa** conformation (*ap, ap* conformer) will also be considered. Analysis of the final wavefunctions will be presented within the Natural Bond Orbital (NBO) basis of Weinhold and coworkers (*20,21*), which we find uniquely suited for calculations on carbohydrate and related systems. Combining *ab initio* calculations and the NBO analysis allows one to probe a molecule's energetic makeup in a manner not directly available from experiment.

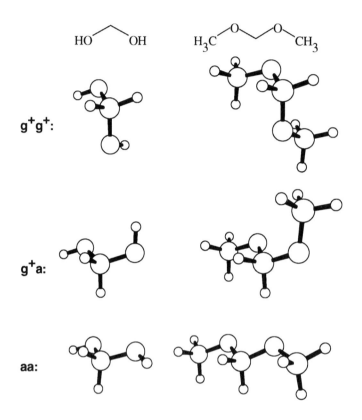

Figure 2: The **g⁺g⁺**, **g⁺a** and **aa** conformers of **1** and **2** considered in this work.

The Origin of the Anomeric Effect

The localized, orthogonal nature of NBOs is central to this study, allowing us to demonstrate that the origin of the anomeric effect is due to $n(O) \rightarrow \sigma^*_{CO}$ (non-Lewis) interactions, as shown in Figure 3. In essence, the anomeric effect can be considered as a departure from an idealized Lewis structure (*22,23,24*). Since the anomeric effect is dependent upon the acetal lone pairs, a method that isolates lone pair interactions is desirable. In a non-orthogonal (delocalized) basis, discerning the effects of the individual lone pairs is complicated by electron density from other atomic centers within the structure. For example, the RHF/6-31G* HOMO of **2** in the canonical MO

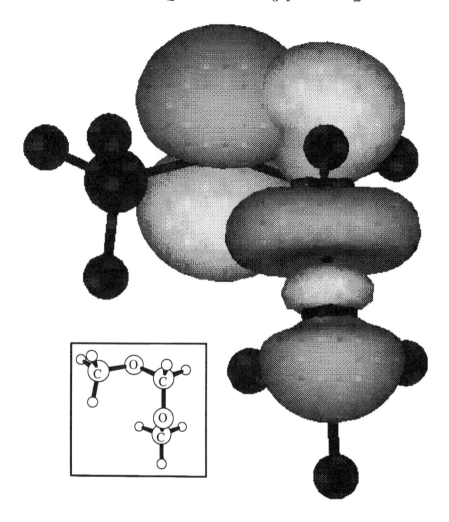

Figure 3: Oxygen centered pure-p lone pair interacting with an antiperiplanar σ^*_{CO} in **2**.

basis has delocalized electron density due to all four oxygen lone pairs in addition to other atomic centers, whereas the equivalent energy NBO is simply an oxygen centered pure p-lone pair (Figure 4). Figures 3 and 4 provide a striking contrast with the uncomplicated NBO picture clearly easier to interpret for anomeric interactions.

The NBO method has been described in detail elsewhere (*23-29*). Here, to set the stage for its application to the anomeric effect, we review its underlying concepts. The NBO procedure works on any electronic wavefunction, ψ, for which a first order density matrix is available. The method is quite economical, requiring only a

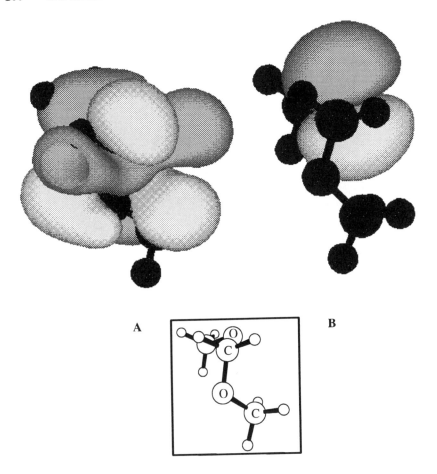

Figure 4: HOMO of **2** in the canonical (**A**) and NBO (**B**) basis.

modest fraction of the total computational effort needed to obtain ψ. The resulting localized orbitals are given in terms of hybrids which form the chemist's traditional language of bonding (*30*). In simple terms, the canonical delocalized MOs comprising an *ab initio* wavefunction are converted into NBOs through a series of reversible matrix transformations. *NBOs are intrinsic to, not imposed upon, a wavefunction.* This point is especially important in that the NBO analysis does not modify or impose any constraints on ψ as a result of the procedure. Instead, intra- and intermolecular interactions are unambiguously shown as a series of stabilizing 2e- interactions. The decomposition of ψ into stabilizing energetic interactions violates no quantum mechanical principle and the results are variationally correct. NBOs for any system are essentially basis set independent, converge as ψ becomes more accurate and appear to be transferable in structurally related molecules, making it the ideal framework for model studies of larger systems (*25*).

Natural orbitals (NOs), whose conception can be traced back to Löwdin, are distinctive in that they are maximally occupied (*31*). Natural atomic orbitals (NAOs) are related to NOs in that they are equivalent for an isolated atom. NAOs form a complete orthonormal set of orbitals ($\langle\theta_i | \theta_j\rangle = \delta_{ij}$) and result from the diagonalization of the one-center blocks of the density matrix followed by an occupancy-weighted symmetric orthogonalization procedure. Natural bond orbitals (NBOs) are generated from NAOs, and can be thought of as a combination of sp^λ or $sp^\lambda d^\mu$ natural hybrid orbitals (NHOs) emanating from atomic centers (*20,27,28*). The NBO analysis breaks down the multicentered or canonical MOs into a series of core orbitals, bonding hybrids, lone pairs, antibonds and Rydberg orbitals. The NBOs are one and two center localized orbitals (lone pairs and bonds, respectively), as opposed to the canonical MOs which are delocalized throughout the entire molecule, and retain the property of maximum occupancy of the NAOs from which they are derived. The Fock matrix $\underline{\underline{F}}$ in the NBO basis is modular, reflecting the localization of electron density in ψ, and is divided into four regions (Figure 5). The localized Lewis structure is contained in Region **I** and represents the bonding interaction between highly occupied NBOs. Regions **II** and **III** contain electron delocalizations (departures from idealized Lewis structures) and most donor-acceptor interactions are found in these regions. Region **IV** corresponds to antibond-antibond interactions which are not important except in unusual situations.

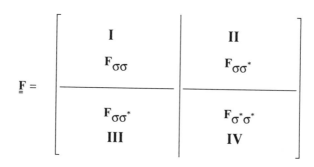

Figure 5: Regions of Fock matrix in the NBO basis.

By definition, the optimal Lewis structure best describing a molecule's electron density is composed of filled, localized one and two center orbitals (collectively referred to as σ orbitals - *20,23*). By contrast, electron density in σ^*, π^* and Rydberg orbitals (referred to generally as σ^* orbitals) correspond to deviations or departures from perfect Lewis structures (*20,23,24*). The antibonds within the NBO basis should not be confused with the virtual MOs from SCF-MO theory. The former have nonzero occupancies which lead to definite energy lowerings, whereas the vitual MOs are completely unoccupied and have no role in the wave function or any observable property. As expected, the occupancy of the σ^* orbitals (within the NBO basis) is generally small in comparison to the occupancy of σ orbitals. The total energy of the system E_{TOTAL} consists of two contributions, the energy of the Lewis structure (E_{LEWIS} or $E_{\sigma\sigma}$) and the energy of the non-Lewis components ($E_{NON\text{-}LEWIS}$ or $E_{\sigma\sigma^*}$):

$$E_{TOTAL} = E_{LEWIS} + E_{NON\text{-}LEWIS} = E_{\sigma\sigma} + E_{\sigma\sigma^*} \qquad (1)$$

The same concept can be applied to a molecule's total electron density:

$$\rho_{TOTAL} = \rho_{LEWIS} + \rho_{NON\text{-}LEWIS} = \rho_{\sigma\sigma} + \rho_{\sigma\sigma^*} \qquad (2)$$

Delineation of the deviations from idealized Lewis structures, which are localized bond pictures or covalent bond models (30), in terms of $E_{\sigma\sigma^*}$ or $\rho_{\sigma\sigma^*}$ is the real power of the NBO method. Typically, these departures comprise less than 1% of E_{TOTAL} and ρ_{TOTAL} for saturated systems (in strongly delocalized systems like benzene, the departure is somewhat larger). The dominance of the Lewis structure in E_{TOTAL} suggests that the localized Lewis structures are good starting points for understanding electronic structure calculations in terms of familiar bonding concepts.

To ascertain how acetal oxygen lone pairs participate in anomeric stabilizations, the relationship of the two apparently different models (stabilizing vs. destabilizing) and Lewis structures must be understood. Although we wish to stress that the two models are in fact numerically equivalent, we cannot take credit for this discovery. Prior to our work, Weinhold carefully examined the origin of the 4e- destabilizing model (lone pair-lone pair repulsions, Figure 1 - 32). He convincingly demonstrated that by including an unfilled σ^*, the 4e- destabilizing model is numerical equivalent to the 2e- stabilizing model (Figure 1 - 32). Under this circumstance, the former cannot be physically distinguished from the latter despite the very different interpretations these models suggest. The problem with the traditional 4e- destabilizing (PMO) treatment is twofold: (i) unjustified application of perturbation theory to a non-Hermitian reference that cannot be physically realized even in principle and (ii) the neglect of overlap integrals of the same numerical magnitude as those used. Orthogonalization and inclusion of all overlap integrals in the 4e- destabilizing basis results in numerical equivalence, both in sign and magnitude, to the 2e- stabilizing model (32). Within an orthogonal basis, $\sigma\rightarrow\sigma^*$ interactions are stabilizing and $\sigma\rightarrow\sigma$ interactions have no energetic contribution (Figure 5). Within a non-orthogonal basis, $\sigma\rightarrow\sigma^*$ interactions are also stabilizing but $\sigma\rightarrow\sigma$ interactions are apparently destabilizing. A non-orthogonal basis can be problematic (26) - AO basis functions are not invariant to unitary transformations, the sum of the coefficients are not unity and the trace of the density matrix does not yield the total number of electrons in the system. Examination of canonical MOs within a delocalized basis can be risky, since the orbitals ignored are of the same energetic magnitude as the orbitals kept.

Weinhold's results do not imply that the lone pairs are isolated from one another (e.g. fail to interact). On the contrary, an acetal's photoelectron spectrum clearly shows splitting between the two p-rich lone pairs (12,33,34). In a non-orthogonal basis, the splitting about an isolated lone pair would appear unequal, with the LP(1)-LP(2) combination displaced more than the LP(1)+LP(2) combination. In an orthogonal basis, two lone pairs of equal energy will split symmetrically about the energy of one of the lone pairs isolated from the other. No net energy gain/loss results from the interaction, just a change in unperturbed lone pair ionization potential in accord with the symmetry combinations LP(1)+LP(2) and LP(1)-LP(2). Thus, regardless of the basis, the observed PE bands will occur at the same ionization potentials, but determining an isolated lone pair's ionization potential will depend upon the basis employed.

To show that the anomeric effect is due primarily to $n(O) \rightarrow \sigma^*_{CO}$ interactions, consider an oxygen centered pure-p lone pair, $n(O)$, and an adjacent σ^*_{CO} orbital. Interaction of NBOs $n(O)$ and σ^*_{CO} is described by the off-diagonal Fock matrix element between the two NBOs ($\langle \varphi_n | \hat{F} | \varphi_\sigma{}^* \rangle$). By deleting specific $\underline{F}$ elements and recomputing ψ, the energetics of interaction are exactly computed. In effect, the electron density normally delocalized from the first to second NBO is forced to remain in the first NBO, resulting in an overall energy increase, equivalent to the interaction energy. If delocalization of $n(O)$ electron density into σ^*_{CO} is truly responsible for the anomeric effect, then deletion of the off-diagonal Fock matrix element between these two NBOs will increase the total energy of the system. If delocalizations are not responsible for anomeric stabilization, deletions will result in no net energy increase.

Deletion energies for **1** and **2** follow the trend expected if the anomeric effect is due primarily to $n(O) \rightarrow \sigma^*_{CO}$ delocalizations (Table I, II). For each conformer considered, the off-diagonal elements between the two pure-p $n(O)$ and the adjacent σ^*_{CO} orbitals were deleted. The resultant $E^{DEL}_{\sigma^*(CO)}$ energies reflect the number of stabilizing $n(O) \rightarrow \sigma^*_{CO}$ interactions present in each conformer. The $\mathbf{g^+g^+}$ conformers show the largest energy increases, reflecting the two $n(O) \rightarrow \sigma^*_{CO}$ interactions historically associated with the anomeric effect. The **aa** conformers show no net increase in energy, indicating the absence of any stabilizing interaction between $n(O)$ and the adjacent σ^*_{CO} orbitals. The $\mathbf{g^+a}$ conformers show an energy increase intermediate between the $\mathbf{g^+g^+}$ and **aa** conformers, consistent with a single stabilizing interaction and Lemeiux's concept of an exo-anomeric effect (*35,36*).

The NBO deletion energies suggest an anomeric stabilization larger than the energy difference between conformers (Table II). However, the net anomeric stabilization is not simply the deletion energy, but the energy difference between the $n(O) \rightarrow \sigma^*_{CO}$ interactions (e.g. $\Delta E^{DEL}_{\sigma^*(CO)} = E^{DEL}_{\sigma^*(CO)}(\mathbf{g^+g^+}) - E^{DEL}_{\sigma^*(CO)}(\mathbf{aa})$) and any other $n(O) \rightarrow \sigma^*$ interactions that arise when $n(O) \rightarrow \sigma^*_{CO}$ interactions are not possible (e.g $\Delta E_{\sigma}{}^* = E(\sigma^*_{CO})(\mathbf{g^+g^+}) - E(\sigma^*_{CO})(\mathbf{aa})$). This difference is termed $\Delta E^{DIF}_{\sigma^*(CO)}$ (Table III). A pure-p $n(O)$ is free to interact with any antibonding or Rydberg acceptor orbital, so in the absence of an appropriately oriented σ^*_{CO} orbital other interactions may be important. While the best acceptors of non-Lewis electron density in **1** and **2** are the strongly polarized σ^*_{CO} orbitals, σ^*_{CH} orbitals are also reasonable acceptors. Thus, while **1**-$\mathbf{g^+g^+}$ has two $n(O) \rightarrow \sigma^*_{CO}$ interactions, **1**-$\mathbf{g^+a}$ has only one $n(O) \rightarrow \sigma^*_{CO}$ and one $n(O) \rightarrow \sigma^*_{CH}$ interaction and the **1**-**aa** conformer has two $n(O) \rightarrow \sigma^*_{CH}$ interactions. Upon deletion of the $n(O) \rightarrow \sigma^*_{CO}$ interactions, **1**-$\mathbf{g^+g^+}$ is destabilized relative to **1**-**aa** by 18.1 kcal/mole. The 18.1 kcal/mole represents interactions of the pure-p $n(O)$ lone pairs with other orbitals, namely $n(O) \rightarrow \sigma^*_{CH}$ interactions (Table II, III).

Table I: Total electronic energies[a,b] for 1 and 2 at different levels of theory

Conformer	RHF/6-31G*	MP2/6-31G*	MP4SDTQ/6-31G*
$1\text{-}g^+g^+$	-189.9006269	-190.3976073	-190.4206080
$1\text{-}g^+a$	-189.8941906 (4.04)	-190.3901042 (4.71)	-190.4131765 (4.66)
$1\text{-}aa$	-189.8867699 (8.70)	-190.3821192 (9.72)	-190.4052157 (9.66)
$2\text{-}g^+g^+$	-267.9539901	-268.6982574	-268.7655990
$2\text{-}g^+a$	-267.9500987 (2.44)	-268.6926909 (3.49)	-268.7600659 (3.47)
$2\text{-}aa$	-267.9450587 (5.60)	-268.6867770 (7.20)	-268.7542161 (7.14)

[a]Electronic energies reported in a.u. [b]Energy differences of conformers are given in parentheses and reported in kcal/mole. The g^+g^+ conformer is taken as the reference.

Table II: Total 6-31G* deletion energies for 1 and 2

Conformer	$E(\sigma^*_{CO})$[a,c]	$E^{DEL}_{\sigma*(CO)}$[b,c]	$E(NS)$[a,d]	E^{DEL}_{NS}[b,d]
$1\text{-}g^+g^+$	-189.8579458	(26.8)	-189.7342016	(104.4)
$1\text{-}g^+a$	-189.8726891	(13.5)	-189.7342101	(100.4)
$1\text{-}aa$	-189.8867699	(0.0)	-189.7319610	(97.1)
$2\text{-}g^+g^+$	-267.9109138	(27.4)	-267.6436214	(195.2)
$2\text{-}g^+a$	-267.9297627	(12.8)	-267.6437144	(192.6)
$2\text{-}aa$	-267.9450587	(0.0)	-268.6378643	(192.8)

[a]Total electronic energy after deletion reported in a.u. [b]The deletion energy (given in parentheses and reported in kcal/mole) is defined as the difference between the RHF electronic energy and the electronic energy after the deletion. [c]$E(\sigma^*_{CO})$, $E^{DEL}_{\sigma*(CO)}$: deletion of vicinal σ^*_{CO} orbitals. [d]$E(NS)$, E^{DEL}_{NS}: NOSTAR deletion e.g. all "starred" orbitals.

Table III: Energy differences[a] for 1 and 2 from Tables I and II

Conformer	$\Delta E(RHF)$[b]	$E^{DEL}_{\sigma*(CO)}$[c]	$\Delta E_\sigma*$[d]	$\Delta E^{DIF}_{\sigma*(CO)}$	E^{DEL}_{NS}[e]	ΔE_{NS}[f]
$1\text{-}g^+g^+$	0.00	26.8	0.0	0.0	104.4	0.0
$1\text{-}g^+a$	4.04	13.5	-9.3	4.0	100.4	0.0
$1\text{-}aa$	8.70	0.0	-18.1	8.7	97.1	1.4
$2\text{-}g^+g^+$	0.00	27.4	0.0	0.0	195.2	0.0
$2\text{-}g^+a$	2.44	12.8	-11.8	2.8	192.3	0.1
$2\text{-}aa$	5.60	0.0	-21.4	3.6	192.8	3.6

[a]All differences reported in kcal/mole; The g^+g^+ conformer is taken as the reference for all differences. [b]The RHF energy difference between conformers. The g^+g^+ conformer is take as the reference. [c]The σ^*_{CO} deletion energy. [d]The difference in $E(\sigma^*_{CO})$ between conformers. [e]The NOSTAR deletion energy. [f]The difference in E^{DEL}_{NS} between conformers.

The difference between the change in $n(O) \rightarrow \sigma^*_{CO}$ interaction energy (26.8 kcal/mole) and the destabilization of the **1-g⁺g⁺** conformer relative to the **1-aa** conformer in the absence of the $n(O) \rightarrow \sigma^*_{CO}$ interactions (18.1 kcal/mole) is 8.7 kcal/mole. This is exactly the difference between **1-g⁺g⁺** and **1-aa** at the SCF level. The $\Delta E^{DIF}_{\sigma^*(CO)}$ differences between **1-g⁺g⁺/1-g⁺a** (4.0 kcal/mole), **2-g⁺g⁺/2-g⁺a** (2.8 kcal/mole) and **2-g⁺g⁺/2-aa** (6.0 kcal/mole) are also comparable to the RHF/6-31G* conformer energy differences.

Optimization of **1-g⁺g⁺** and **1-g⁺a** in the absence of all $n(O)$ delocalizations results in a dramatic lowering of the energy difference between the conformers. At RHF/6-31G*, **1-g⁺g⁺** is more stable than **1-g⁺a** by 4.04 kcal/mole. With all $n(O)$ delocalizations deleted, the difference between conformers drops to 0.80 kcal/mole. The energy difference between **1-g⁺g⁺** and **1-aa** is effectively 0.00 kcal/mole. A change in CO bond lengths also occurs. For **1-g⁺g⁺**, r_{CO} lengthens to 1.4496Å (Δr_{CO} =0.064Å) and for **1-g⁺a**, r_{CO} lengthens to 1.4470Å ($\Delta r_{CO} = 0.053$Å). If the anomeric effect did not arise from $n(O) \rightarrow \sigma^*_{CO}$ interactions, then optimization in the absence of these interactions would not result in either geometry changes or energy changes between conformers.

Specific deletion of Fock matrix elements can be extended to included all off-diagonal interactions, and is termed a NOSTAR calculation (deletion of electron density in all "starred" orbitals - *37,38*). Technically, energies of single deletions are not variationally correct, but NOSTAR calculations give energies consistent with the variational principle and provide the solution to the 1-electron Slater determinant. Naturally, the NOSTAR deletion energies values are larger than the simple $n(O) \rightarrow \sigma^*_{CO}$ deletion energies, reflecting a larger number of deletions (Table II). The larger molecular framework in **2** gives rise to more $\sigma \rightarrow \sigma^*$ interactions, which accounts for the increase in NOSTAR deletion energies compared to **1**.

The NOSTAR deletion energies also support the hyperconjugative model. For **1**, $E^{DEL}_{NS}(g^+g^+) - E^{DEL}_{NS}(g^+a) = 4.00$ kcal/mole, which is about the energy difference of 4.04 kcal/mole of the conformers at RHF/6-31G*. The same analysis for **2** yields a $E^{DEL}_{NS}(g^+g^+) - E^{DEL}_{NS}(g^+a) = 2.60$ kcal/mole, which again compares well with the conformational energy of 2.44 kcal/mole at RHF/6-31G*. The difference in NOSTAR electronic energies, ΔE_{NS}, between the **g⁺g⁺** and **g⁺a** conformers is small (0.00 kcal/mole for **1**; 0.1 kcal/mole for **2**). For **1**, $E^{DEL}_{NS}(g^+g^+) - E^{DEL}_{NS}(aa) = 7.3$ kcal/mole, which is smaller than the 8.7 kcal/mole difference between conformers at RHF/6-31G*. However, $\Delta E_{NS}(g^+g^+\text{-}aa)$ is rather large (1.4 kcal/mole) and must be added to ΔE^{DEL}_{NS} to equal the RHF/6-31G* conformational energy difference. ΔE_{NS} corresponds to the dipole-dipole destabilization of a structure in the absence of all stabilizing 2-electron delocalizations.

NOSTAR optimizations of the three conformations provide final proof that the anomeric effect is due primarily to $n(O) \rightarrow \sigma^*_{CO}$ interactions. An overlay of the optimized RHF/6-31G* and NOSTAR/6-31G* structures of **2-g⁺a** is shown in Figure 6. Clearly, large geometry changes are calculated (Tables IV, V). Of particular note is the change in r_{CO} of roughly 0.13Å. CO bonds directly involved in $n(O) \rightarrow \sigma^*_{CO}$ interactions lengthen up to 0.03Å more than the CO bonds not involved with anomeric stabilization. Other changes in geometry also take place, such as the OCO bond angle. To maximize the overlap between the oxygen lone pair and the adjacent σ^*_{CO} orbital, $\angle_{OCO}$ enlarges. As the stabilizing $n(O) \rightarrow \sigma^*_{CO}$ interactions are removed, the enlarged OCO bond angle is free to relax (shrink). For **2-g⁺g⁺**, **2-g⁺a** and **2-aa**, $\Delta\angle_{OCO}$ = 6.0°, 3.9° and 1.5° respectively. The sum of incremental $\Delta\Delta\angle_{OCO}$ for **2-g⁺g⁺→2-g⁺a** = 2.1° and **2-g⁺a→2-aa** = 2.4° is equal to the $\Delta\Delta\angle_{OCO}$ of **2-g⁺g⁺→2-aa** = 4.5°. The larger bond angle changes upon NOSTAR optimization also indicate the number of $n(O) \rightarrow \sigma^*_{CO}$ interactions. The NOSTAR optimizations of **1** produce larger geometry changes than optimization in the absence of just $n(O) \rightarrow \sigma^*_{CO}$ interactions. The largest stabilizing donor-acceptor interactions in **1** and **2** are clearly the $n(O) \rightarrow \sigma^*_{CO}$ interactions, but a host of other delocalizations exist (e.g. $n(O) \rightarrow \sigma^*_{CH}$, $\sigma_{OC} \rightarrow \sigma^*_{CO}$, $\sigma_{OC} \rightarrow \sigma^*_{CO}$). The larger geometry change results from the loss of all $\sigma \rightarrow \sigma^*$ interactions, which include the $n(O) \rightarrow \sigma^*_{CO}$ interactions.

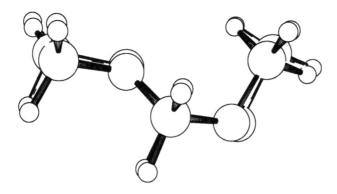

Figure 6: Overlay of RHF/6-31G* (foreground) and NOSTAR/6-31G* (background) optimized geometries of **2-g⁺a**.

If the $n(O) \rightarrow \sigma^*_{CO}$ were not at the heart of the anomeric effect, then reoptimization would fail to produce such dramatic changes in molecular geometry (Tables IV, V). The persuasive deletion optimizations are unique; no other method provides direct evidence for or against the origin of the anomeric effect by determining geometries and conformational energies in its absence. One reasonable objection to the deletion calculations is that electron correlation effects were not considered. Typically, an effective one-electron Hamiltonian (Fock) operator is not available for correlated methods, making the deletion calculations impossible at higher levels of theory.

Table IVa: Optimized 1-g⁺g⁺ 6-31G* RHF and NOSTAR geometries

	RHF	NOSTAR	Δ^a
$r_{CO}{}^b$	1.3859	1.5103	0.1244
$r_{OH}{}^b$	0.9486	0.9457	0.0029
$\angle_{OCO}{}^c$	112.4	104.1	8.3
$\angle_{HOC}{}^c$	108.8	105.7	3.1

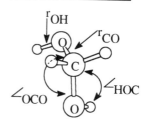

$^a\Delta = |\text{NOSTAR-RHF}|$. ^{b}Bond lengths given in Angströms. ^{c}Bond angles given in degrees.

Table IVb: Optimized 1-g⁺a 6-31G* RHF and NOSTAR geometries

	RHF	NOSTAR	Δ^a
$r_{CO1}{}^b$	1.3736	1.5063	0.1327
$r_{CO2}{}^b$	1.3941	1.5071	0.1130
$r_{OH1}{}^b$	0.9477	0.9455	0.0022
$r_{OH2}{}^b$	0.9485	0.9445	0.0040
$\angle_{OCO}{}^c$	108.8	104.7	4.1
$\angle_{HOC1}{}^c$	108.9	105.8	3.1
$\angle_{HOC2}{}^c$	110.1	107.4	2.7

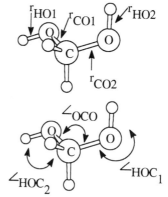

$^a\Delta = |\text{NOSTAR-RHF}|$. ^{b}Bond lengths given in Angströms. ^{c}Bond angles given in degrees.

Table IVc: Optimized 1-aa 6-31G* RHF and NOSTAR geometries

	RHF	NOSTAR	Δ^a
$r_{CO}{}^b$	1.3824	1.5046	0.1222
$r_{OH}{}^b$	0.9467	0.9444	0.0013
$\angle_{OCO}{}^c$	105.4	105.3	0.1
$\angle_{HOC}{}^c$	109.6	107.1	2.5

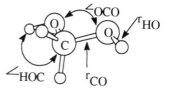

$^a\Delta = |\text{NOSTAR-RHF}|$. ^{b}Bond lengths given in Angströms. ^{c}Bond angles given in degrees.

Although electron correlation is required for quantitative modeling of the glycosidic linkage in carbohydrates (*vide infra*), the RHF/6-31G* calculations are qualitatively the same as calculations at higher levels of theory, e.g. MP4SDTQ/6-31G*.

The contributions of the CO bond dipole-dipole interactions can be ascertained within the NOSTAR basis. Within the NBO basis, bond dipoles arise from

Table Va: Optimized 2-g⁺g⁺ 6-31G* RHF and NOSTAR geometries

	RHF	NOSTAR	Δ^a
$r_{CO}{}^b$	1.3800	1.5097	0.1297
$r_{OC}{}^b$	1.4007	1.5272	0.1265
$\angle_{OCO}{}^c$	114.0	108.0	6.0
$\angle_{COC}{}^c$	115.1	112.8	2.3

$^a\Delta = |\text{NOSTAR-RHF}|$. ^{b}Bond lengths given in Angströms. ^{c}Bond angles given in degrees.

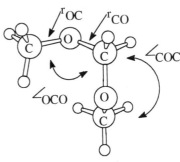

Table Vb: Optimized 2-g⁺a 6-31G* RHF and NOSTAR geometries

	RHF	NOSTAR	Δ^a
$r_{CO1}{}^b$	1.3686	1.5108	0.1422
$r_{CO2}{}^b$	1.3882	1.5064	0.1182
$r_{OC1}{}^b$	1.4014	1.5257	0.1243
$r_{OC2}{}^b$	1.3930	1.5242	0.1312
$\angle_{OCO}{}^c$	110.3	106.4	3.9
$\angle_{COC1}{}^c$	115.5	112.7	2.8
$\angle_{COC2}{}^c$	114.0	111.6	2.4

$^a\Delta = |\text{NOSTAR-RHF}|$. ^{b}Bond lengths given in Angströms. ^{c}Bond angles given in degrees.

Table Vc: Optimized 2-aa 6-31G* RHF and NOSTAR geometries

	RHF	NOSTAR	Δ^a
$r_{CO}{}^b$	1.3756	1.5077	0.1321
$r_{OC}{}^b$	1.3933	1.5254	0.1321
$\angle_{OCO}{}^c$	106.6	105.1	1.5
$\angle_{COC}{}^c$	114.0	111.7	2.3

$^a\Delta = |\text{NOSTAR-RHF}|$. ^{b}Bond lengths given in Angströms. ^{c}Bond angles given in degrees.

2-electron delocalizations. An optimized structure within the NOSTAR basis is a "pure" Lewis structure. In the limiting case of a pure Lewis structure with no electrostatic contribution, the bond lengths and bond angles of different conformers would be identical (22,23,30). Once all delocalizations have been removed, the remaining effects must be electrostatic in nature and reflect the true dipole-dipole contributions.

The NOSTAR geometries of **1** and **2** show slight differences, which are due to the bond dipoles. However, the NOSTAR geometry differences are the dramatically smaller than the RHF/6-31G* level. These data suggest that while dipole-dipole interactions do exist, their contribution to the anomeric effect appear to be about an order of magnitude less important than the $n(O) \rightarrow \sigma^*_{CO}$.

The major objection to the hyperconjugative model is its purported failure to explain the apparent lessening of the anomeric effect in more polar solvents.[12] One must, however, consider *all* intra- and intermolecular donor-acceptor interactions when examining solvation effects. Solvation and other intermolecular interactions share a common origin with the anomeric effect in that 2-electron delocalizations dictate the observed chemistry. As a solvent becomes more polar, the solvent's bonds and antibonds must necessarily become more polarized. More polarized bonds are better acceptors of electron density and provide more stabilization when interacting with polarized solutes. Since the $n(O)$ is an effective donor orbital, interactions with polarized solvent σ^* orbitals can become competitive with intramolecular σ^*_{CO} overlap. The more polar the solvent, the more polarized the σ^* orbitals and the greater the competition between intramolecular vs. intermolecular $n(O) \rightarrow \sigma^*$ overlap. In the limiting case, the solvent σ^* orbitals would provide better overlap than intramolecular σ^* orbitals. While such a scenario should prove unlikely, the realization that solvation is also controlled by 2-electron stabilizing interactions allows rationalization of the experimentally observed solvation effects.

Quantitative Modeling of the Anomeric Effect

The quantitative modeling of the anomeric effect and glycosidic linkage at the *ab initio* level reduces to fully correlating the four lone pair electrons (*37,39*). This conclusion follows from the observation that the bond lengths and bond angles of **1** and **2** approach limiting values at higher levels of theory. The rotational potential surfaces of **1** and **2** have been optimized up through the MP4SDTQ/6-31G* level of theory. While the complete details of these calculations will be communicated elsewhere, two aspects of the work are relevant to this discussion: (*i*) how correlation of the lone pairs affect geometries and (*ii*) why quantitative modeling is important. Recently, we have shown that molecules with two vicinally oriented lone pairs (e.g. hydrazine) converge in a basis set and correlation method independent manner provided that all four lone pair electrons are correlated (*37,39*). The geminal oxygen lone pairs follow the same computational trends as vicinal lone pairs, suggesting that geminal interactions can also be effectively modeled at the MP4SDTQ level of theory.

Recently, a report modeling the conformations of **1** optimized up through MP2/6-311G** has appeared (*40*). Two landmark *ab initio* studies on the rotational potential surface of **2** exist. In the first, Pople and co-workers examined the potential surface at RHF/4-31G (*41*). Murcko and Wiberg optimized the surface of **2** at RHF/6-31G* and performed single point calculations at MP2/6-31G* (*42*). Despite the detail of these studies, none included enough correlation, if at all, during geometry optimization. The study on vicinal lone pairs concluded that lone pairs from RHF geometry optimizations emerge too diffuse, overestimating $\sigma \rightarrow \sigma^*$ interactions. Shortened CX bond lengths (X = N,O,P,S) and larger CXC, CXX, XCX bond angles result. In the case of **1** and **2**, the RHF geometries underestimate the two CO bond lengths and overestimate the OCO and COH/COC bond angles. To maximize overlap between the donor and acceptor, the CO_1 bond shortens and the OCO bond angle increases. The diffuse RHF lone pairs cause distortion in ground state geometries by maximizing these interactions, causing geometry changes which reinforce the effect.

For quantitative geometric modeling of the glycosidic linkage, correlation effects prove important. Optimization of **1** and **2** including correlation produces substantial geometry changes relative to RHF geometries (Tables VI, VII). Calculations including correlation effects at the MP2 and MP4SDTQ levels compact the lone pairs, in effect reducing the magnitude of the $\sigma \rightarrow \sigma^*$ interactions, which are, in this case, $n(O) \rightarrow \sigma^*_{CO}$ interactions. Cole and Bartlett recently suggested that in atomic systems MBPT(2) and MBPT(4) wavefunctions (equivalent to MP2 and MP4SDTQ corrections) account for roughly 90% and 100% of the total electron correlation effects for a particular basis set (*43*). As the correlation level increases from MP2 to MP4SDTQ, a larger geometry change is observed relative to the RHF reference structure, consistent with the notion of increased electron correlation. The RHF→MP2 geometry change is larger in magnitude than the MP2→MP4SDTQ change, suggesting that MP4SDTQ geometries are approaching limiting values. These changes are internally consistent with calculations on other systems, e.g. hydrazine. In the absence of the other data, the idea of compacted lone pairs reducing the magnitude of $n(O) \rightarrow \sigma^*_{CO}$ is reasonable.

The need for correlation to accurately model the magnitudes of $n(O) \rightarrow \sigma^*_{CO}$ interactions does not invalidate the conclusions of the previous section. As before, the existence of $n(O) \rightarrow \sigma^*_{CO}$ interactions is consistent with interacting lone pair electrons that are not repulsive. Though dramatic geometry changes are observed when **1** and **2** are optimized in the presence of correlation effects, the resulting geometries are still different than the NOSTAR geometries. Thus, even though r_{CO} of **2-g⁺g⁺** at MP4SDTQ/6-31G* is longer than at RHF/6-31G* (1.4106Å vs. 1.3800Å), the former is still substantially shorter than the corresponding NOSTAR/6-31G* value of 1.5097Å. Therefore, while the MP4SDTQ/6-31G* models the geometry of **1** and **2** more quantitatively than other methods, $n(O) \rightarrow \sigma^*_{CO}$ interactions still exert an anomeric conformational effect.

Changes in bond lengths and bond angles as a function of dihedral angle, the so-called "cross-terms" in empirical force field terminology, is a recognized but largely unimplemented aspect of current molecular mechanic programs (*44*). This study actually started as an effort to develop better molecular mechanics parameters for molecular dynamics (MD) modeling of carbohydrate structures. Within a MD/MM framework, the results are only as good as the parameters and approximations used within the empirical force field. *Ab initio* potentials have been used to obtain terms for MM parameterization not available from experiment (*45*). While the current generation of force fields, namely MM3, have made provisions within their frame work for cross terms, the parameterization is incomplete (*46-50*).

Table VI: Optimized 6-31G* geometries of **1** at different levels of correlation [a]

	1-g^+a				1-g^+g^+		
	RHF	MP2	MP4SDTQ		RHF	MP2	MP4SDTQ
r_{CO1}^b	1.3736	1.3929	1.3972	r_{CO}^b	1.3859	1.4084	1.4128
r_{CO2}^b	1.3941	1.4184	1.4230	r_{OH}^b	0.9486	0.9731	0.9748
r_{OH1}^b	0.9477	0.9727	0.9741	$\angle_{OCO}^c$	112.4	112.5	112.5
r_{OH2}^b	0.9485	0.9723	0.9725	$\angle_{HOC}^c$	108.8	106.5	106.4
$\angle_{OCO}^c$	108.8	108.1	108.1				
$\angle_{HOC1}^c$	108.9	106.8	106.8				
$\angle_{HOC2}^c$	110.1	108.2	108.7				

[a]See Table IV for bond length and bond angle definitions. [b]Bond lengths reported Angströms. [c]Bond angles reported in degrees.

Table VII: Optimized 6-31G* geometries of **2** at different levels of correlation [a]

	2-g^+a				2-g^+g^+		
	RHF	MP2	MP4SDTQ		RHF	MP2	MP4SDTQ
r_{CO1}^b	1.3686	1.3893	1.3921	r_{CO}^b	1.3800	1.4058	1.4106
r_{CO2}^b	1.3882	1.4142	1.4183	r_{OC}^b	1.4007	1.4266	1.4310
r_{OC1}^b	1.3930	1.4142	1.4222	$\angle_{OCO}^c$	114.0	113.6	114.0
r_{OC2}^b	1.4014	1.5272	1.4317	$\angle_{COC}^c$	115.1	111.0	111.0
$\angle_{OCO}^c$	110.3	109.9	109.6				
$\angle_{COC1}^c$	115.5	112.4	112.2				
$\angle_{COC2}^c$	114.0	111.2	111.0				

[a]See Table V for bond gength and bond angle definitions. [b]Bond lengths reported Angströms. [c]Bond angles reported in degrees.

Cross terms involving torsional motions in force fields require accurate geometries for accurate potentials. Figures 7 and 8 demonstrate the existence of cross terms or the coupling of motions, where changes in CO bond lengths and OCO bond angles are observed as a function of HOCO dihedral angle. The origin of these effects is tied to the anomeric effect - namely $n(O) \rightarrow \sigma^*$ interactions. In the case of **1** and **2**, CO bond rotation changes the orientation of the $n(O)$ donor orbital relative to the σ^*_{CO} acceptor orbital. Although maximum orbital overlap takes place when the donor and acceptor are antiperiplanar, partial donation takes place over most of the CO bond rotational profile. To maximize overlap between the donor and acceptor, the CO_1 bond length shortens and the OCO bond angle enlarges. At a rotational transition state, stabilizing $\sigma \rightarrow \sigma^*$ donations ($n(O) \rightarrow \sigma^*_{CO}$ donations) are lost. In the absence of these donations, the bonds lengthen and the bond angles decrease. The g^+g^+ and g^+a geometries of the of **1** and **2** observe these general trends (Figures 7, 8, Tables VI, VII). The g^+g^+ conformations have the maximum number of $n(O) \rightarrow \sigma^*_{CO}$

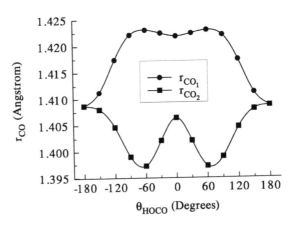

Figure 7: Plot of changes in MP4SDTQ/6-31G* optimized CO bond lengths upon CO bond rotation in **1** .

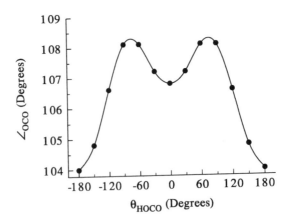

Figure 8: Plot of changes in MP4SDTQ/6-31G* optimized OCO bond angle upon CO bond rotation in **1** .

interactions and both HOCO dihedral angles are at their minimum energy values on the CO bond rotational surface. For the $\mathbf{g^+a}$ conformations, the two different HOCO dihedral angles have different bond lengths and bond angles due to different acceptors for $n(O)$ delocalizations. This, coupled with the dramatic change in stabilization energy of the two different conformers (Table I), provides strong evidence that the $n(O) \rightarrow \sigma^*_{CO}$ interactions are responsible for all the effects discussed.

Methods

All calculations were performed on a Stardent-3000, an IBM RS6000/530, an IBM RS6000/560 or a Cray Y-MP4/464 with Gaussian-90 or Gaussian-92 (*16,17*). Standard Pople style basis sets were used as implemented in Gaussian-90 and/or Gaussian-92. All correlation methods were used as implemented in Gaussian 90/Gaussian-92. Abbreviations for correlated methods are as follows: MP2 - second order Moller-Plesset correlation energy correction; MP4(SDTQ) - fourth order Moller-Plesset correlation energy correction including triple substitutions (*18,19*). The NBO analysis was used as implemented in Gaussian-90 and Gaussian-92 (*16,17,20,21*).

Conclusions

The anomeric effect represents a specific example of general 2-electron $\sigma \rightarrow \sigma^*$ stabilizing interactions, and can be thought of as a departure from simple Lewis-type bonding. The important conclusions of this study are threefold: (*i*) The anomeric effect is due primarily to stabilizing 2-electron $n(O) \rightarrow \sigma^*_{CO}$ interactions, (*ii*) quantitative modeling of the anomeric effect can be achieved at MP4SDTQ/6-31G* and, (*iii*) quantitative modeling of cross terms in empirical force fields are shown by these calculations to be important. Central to these conclusions is the NBO analysis, which provided the framework in which these questions were phrased. Also of note were Weinhold's observations that the two models (4-electron repulsive versus 2-electron attractive) actually arise from the same details of the total electron density when all interactions are included.

Acknowledgments

The authors would like to express their gratitude to Frank A. Weinhold for his insights during this work. The authors also benefited from stimulating conversations with Eric D. Glendening, Gary R. Weisman and Mary Beth Carter during the course of this work. We are grateful for financial support from the Whitaker Foundation. Calculations were performed on computers purchased with NSF support (CHE-9007850). Cray Y-MP4/464 supercomputer calculations were performed at the NCSA (CHE-920025N).

Literature Cited

1. Jungius, C. L. *Z. Phys. Chem.*, **1950**, *52*, 97.
2. Lemieux, R. U. *Adv. Carb. Chem.*, **1954**, *9*, 1.
3. Haynes, L. J.; Newth, F. H. *Adv. Carb. Chem.*, **1955**, *10*, 207.
4. Edward, J. T. *Chem. Ind*, (London), **1955**, 1102.
5. Lucken, E. A. C. *J. Chem. Soc.*, **1959**, 2954.
6. de Hoog, A. J.; Buys, H. R.; Altona, C.; Havinga, E. *Tetrahedron*, **1969**, *25*, 3365.
7. Durette, P. L.; Horton, D. in *Advances in Carbohydrate Chemistry and Biochemistry*; Tipson, R. S.; Horton, D., Eds.; Academic Press: New York, NY, 1971, Vol 26; p 49.
8. Eliel, E. L. *Angew. Chem.*, **1972**, *17*, 779 and references therein.
9. *Anomeric Effect: Origin and Consequence*; Szarek, W. A.; Horton, D., Eds.; ACS Symposium Series, Vol. 87; ACS publications: Washington, D.C., 1979.
10. Kirby, A. J. *The Anomeric Effect and Related Stereoelectronic Effects at Oxygen*; Springer-Verlag: Berlin, 1983.
11. Deslongchamps, P. *Stereoelectronic Effects in Organic Chemistry*; Pergamon Press: Exeter, England, 1983.
12. Tvaroska, I.; Bleha, T. in *Advances in Carbohydrate Chemistry and Biochemistry*; Tipson, R. S.; Horton, D., Eds.; Academic Press: San Diego, CA, 1989, Vol. 47; p 45.
13. Box, V. G. S. *Heterocycles*, **1984**, *22*, 891.
14. Inagaki, S.; Iwase, K.; Mori, Y. *Chem. Lett.*, **1986**, 417.
15. Juaristi, E.; Cuevas, G. *Tetrahedron*, **1992**, *48*, 5019.
16. *Gaussian 90*, Revision H, Frisch, M. J.; Head-Gordon, M.; Trucks, G. W.; Foresman, J. B.; Schlegel, H. B.; Raghavachari, K; Robb, M; Binkley, J. S.; Gonzalez, C.; Defrees, D. J.; Fox, D. J.; Whiteside, R. A.; Seeger, R.; C. F. Melius, C. F.; J. Baker, J.; Martin R. L.; Kahn, L. R.; Stewart, J. J. P.; Topiol, S.; Pople, J. A.; Gaussian, Inc., Pittsburgh PA, 1990.
17. *Gaussian 92*, Revision A, Frisch, M. J.; Trucks, G. W.; Head-Gordon, M.; Gill, P. M. W.; Wong, M. W.; Foresman, J. B.; Johnson, B. G.; Schlegel, H. B.; Robb, M. A.; Replogle, E. S.; Gomperts, R.; Andres, J. L.; Raghavachari, K.; Binkley, J. S.; Gonzalez, C.; Martin, R. L.; Fox, D. J.; Defrees, D. J.; Baker, J.; Stewart, J. J. P.; Pople, J. A. Gaussian, Inc., Pittsburgh PA, 1992.
18. Hehre, W. J.; Radom, L.; Schleyer, P. v.R.; Pople, J. A. *Ab Initio Molecular Orbital Theory*; Wiley-Interscience: New York, NY, 1986.
19. Clark, T. *A Handbook of Computational Chemistry*; Wiley-Interscience: New York, NY, 1985.
20. Reed, A. E.; Curtiss, L. A.; Weinhold, F. *Chem. Rev.*, **1988**, *88*, 899 - 926.
21. Glendening, E. D.; Reed, A. E.; Carpenter, J. E.; Weinhold, F. *NBO 3.0 Program Manual Technical Notes*; WIS-TCI-756, Madison, 1990.
22. Lewis, G. N. *Valence and the Structure of Atoms and Molecules*; The Chemical Catalogue Company: 1923.

23. Weinhold, F.; Carpenter, J. E. *The Structure of Small Molecules and Ions*; Naaman, R.; Vager, Z., Eds; Pelnum: New York, NY, 1988; p 227.
24. Reed, A. E.; Weinhold, F. *Isr. J. Chem.*, **1991**, *31*, 277.
25. Carpenter, J. E.; Weinhold, F. *J. Am. Chem. Soc.*, **1988**, *110*, 368 - 372.
26. Weinhold, F. A.; Carpenter, J. E. *Journal of Molecular Structure*, **1988**, *165*, 189.
27. Foster, J. P.; Weinhold, F. *J. Am. Chem. Soc.*, **1980**, *102*, 7211 - 7218.
28. Reed, A. E.; Weinstock, R. B.; Weinhold, F. *J. Chem. Phys.* **1985**, *83*, 735 - 746.
29. Reed, A. E.; Weinhold, F. *J. Chem. Phys.* **1985**, *83(4)*, 1736 - 1740.
30. Pauling, L. "The Nature of the Chemical Bond", 3rd Edition; Cornell University Press, Ithica, New York: 1960.
31. Löwdin, P.-O. *Phys. Rev.*, **1955**, *97*, 1474 - 1489.
32. Weinhold, F. "Are Four-Electron Interactions Destabilizing?", WISC-TCI-789, Madison, WI.; 1992.
33. Jorgensen, F. S.; Norskov-Laurisen, L. *Tetrahedron Lett.*, **1982**, 5221.
34. Tvaroska, I.; Bleha, T. *Can. J. Chem.*, **1979**, *57*, 424.
35. Thogersen, H.; Lemieux, R. U.; Bock, K.; Meyer, B. *Can. J. Chem.*, **1982**, *60*, 44.
36. Lemieux, R. U.; Koto, S.; Voisin, D. in *Anomeric Effect: Origin and Consequence*; Szarek, W. A.; Horton, D., Eds.; ACS Symposium Series, Vol. 87; ACS publications: Washington, DC, 1979. p 17.
37. Petillo, P. A., Ph.D. Thesis, University of Wisconsin, 1991.
38. Glendening, E. D., Ph.D. Thesis, University of Wisconsin, 1991.
39. Petillo, P. A.; Nelsen, S. F. *J. Am. Chem. Soc.*, submitted for publication.
40. Woods, R. J.; Szarek, W. A.; Smith, V. H. *J. Chem. Soc. Chem. Commun.*, **1991**, 334.
41. Jeffrey, G. A.; Pople, J. A.; Binkley, J. S.; Vishveshwara, S. *J. Am. Chem. Soc.*, **1978**, *100*, 373.
42. Wiberg, K. B.; Murcko, M. A. *J. Am. Chem. Soc.*, **1989**, *111*, 4821.
43. Cole, S. J.; Bartlett, R. J. *J. Chem. Phys.*, **1987**, *86*, 873.
44. Burkert, U.; Allinger, N. L *Molecular Mechanics*; ACS Monograph, Vol. 177; ACS Publications: Washinton, DC, 1982.
45. Schmitz, L. R.; Allinger, N. L.; Profeta, S., Jr. *J. Comp. Chem.*, **1988**, *9*, 460-464.
46. Allinger, N. L.; Yuh, Y. H.; Lii, J.-H. *J. Am. Chem. Soc.*, **1989**, *111*, 8551.
47. Lii, J.-H.; Allinger, N. L. *J. Am. Chem. Soc.*, **1989**, *111*, 8566.
48. Lii, J.-H.; Allinger, N. L.; *J. Am. Chem. Soc.*, **1989**, *111*, 8576.
49. Dowd, M. K.; Zeng, J.; French, A. D.; Reilly, P. J. *Carbohydr. Res.*, **1992**, *230*, 223.
50. Stuike-Prill, R.; Meyer, B. *Eur. J. Biochem*, **1990**, *194*, 903.

RECEIVED May 12, 1993

Chapter 10

Application of Quantum Theory of Atoms in Molecules to Study of Anomeric Effect in Dimethoxymethane

N. H. Werstiuk, K. E. Laidig[1], and J. Ma

Department of Chemistry, McMaster University, Hamilton, Ontario L8S 4M1, Canada

The quantum theory of atoms in molecules (AIM) has been applied to a study of four conformations of dimethoxymethane (**gg-1** , **ag-1**, **aa-1**, and **90,90-1**), three conformations of methyl vinyl ether (**sc-2**, **tsc-2**, and **tsc2-2**), and two conformations of methyl n-propyl ether (**aa-3** and **ga-3**) to establish how the anomeric effect in dimethoxymethane is made manifest in the topology of the charge density. The AIMPAC set of programs and DOMAIN were used to compute the atom electron populations, the atom energies and to analyze the topology of the charge density. By studying the s-cis planar (**sc-2**) and the twist conformations **tsc-2** and **tsc2-2** of methyl vinyl ether we have established that the Laplacian and its ellipticity are sensitive probes of the differential delocalization/localization of geminal lone pairs on oxygen. That the Laplacians of the nonbonded charge concentrations are virtually identical in the case of **gg-1** is surprizing and suggests that the classical picture of the anomeric effect in which there is a differential delocalization of a lone pair of electrons antiperiplanar to a C-O bond does not hold for this acetal. Even so, the nbccs of **aa-1** and **aa-3** are larger and have a smaller ellipticity than in the case of **gg-1** indicating that the localization of the lone pairs is greater in **aa-1** than in **gg-1**.

The anomeric effect, first uncovered in the chemistry of α-methylglycosides has played a central role in many attempts to rationalize and predict changes in geometries and reactivities of compounds which have nonbonding electrons antiperiplanar to polar sigma bonds. Its role in inducing changes in geometries and reactivities of a wide range of compounds has been probed experimentally and computationally. Much of the experimental work is documented in two excellent monographs (*1,2*) and a review

---

[1]Current address: Department of Chemistry, University of California—Berkeley, Berkeley, CA 94720

0097–6156/93/0539–0176$08.25/0

(*3*). Computational studies have been carried out with ab initio and semiempirical methods (see reference *4)* and to this point, it is safe to say that the stabilizing component of the anomeric effect has been described on the basis of frontier molecular orbital theory. The results of one of the most recent ab initio computational studies of dimethoxymethane (**1**) and methyl n-propyl ether (**3**) have been published by Wiberg and Murcko (*4*); 6-31G* and 6-31+G* basis sets were used to compute the geometries and total energies of the gauche-gauche (**gg**), anti-gauche (**ag**) and anti-anti (**aa**) conformers of **1** and the analogous conformers of **3**. The **gg** geometrical structure of **1** was computed to be the lowest energy conformation and the authors deduced that the anomeric stabilization of **gg-1** (the 67,67-degree conformation) which formally has two anomeric interactions is 4.6 kcal/mol relative to **aa-1** (the 180,180-degree conformation). Consequently, **gg-1** should be preferentially populated in the gas phase. That this is the case has been established by gas-phase electron diffraction (*5*) and ultraviolet photoelectron spectroscopic studies (*6*). A number of energy decomposition schemes (see reference *7)* have been used in attempts to disect the anomeric stabilization energy as obtained from ab initio calculations into steric, electrostatic, and electronic energies. In the most recent attempt, Grein (*7*) carried out a Fourier analysis of the ΔEs between dihydroxymethane conformers to evaluate constants V_1, V_2, and V_3 corresponding to the dipolar interaction, bonding and repulsion potentials, respectively. It was concluded that the anomeric stabilization is due to partial π bonding in the CH_2-OH fragment of dihydroxymethane and that the "antiperiplanar alignment of a lone pair with a C-OH bond was incidental and of no immediate concern; the important factor was the coplanar alignment of a p-type lone pair with the C-OH bond and that this was maximized in the 90,90-degree conformation". This rationale shows the arbitrary nature of orbital models which continue to be used to define the stabilizing component of the anomeric effect.

We have applied the quantum theory of atoms in molecules (AIM) (*8*) to a study of **1**, methyl vinyl ether (**2**), methyl n-propyl ether (**3**), O-protonated **1**, and O-protonated **3** to establish how the anomeric effect is made manifest in the topology of the charge density of **1** and its protonated analogue. In this approach, no orbital model is assumed. Wavefunctions of optimized geometries are used with the AIMPAC set of programs (*8*) to calculate the properties of the critical points (cps) in the charge densities (ρ) and Laplacians ($-\nabla^2\rho(\mathbf{r})$) of bonds and nonbonded charge concentrations (nbccs). These calculations provide the only way to characterize quantitatively nbccs which correspond to the lone pairs defined in the VSEPR model (*8*). Because this is done by computing their size, their shape, the location of the maxima in the nonbonded density and the orientation of the maxima with respect to adjacent bonds, it is possible to gain information on the delocalization/localization of nonbonding electrons. Acetal **1** was chosen over dihydroxymethane because potential problems with hydrogen bonding are obviated and it should be the best model for simple dialkyl acetals. Of prime interest were the **gg**, **ag**, and **aa** conformations of **1**. There are two anomeric interactions (two lone pairs aligned antiperiplanar with two C-O bonds) in **gg-1**, one in **ag-1**, and none in **aa-1.** Vinyl ether **2** was studied because it represents a case where the interaction of lone pairs on oxygen with a double bond should show an angular dependency and this should be reflected in the topology of the charge density. Because **3** does not have a C-O bond vicinal to an oxygen lone pair, it was studied as a case where anomeric stabilization is small or negligible. In this paper we

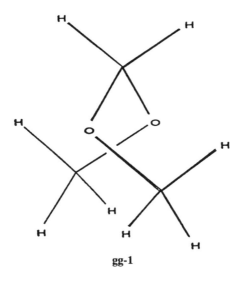

gg-1

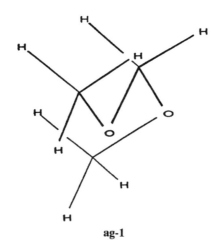

ag-1

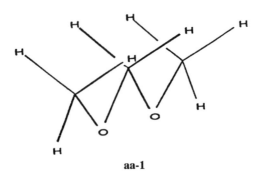

aa-1

present and discuss the results of the AIMPAC computational studies on **1**, **2**, and **3**. The results of the study on the O-protonated analogues of **1** and **3** will be presented in a following paper.

RESULTS AND DISCUSSION

Dimethoxymethane and Methyl Vinyl Ether Conformers

The ab initio computational packages Gaussian 90 (9), Gaussian 92,(10) and Gamess (11) were used to compute the equilibrium geometries (Table 1), the total

Table 1. Geometric Bond Lengths of Dimethoxymethane Conformers

	gg-1[a]	gg-1[b]	ag-1	aa-1	aa-1[b]	90,90-1
C(1) - O(2)	1.4005[a]	1.4018	1.3933	1.3935	1.3951	1.3985
O(2) - C(3)	1.3816	1.3826	1.3887	1.3756	1.3763	1.3823
C(3) - O(4)	1.3816	1.3826	1.3678[c]	1.3756	1.3763	1.3823
O(4) - C(5)	1.4005	1.4018	1.4013[c]	1.3935	1.3951	1.3985
C(1) H$_3$	1.0885	1.0882	1.0893	1.0807	1.0805	1.0815
	1.0815	1.0811	1.0811	1.0891	1.0889	1.0881
	1.0844	1.0846	1.0890	1.0891	1.0899	1.0863
C(3) H$_2$	1.0859	1.0852	1.0945	1.0954	1.0945	1.0854
	1.0859	1.0852	1.0866	1.0954	1.0945	1.0854
C(5) H$_3$	1.0885	1.0882	1.0888	1.0807	1.0805	1.0815
	1.0815	1.0811	1.0812	1.0891	1.0889	1.0881
	1.0844	1.0846	1.0831	1.0891	1.0889	1.0863

[a]The bond distances are given in angstroms.
[b]These are the values obtained with the 6-31++G** basis set.
[c]This is the gauche fragment of **ag-1**.

energies and the wavefunctions of **gg-1**, **ag-1**, **aa-1** and **90,90-1** (the structural formula of this conformation is not shown) with the 6-31G** basis set. There are two notable differences in the bond lengths computed for **gg-1** and **aa-1**; the CH$_3$-O and O-CH$_2$ bonds of **gg-1** are longer and the methylene C-H bonds are shorter than the corresponding bonds of **aa-1**. It appears, based on fact that the O-CH$_2$ bonds of **gg-1** are longer than the bonds of **aa-1**, that the proposal (7) that the stabilizing component of the anomeric effect is due to an increase in π-bonding between the p-type lone pair and the adjacent C-O bond does not hold for **1**. The component energies of **gg-1**, **ag-1**, and **aa-1** are given in Table 2. It is seen that the the difference in energy between **aa-1** and **gg-1** (Table 3) is dominated by a decrease in V_{ne} (this corresponds to an increase in the stabilizing attractive interaction) even though V_{ee} and V_{nn} (destabilizing repulsive interactions) increase in going from **aa-1** to **gg-1**. PROAIMV of AIMPAC was used to compute the energies of the atoms; it is the decreases in V_{ne} of the carbons - C(1) and C(5) - and the oxygens - O(2) and C(4) - which dominate the large change in V_{ne} (see Table 3) when **aa-1** is transformed into **gg-1**, but the change in V_{ne} of C(3), the methylene carbon, is relatively small. Because an increase

Table 2. Component Energies of Dimethoxymethane Conformers

	gg-1	ag-1	aa-1
E^a	-267.967024	-267.962781	-267.957567
V_{ne}	-1031.517800	-1026.092212	-1020.354196
V_{ee}	295.049037	292.316103	289.405797
V_{nn}	200.534781	197.850536	195.033170
V	-535.933982	-535.925572	-535.915123
T	267.966957	267.962791	267.957662

^{a}Values are given in hartrees (1 hartree = 627.51 kcal).

Table 3. Differences in Component Energies

	Δ(gg-1 - aa-1)	Δ(gg-1 - aa-1)	Δ(ag-1 - aa-1)
E^a	-0.009458 (-5.93)^{b,c}	-0.004242 (-2.66)^b	-0.005215 (-3.27)^b
V_{ne}	-11.163604^d	-5.425588	-5.738016
V_{ee}	5.643240	2.732934	2.910306
V_{nn}	5.501611	2.684245	2.817366
V	-0.018859	-0.008410	-0.010449
T	0.009295	0.004166	0.005129

^{a}Values are in hartrees.
^{b}The values in parentheses are in kilocalories.
^{c}Conformer **90,90-1** is computed to be 2.69 kcal/mol higher in energy than **gg-1** with the 6-31G** basis set.
^{d}The V_{ne}s of C1(5) and O2(4) decrease by 2.636534 and 2.042095 hartrees, respectively, relative to **aa-1**. V_{ne} of C3 decreased by 0.230689 hartrees.

in bond length usually results in a decrease in the stabilizing attractive and destabilizing repulsive interactions, the increases found in these values when **aa-1** is transformed into **gg-1** is atypical behaviour and establishes that **gg-1** is a special case.

Analysis of Atom Electron Populations, Atom Energies and Bond Charge Densities. AIMPAC was used to compute the properties of the charge density. Table 4 lists the atom electron populations of the four conformers of **1** obtained with PROAIMV. In going from **aa-1** to **gg-1**, overall, the carbons and the oxygens gain electrons at the expense of four hydrogens - the methylene hydrogens and one on each methyl group. As far as the group electron populations are concerned (Table 4), the methyls and the methylene group lose electrons while the oxygens gain electrons. In terms of the electron populations, conformer **90,90-1** resembles **gg-1** but the gain in electron populations by C(1), C(3) and C(5) and the oxygens is not as great as it is in the case of **gg-1**. As far as the energies of the atoms of **aa-1** and **gg-1** are concerned, the methyl and methylene carbons along with two hydrogens - one on each methyl group - are stabilized, but the oxygens and remaining six hydrogens are destabilized; in fact, it is the methylene carbon C(3) that is stabilized to the largest degree (Table 5) simply because it gains the most charge. This indicates that the methylene group plays an important role in determining the relative energies of **gg-1** and **aa-1**. The differences in the energies of the atoms computed for transforming **aa-1** into **ag-1** are roughly half the differences found for the conversion of **aa-1** into **gg-1**.

It is interesting to note that the methylene hydrogen that is antiperiplanar to the oxygen nbcc of the gauche fragment of **ag-1** is similar in energy to the hydrogens of the methylene group of **gg-1**; the other hydrogen is similar in energy to the methylene hydrogens of **aa-1**. These relationships to **gg-1** and **aa-1** are also seen in the methylene C-H bond lengths (Table 1). In the case of **90,90-1**, the carbons are stabilized and the oxygens destabilized relative to **gg-1**. Conformers **gg-1** and **aa-1** were also studied with the 6-31++G** basis set to determine how the geometries, the

Table 4. Electron Populations of Atoms

	gg-1	ag-1	aa-1	90,90-1
Heavy Atoms				
C(1)	5.1792	5.1654	5.1664	5.1707
O(2)	9.3265	9.3180	9.3078	9.3226
C(3)	4.5717	4.5475	4.5269	4.5679
O(4)	9.3265	9.3208[a]	9.3078	9.3226
C(5)	5.1792	5.1826[a]	5.1664	5.1707
Hydrogens				
C(1) H$_3$	1.0711	1.0735	1.0315	1.0744
	1.0426	1.0346	1.0722	1.0409
	1.0477	1.0706	1.0722	1.0541
C(3) H$_2$	1.0501	1.0840	1.0843	1.0535
	1.0501	1.0495	1.0843	1.0536
C(5) H$_3$	1.0711	1.0753[a]	1.0315	1.0744
	1.0426	1.0422[a]	1.0722	1.0409
	1.0477	1.0365[a]	1.0722	1.0541
Total Electrons	42.0061[b]	42.0005[b]	41.9957[b]	42.0004[b]

[a]These values are for the gauche fragment of **ag-1**.
[b]Group populations for **gg-1**; CH$_3$(8.3406), O(9.3265), CH$_2$(6.6719): for **ag-1**; CH$_3$(8.3441), O(9.3180), CH$_2$(6.6955), O(9.3208(gauche fragment)), CH$_3$(8.3368, gauche fragment): for **aa-1**; CH$_3$(8.3423), O(9.3078), CH$_2$(6.6807): for **90,90-1**; CH$_3$(8.3701), O(9.3226), CH$_3$(6.6750).

component energies, the atom electron populations, and the atom energies are affected by the addition of diffuse functions to the 6-31G** basis. The CH$_3$-O and O-CH$_2$ bonds (Table 1) are slightly longer than those computed with the 6-31G** basis set, but overall the changes in the other parameters were found to be small; the difference in energy between **gg-1** and **aa-1** was still dominated by V_{ne} (data not shown). The properties of the critical points in the charge densities (ρ), the Laplacians ($-\nabla^2\rho$) and the ellipticities (ε) of the C-O bonds of gg-1, **ag-1**, **aa-1** and **90,90-1** (Table 6) show

Table 5. Atom Energies of Dimethoxymethane Conformers

	gg-1	ag-1	aa-1	Δ(gg-1 - aa-1)	Δ(90,90-1 - gg-1)
Atom					
C(1)	-37.316840[a]	-37.301574	-37.302771	-0.014069	-0.006412
O(2)	-75.552383	-75.558960	-75.565583	0.006577	0.003000
C(3)	-36.910041	-36.894246	-36.880156	-0.015795	-0.004405
O(4)	-75.552383	-75.560812	-75.565583	0.006577	0.003000
C(5)	-36.910041	-37.319548	-37.302771	-0.014069	-0.006412
H(6)	-0.672936	-0.683404	-0.684153	0.011217	-0.000703
H(7)	-0.672936	-0.673110	-0.684153	0.011217	-0.000703
H(8)	-0.666208	-0.666993	-0.666633	0.000425	0.002773
H(10)[b]	-0.657842	-0.654525	-0.653437	-0.004405	-0.000648
H(12)	-0.661453	-0.666082	-0.666633	0.005080	0.000271
H(9)	-0.666208	-0.667725	-0.666633	0.000425	0.002773
H(11)[b]	-0.657842	-0.657967	-0.653437	-0.004405	-0.000698
H(13)	-0.661453	-0.657469	-0.666633	0.000271	0.000271

[a]Values given in hartrees (1 hartree = 627.51 kcal/mol).
[b]This C-H is antiperiplanar to the adjacent C-O bond.

that the CH_3-O and O-CH_2 bonds of **1** differ in terms of the topology of the charge density; the Laplacians of the O-CH_2 bonds of **gg-1** and **aa-1** are larger than the Laplacians of the CH_3-O bonds, as are the values of ρ, even though the cps are located at virtually indentical points along the geometric bond vectors. Moreover, the ellipticities of ρ at the cps of the O-CH_2 bonds of **gg-1**, **ag-1**, and **aa-1** (0.1785, 0.2154/0.1583, and 0.1955) are larger than the ellipticities of the CH_3-O bonds (0.0095, 0.0037/0.0135, and 0.0039). Contour plots of ρ (not shown) in a plane located at the bond critical point perpendicular to the vector connecting the bond cps to C3 for **aa-1** and **gg-1** showed that the soft axis of curvature is perpendicular to O-CH_2-O plane in both cases. If the antiperiplanar lone pair interacted selectively with the C-O bond, then the soft axis would be expected to be coplanar with the three-atom plane in the case of **gg-1**. These results indicate that the methylene groups of **aa-1** and **gg-1** have π character, but on the basis of the ellipticities of the O-CH_2 bonds, the methylene group of **aa-1** has more π character than methylene of **gg-1**. In the case of **ag-1**, the ellipticity (0.2154) of the O-CH_2 bond of the anti fragment is larger than ellipticity (0.1955) of the O-CH_2 bond of **aa-1**. On the other hand, the ellipticity (0.1583) of the O-CH_2 bond of the gauche fragment is smaller than the ellipticity (0.1785) of the O-CH_2 bond of **gg-1** . While care should be exercized in interpreting differences in ellipticities of polar heteronuclear bonds, the fact that the bond critical points of the CH_3-O and O-CH_2 bonds are located at virtually identical positions along the bond vectors relative to the C and O atoms validates the comparison in the case of **1**. These results reveal the unique character of the acetal linkage and establish that the anti and gauche fragments of **ag-1** are not identical to the corresponding fragments of **aa-1** and **gg-1**. Table 6 also gives the same sets of data for **gg-1** and **aa-1** obtained the 6-31++G** basis set. The data correlate closely with the values computed with the

Table 6. Properties of Bond Critical Points of Dimethoxymethane Conformers

Bond and Property	gg-1	ag-1	aa-1	90,90-1
CH₃ - O(2)-				
ρ	0.2615 (0.4499, 0.9561)[b] [0.2594][b]	0.2657 (0.4430, 0.9509) [0.2632]	0.2653 (0.4427, 0.9519)	0.2625(0.4426, 0.9548)
$-\nabla^2 \rho$	0.1115 [0.0876]	0.1070	0.0986 [0.0722]	0.1093
$\in$	0.0095 [0.0082]	0.0037	0.0039 [0.0028]	0.0101
-O(2) - CH₂-				
ρ	0.2795 (0.9425, 0.4397) [0.2782]	0.2762 (0.9461, 0.4429)	0.2870 (0.9370, 0.4389) [0.2859]	0.2794(0.9427, 0.4402)
$-\nabla^2 \rho$	0.2109 [0.2031]	0.2544	0.2688 [0.2751]	0.2235
$\in$	0.1785 [0.1855]	0.2154	0.1955 [0.2022]	
-CH₂ -O(4)[c] -				
ρ		0.2904 (0.4357, 0.9329)[d]		
$-\nabla^2 \rho$		0.2373		
$\in$		0.1583		
-O(4)-CH₃[c]				
ρ		0.2598 (0.9572, 0.4447)[d]		
$-\nabla^2 \rho$		0.0898		
$\in$		0.0135		

[a]The values in parenthesis correspond to the distances from the bond critical points to C(O) and O(C) in angstroms. The sum corresponds to the geometric bond distance.
[b]The values in brackets were obtained from the wavefunction computed with the 6-31++G** basis set.
[c]Identical by symmetry to the other CH₂-O and CH₃-O bonds.
[d]These values are for the gauche fragment of **ag-1**.

6-31G** basis set. There is a small increase in the ellipticity of the O-CH$_2$ bond and a decrease in the ellipticity of the CH$_3$-O bond, but on the whole, the data show that the addition of diffuse functions produces only small systematic changes in these properties; ΔE increased only marginally to 6.17 from 5.93 kcal/mol.

Analysis of Nonbonded Charge Concentrations. Because the Laplacian is sensitive to small changes in ρ near a maximium or minimum, local variations in the nbccs arising from differences in hyperconjugative and conjugative interactions should be reflected in differences not only in the magnitude of $-\nabla^2\rho$ at the (3,-3) cps but in the size and shape of the nbccs and the spacial orientation of the maxima relative to adjacent bond vectors. To establish whether the Laplacian can be used to quantify the differential delocalization/localizaton of geminal lone pairs (lps) on oxygen, several conformations of methyl vinyl ether (**2**) were studied with AIMPAC, the wavefunctions being obtained with the 6-31G** basis set. Table 7 lists data for planar s-cis methyl vinyl ether (**sc-2**), the s-cis conformer twisted by 22.5° (**tsc-2**) and

Table 7. Properties of the Laplacian at Critical Points of Nonbonded Charge Concentrations of Methyl Vinyl Ether

Conformer	$-\nabla^2\rho$	distance/Å	ε	cp, cp angle/deg	cp, C=C dihedral angle/deg
sc-2	6.3319	0.3358	2.98	115.4	118.7
	6.3319	0.3358	2.98		
tsc-2	6.5765	0.3348	2.29	121.2	139.5
	6.1812	0.3362	2.88		92.3
tsc2-2	6.8922	0.3336	1.36	139.3	173.9
	6.5721	0.3346	1.38		40.6

the s-cis conformer twisted by 65° (**tsc2-2**). The planar s-trans comformer was not used in this study because calculations showed that it was 1.93 kcal/mol higher in energy **sc-2**. A twist angle of 22.5° was used in attempt to align one nbcc - on the basis of the coordinates of its critical point - perpendicular to the plane of the vinyl group. This gave a dihedral angle of 92.3° between the (3,-3) cp of the conjugated nbcc and the plane defined by the vinyl group. In the case of **tsc2-2,** the dihedral angles between the nbcc cps and the plane of the vinyl group were found to be 173.9° (this nbcc virtually lies in the plane of the vinyl group) and 40.6°. For **sc-2**, the nbcc cps are located in the range of 0.33 Å from the oxygen nucleus and the Laplacians have values of 6.3319. The angle between the nbcc cps is 115.3°. It is interesting to note that the ellipticity of the double bond (0.5499) is greater than the ellipticity of the double bond of ethene (0.4458)(8). That the double bond of **2** has a higher ellipticity than ethene is one of the reasons why it is more reactive towards electrophiles than ethene. When the methyl group was rotated 22.5° out of plane, the energy of **2** increased (ΔE = 0.93 kcal/mol) and two different nbccs developed. The Laplacian of the nbcc cp which is nearly coplanar with the π-system (perpendicular to the plane of

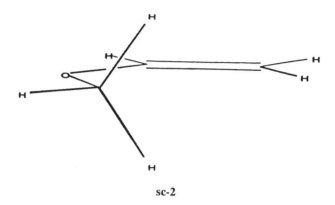

sc-2

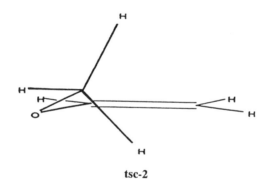

tsc-2

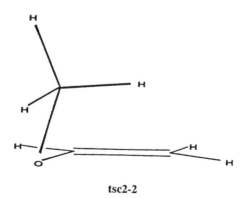

tsc2-2

the vinyl group) decreased to 6.1872 and the other one increased to 6.5765 (Δ = 0.3893) from a value of 6.3319 computed for **sc-2**. That the average of the Laplacians (6.3818) is a slightly greater (Δ = 0.0499) than the Laplacians of **sc-2** and the angle between the nbccs increases to 121.2°, indicates that overall there is a small increase in the localization of the lps accompanying the conversion of **sc-2** to **tsc-2**. This result is in keeping with the small increase in energy of **2**. Furthermore, the ellipticity of the double bond decreases to 0.5440. When the dihedral angle is increased to 65° there is a substantial increase in energy (ΔE (**sc-2** - **tsc2-2**) = 4.69 kcal/mol) and the magnitudes of the Laplacians of both nbccs. The nbcc which lies in the plane of the vinyl group has a $-\nabla^2\rho$ value of 6.8922 and the other one a value of 6.5721. That the average is 6.7321 (the difference between this value and the Laplacians of **sc-2** is 0.4020) shows there is a significant increase in the localization of the lps relative to **sc-2**. Furthermore, the angle between the (3,-3) cps increased to 139.3° and the ellipticity of the double bond decreased to 0.5345. The increase in the the magnitudes of the Laplacians of the nbccs is also accompanied by a shortening of the carbon-carbon double bond through the series **sc-2**(1.3205), **tcs-2**(1.3196), and **tsc2-2**(1.3157). *The study of 2 clearly shows that Laplacians can be used as sensitive probes to detect the differential delocalization/localization of geminal lps on oxygen.*

If the classical picture of the anomeric effect holds for **gg-1**, two different nbccs, one antiperiplanar to the C-O bond and the other antiperiplanar to C-H, should be computed for **gg-1**; four different nbccs are expected for **ag-1** and the nbccs of **aa-1** should be identical. The values of $-\nabla^2\rho$ at the (3,-3) critical points of the nbccs of **gg-1**, **ag-1**, **aa-1**, and **90,90-1** as well as the distances of the cps from the oxygen nuclei are given in Table 8. The Laplacians of the nbccs of **gg-1** (6.5816 for the nbcc antiperiplanar to the $C(3)H_2$-O bond and 6.5832 for the one antiperiplanar to the C(3)-H bond) are virtually identical ($\Delta(-\nabla^2\rho)$ = 0.0016). This is the case even when the basis set is extended to include diffuse functions. This unexpected result establishes that the classical picture of the anomeric effect in which a lone pair selectively interacts with a C-O bond antiperiplanar to it appears not hold in the case of **gg-1**. Four different Laplacians are found for **ag-1**, two identical ones for **aa-1**, and *two different* ones for **90,90-1**. But the Laplacians of **gg-1** (6.5816) are smaller than the Laplacians of **aa-1** (6.8222) indicating that the lps are less localized in **gg-1** than in **aa-1**. It is noteworthy that an *increase* in the average of the two the Laplacians of **2** by 0.4020 results in an *increase* in ΔE to 4.69 kcal/mol. That the increase in energy (ΔE = 5.93 kcal/mol) for the transformation of **gg-1** to **aa-1** is accompanied by an increase of 0.4796 (2 x 0.2398) in the average of the Laplacians for the two sets of lps correlates remarkably well with the data obtained for **2**. In the case of **ag-1**, the Laplacians of the anti fragment (6.7794 and 6.7153) are smaller - the average being 6.7474 - than the Laplacians of **aa-1** (6.8222) indicating that the lps in this fragment less localized than in **aa-1**; the ellipticity of the O-CH_2 bond of the anti fragment is greater than the ellipticity of the O-CH_2 bond of **aa-1**. As far as the gauche fragment of **ag-1** is concerned, the Laplacian of the nbcc antiperiplanar to the adjacent CH_2-O is only marginally smaller (6.6296) than the Laplacian (6.6982) of the nbcc that is antiperiplanar to the C(3)-H bond (Δ = 0.0684). But the Laplacians are larger (the average is 6.6639) than the Laplacians of **gg-1**. This indicates that the lone pairs of the gauche fragment are more localized in **ag-1** than in **gg-1**. In this case, the elliptcity of the O-CH$_2$ is smaller than the ellipticity of the bond of **gg-1**. The results obtained for **ag-1** suggest that the stabilization of **ag-1** relative to **aa-1** derives from an decrease in the localization of the lps of the gauche fragment *and the anti fragment*. In the case of **90,90-1**, the nbccs have substantially different Laplacians ($\Delta = 0.1378$) with the Laplacian of the syn nbcc (the dihedral angle between its (3,-3) cp and the adjacent CH$_2$-O bond vector is 19.7°) being smaller than the Laplacian of the anti nbcc (the dihedral angle is 160.6°). This shows that the anti lone pair is more localized than the syn one. That the average of the Laplacians is (6.5955) is larger than the average of the nbccs of **gg-1** (6.5824) indicates that the lone pairs are more localized than in the case of **gg-1**. These results do not support Grein's proposal.

All cps in the Laplacian are located around 0.33 Å from the oxygen nuclei. Table 8 also lists the angles between the (3,-3) cps of the nbccs of **gg-1**, **ag-1**, **aa-1**, and **90,90-1** which are found to be much larger (133.5° and 134.4° for gauche fragments of **gg-1** and **ga-1**, and 140.2° and 139.9° for and the anti fragments of **ag-1** and **aa-1**) than the tetrahedral angle of 109.5° and the angles found for **sc-2** (115.3°) and **tsc2** (121.2°). The angle for **90,90-1** is 134.5°. The same properties were computed for **aa-1** and **gg-1** with the 6-31++G** basis set (Table 8). It is clear that adding diffuse functions produces only small changes in the Laplacians and the angles between the nbcc cps. Displays of selected contours of -∇^{2}ρ in a plane defined by the (3,-3) nbcc cps and the oxygen nuclei of **gg-1** and **aa-1** are given in Figure 1. It is noteworthy that a contour plot of ρ in the cp-cp-O2 plane (Figure 2) does not show the rabbit ears usually written for lone pair electrons; nbccs can only defined by the Laplacian. That the angle between the cps of the nbccs of **aa-1** is greater than the angle for **gg-1** is consistent with greater localization of the nonbonding electrons in in the anti-anti conformation than in the gauche-gauche conformation. In the extreme case when one pair of electrons is completely delocalized or forms a π bond, the nbccs should coalesce into one lump characterized by a single critical point. Calculational results obtained for **sc-2**, **tsc-2**, and **tsc2-2** (*vide supra*) and O-protonated **gg-1** (data not shown) are certainly in accord with this expectation. For **gg-1** and **ag-1** the dihedral angles between the nbcc cps and the adjacent C-O bond vectors (176.2° and 173.8°) deviate from the ideal antiperiplanar angle of 180°.

That the nbccs of **gg-1** and **aa-1** have different shapes is seen in the ellipticities (Table 8) of the Laplacian ($\varepsilon = \lambda_2/\lambda_3 - 1$ where λ_2 and λ_3 are the negative curvatures perpendicular to the cp-nuclear vector at the (3,-3) cp). A spherical nbcc would have an ε of zero. The small difference in the ellipticities of the nbccs of **gg-1** (1.68 and 1.65) indicates that the nbccs are virtually identical. Given that the 'delocalized' nbccs of **sc-2** and **tsc-2** exhibit εs of 2.98, and 2.88 and 2.29 (Table 8), the difference in the ellipticities of the nbccs of **aa-1** (1.19) and **gg-1** (1.68 and 1.65) shows, as did the magnitudes of the Laplacians, that the lps of **aa-1** are more localized than the lps of **gg-1**. As far as the the anti fragment of **ag-1** is concerned the εs are the same as the εs of **aa-1**; for the gauche fragment the εs are smaller than the εs of **gg-1**. This result is in accord with the conclusion reached on the basis of the magnitudes of the Laplacians; the lps of the gauche fragment are more localized than in the **gg-1**. That the anti nbcc of **90,90-1** exhibits an ε (1.60) that is smaller than the ellipticity of the syn nbcc (1.72) and the nbccs of **gg-1** (1.68 and 1.65) shows that the anti lone pair is more localized than the syn one; the syn lp is less localized than the lps of **gg-1**. This

Table 8. Properties of the Laplacian at Critical Points of Nonbonded Charge Concentrations

Conformer	atom and nbcc type	$-\nabla^2\rho$	distance Å	ϵ	cp, cp angle/deg	cp, C-O dihedral angle/deg
gg-1	O(2) anti to C(3) - O(4)	6.5816 [6.5145][b]	0.3349 [0.3351]	1.68 [1.68]	133.5 [134.0]	176.2[a] [176.9]
	O(2) anti to C(3) - H	6.5832 [6.5168]	0.3385 [0.3350]	1.65 [1.65]		
ag-1	O(2) anti to C(3) - H	6.7774	0.3341	1.19	140.2	
	O(2) anti to C(3) - H	6.7153	0.3344	1.19		
	O(4) anti to C(3) - O(2)	6.6296[c]	0.3347	1.54	134.4	173.8[a]
	O(4) anti to C(3) - H	6.6982[c]	0.3343	1.59		
aa-1	O(2) anti to C(3) - H	6.8222 [6.7545]	0.3339 [0.3440]	1.19 [1.19]	139.9 [140.1]	
90,90-1	0(2) anti to C(3) - O(4)	6.6644	0.3346	1.60		160.6
	syn to C(3)-O(4)	5.5266	0.3349	1.72	134.5	19.7

[a] This is the cp of the nbcc which is antiperiplanar to the adjacent C-O bond.
[b] Computed with the 6-31++G** wavefunction.
[c] This is the gauche fragment of ag-1.

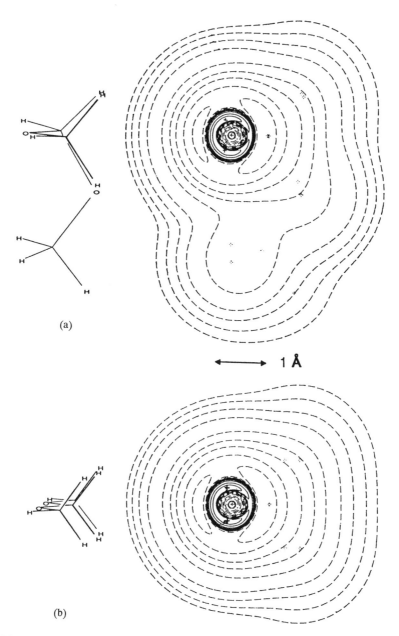

Figure 1. Displays of contours of the Laplacian of ρ in a plane defined by the (3,-3) critical points and the oxygen nuclei (+). The contours (dashed lines > 0, solid lines < 0) increase and decrease from a zero contour in steps of $\pm2 \times 10^n$, $\pm4 \times 10^n$ $\pm8 \times 10^n$ beginning with $n = -3$ and increasing in steps of unity: (a), **gg-1**; (b), **aa-1**.

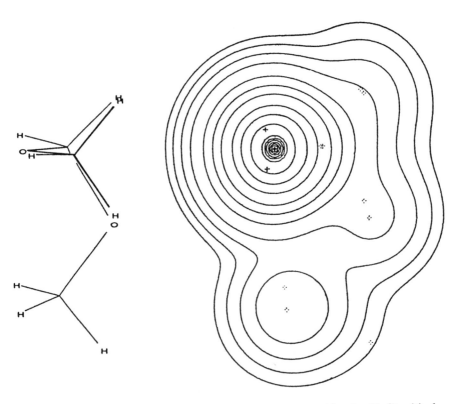

Figure 2. Display of the contours of ρ in a plane defined by the (3,-3) critical points and the oxygen nucleus (+) of **gg-1**.

is in accord with the relative magnitudes of the Laplacians. Figure 3 shows dislpays of selected contours of $-\nabla^2\rho$ for the nbcc of **gg-1** that is anti to C(3)-O(4)(Figure 3(a)) and the one antiperiplanar to C(3)-H(Figure 3(b)) in a planes perpendicular to the nbcc cp - oxygen nucleus bond vectors. Figure 3(c) shows a similar display of selected contours of $-\nabla^2\rho$ for the nbcc of **aa-1**. The differences in the ellipticities of the nbccs are seen in the displays. It is clear that the magnitude *and the ellipticity* of the Laplacian can be used to gain information about the differential delocalization/localization of geminal lone pairs.

Nevertheless, it was still necesary to obviate the potential problem of using the Laplacian at a single point - the critical point - to analyze nbccs of different shapes. To this end a program DOMAIN (*11*) was used to integrate sections of the nbccs. DOMAIN uses an algorithm based on lattice growth to quantify a region of charge concentration. From a seed point taken to be the local maximum in $-\nabla^2\rho$ the ((3,-3) critical point), a lattice is grown in three dimensions on the condition that at each new point in $-\nabla^2\rho$ has a larger value than some predetermined boundary value of $-\nabla^2\rho$. Clearly, the choice of the boundary value is important in order to obtain meaningful results. For molecules containing regions of charge concentration which are bounded only by a regions of charge depletion, it is possible to use a contour value of zero (this ideal situation is found in the 'lone pair' concentration of PF_4^-). In other molecules where this is not the case, some other boundary condition must be used. In these situations, the value of $-\nabla^2\rho$ at a (3,-1) or (3,+1) critical point can be used to define the domain. DOMAIN yields two measures of the size of the nbcc based on the sum of the $-\nabla^2\rho$ values at each of the lattice points and the number of lattice points; the volume V in atomic units ($V = n^3 i$, where i is the number of lattice points in the bounded region and n is the length of the unit cell) and the amount of charge concentration in the region L ($L = \Sigma(-\nabla^2\rho)_i . n^3$ and has the units of charge per unit area). In addition to calculating the values of L and V for a region of charge concentration, it is also possible to calculate the number of electrons R in the volume bounded by the chosen boundary in $-\nabla^2\rho$. The values of $-\nabla^2\rho$ at the (3,-1) cps between the (3,-3) cps were chosen as the boundaries defining the domains for the four conformers of **1**, three conformers of **2** (**sc-2**, **tsc-2**, and **tsc2-2**) and two conformers of **3** (**aa-3** and **ga-3**). Although these contour values do not define the whole nbcc, the sections selected are considered to accurate relative measures of the whole nbccs because the maxima are located at virtually identical distances from the oxygen nuclei. DOMAIN was first used to integrate the nbcc domains of **sc-2**, **tsc-2**, and **tsc2-2** and the data are given in Table 9. The conjugated nbcc of **tsc-2** has smaller V, L, and R values than the nonconjugated nbcc. It also has substantially smaller V, L, and R values than the nbccs of **sc-2**. Displays of selected contours of the integrated domains of **sc-2** and **tsc-2** are given in Figure 4(a) and (b), respectively, and provide for a visual comparison to be made with the plots obtained for **gg-1** and **aa-1**. In keeping with the magnitude of $-\nabla^2\rho$, the nbcc which lies in the plane of of the vinyl group of **tsc2-2** has the largest V, L, and R values. Table 9 gives V, L, and R for the nbccs of **gg-1**, **ag-1**, **aa-1**, **90,90-1**, **aa-3**, and **ga-3** obtained with DOMAIN. It is seen that V, L, and R of the nbccs of **gg-1** are smaller than the corresponding values for **aa-1** in keeping with the values of $-\nabla^2\rho$ at the (3,-3) cps. Displays of selected contours of $-\nabla^2\rho$ in a plane defined by the oxygen nucleus, the (3,-3) and (3,-1) critical points for **gg-1** and **aa-1** are given in Figure 5. The differences in the sizes

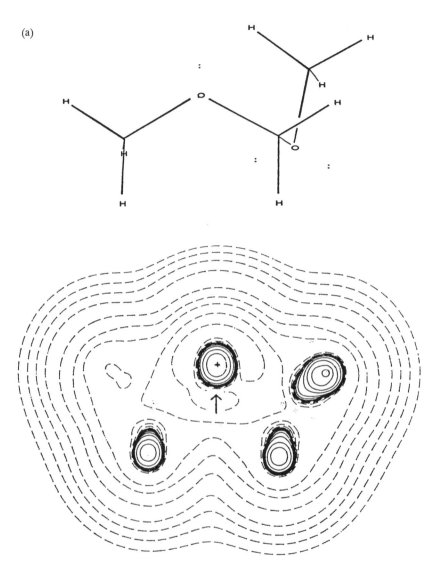

Figure 3(a). Display of the contours of the Laplacian of ρ in a plane perpendicular to a vector connecting the $(3,-3)$ critical point ($\uparrow$) and the oxygen nucleus at the critical point of the nbcc of **gg-1** that is antiperiplanar to the C(3)-O(4) bond.

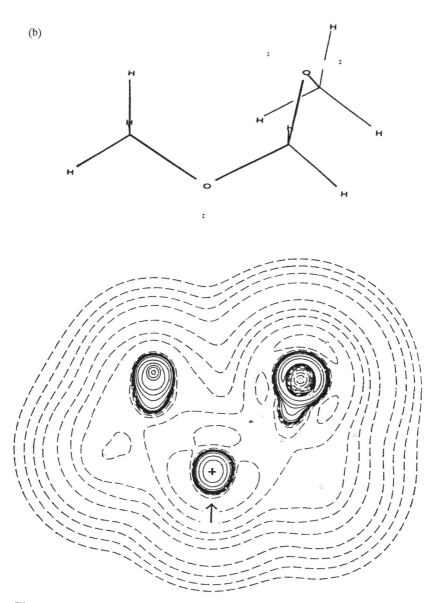

Figure 3(b). Display of the contours of the Laplacian of ρ in a plane perpendicular to a vector connecting the (3,-3) critical point (↑) and the oxygen nucleus at the critical point of the nbcc of **gg-1** that is antiperiplanar to the C(3)-H bond.

Continued on next page.

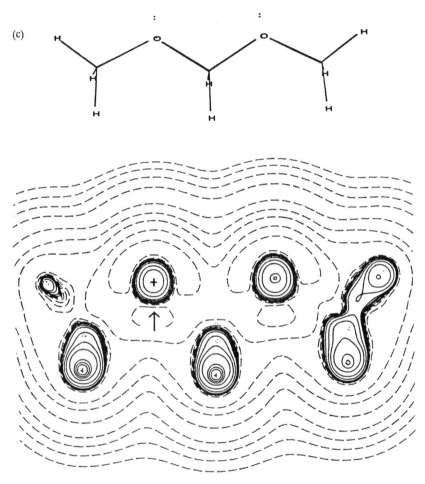

Figure 3(c). Display of the contours of the Laplacian of ρ in a plane perpendicular to a vector connecting the (3,-3) critical point ($\uparrow$) and the oxygen nucleus at the critical point of the nbcc of **aa-1**.

Table 9. Properties of the Nonbonded Charge Concentrations of Dimethoxyethane, Methyl Vinyl Ether and Methyl n-Propyl Ether Conformers

Conformer	$-\nabla^2\rho^a$	V	L	R
gg-1 nbcc anti to C(3)-H	5.5366	0.0289	0.1792	0.0303
gg-1 nbcc anti to C(3)-O(4)	5.5366	0.0287	0.1783	0.0301
aa-1	5.3466	0.0453	0.2667	0.0432
90,90-1 nbcc syn to C(3)-O(4)	5.5402	0.0284	0.1677	0.0271
90,90-1 nbcc anti to C(3)-O(4)	5.5402	0.0325	0.1938	0.0312
sc-2	5.8799	0.0108	0.0651	0.0104
tsc-2 conjugated nbcc	5.7812	0.0091	0.0543	0.0087
tsc-2 non-conjugated nbcc	5.7812	0.0219	0.1331	0.0212
tsc2-2 in-plane nbcc	5.3959	0.0465	0.2761	0.0445
40.6° nbcc	5.3959	0.0346	0.2009	0.0327
aa-3	5.3881	0.0419	0.2468	0.0399
ga-3 nbcc anti to C(3)-H	5.4996	0.0356	0.2117	0.0341
ga-3 nbcc anti to C(3) - C(4)	5.4896	0.0383	0.2286	0.0367

^{a}The value of $-\nabla^2\rho$ at the (3, -1) cp between the nbcc's on oxygen used as the contour to define the boundary of the domain.

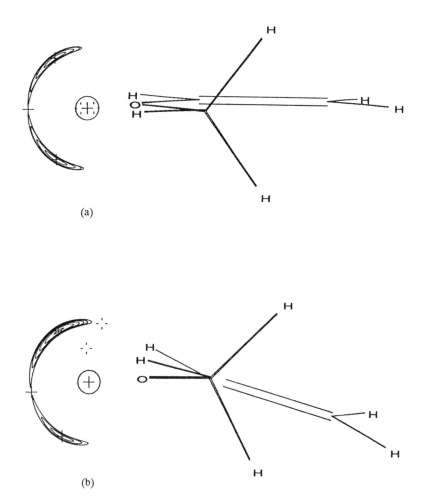

Figure 4. Displays of the contours of the sections of the Laplacian of ρ integrated with DOMAIN in a plane defined by the (3,-3) critical points and the oxygen nuclei (+), along with the corresponding conformations. The markers (+) show the positions of four nuclei not in the plane: (a), **sc-2**; (b), **tsc-2**. The contours start at the value of $-\nabla^2\rho$ at the (3,-1) critical point and increase in increments of 0.15.

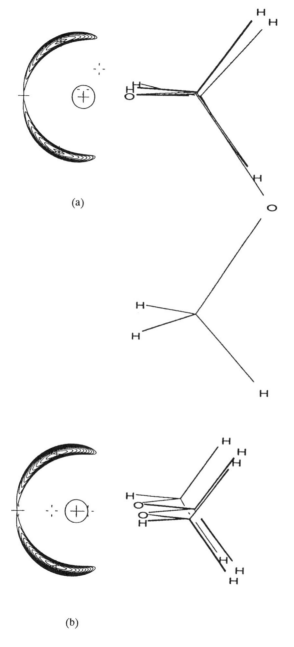

(a)

(b)

Figure 5. Displays of the contours of the sections of the Laplacian of ρ integrated with DOMAIN in a plane defined by the (3,-3) critical points and the oxygen nuclei (+), along with the corresponding conformations. The markers (+) show the positions of four nuclei not in the plane: (a), **gg-1**; (b), **aa-1**.

of the nbccs defined by the contour value at the (3,-1) critical point between the nbccs are clearly seen in these displays. *The results obtained with DOMAIN show that the Laplacians at nbcc cps can be used to gain information on differential delocalization/localization of lps even though the shapes of the nbccs differ.*

Currently, charge densities are obtained with high precision x-ray crystallographic structure determinations, and density difference maps derived from the densities are used to identify regions of charge concentration or depletion relative to the unperturbed atom. As a result, there is experimental evidence to support the AIMPAC calculational results that the maxima of oxygen nbccs are located in the range of 0.3 Å from the nucleus and that the angles between the maxima can fall in the range of 130°. Dunitz and Sieler obtained the density difference maps of the oxygens in the bisperoxide **4** with a high precision x-ray crystallographic structure determination (*13*) and estimated that the maxima in the nbccs are located between 0.3 to 0.4 Å from the nucleus and that the angle between the maxima is between 120° and 130°. That the maxima in the density difference map (the contour spacing is 0.075 eÅ^{-3}) of the oxygen of **4** (Figure 6) and the nbccs cps of **gg-1** are virtually superposable, establishes the validity of the AIMPAC calculations results.

Methyl Propyl Ether Conformations

Equilibrium geometries of **aa-3** and **ga-3** were computed with the 6-31G** basis set and the geometric bond lengths are given in Table 10. In keeping with the

Table 10. Geometric Bond Lengths of Methyl Propyl Ether Conformers

	aa-3	ga-3
C(1) - O(2)	1.3919	1.3911[a]
O(2) - C(3)	1.3955	1.3994[a]
C(3) - C(4)	1.5176	1.5251
C(4) - C(5)	1.5274	1.5281
C(1) - H$_3$	1.0822	1.0821
	1.0892	1.0876
	1.0892	1.0893
C(3) - H$_2$	1.0919	1.0848
	1.0914	1.0911
C(4) - H$_2$	1.0861	1.0863
	1.0861	1.0867
C(5) - H$_3$	1.0844	1.0849
	1.0868	1.0867
	1.0869	1.0867

[a]This is the gauche fragment of **ga-3**.

results of experimental studies (see reference *4*), **aa-3** was found to be marginally lower in energy than **ga-3** (*13*). For **3**, unlike the case of **1**, increases in the repulsive terms V_{ee} and V_{nn} (Table 11) outweigh the increase in the stabilizing attractive interaction V_{ne}. This means that steric effects outweigh stabilizing electronic effects.

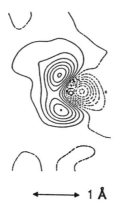

Figure 6. Display of the contours of the density difference map of the oxygen of bisperoxide **4.** Taken from the publication by J.D. Dunitz and P. Seiler. *J. Am. Chem. Soc.* **1983**, *105*, 7056.

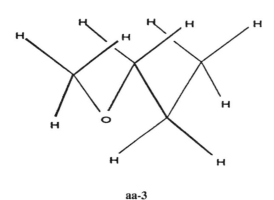

aa-3

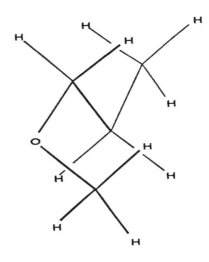

ga-3

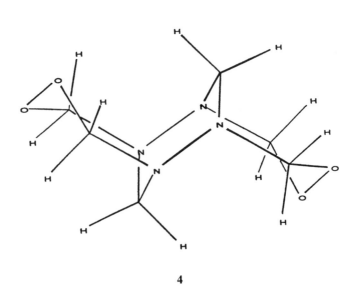

4

Table 11. Component Energies of Methyl Propyl Ether Conformers

	ga-3	aa-3	Δ(ga-3 - aa-3)
E	-232.152393	-232.155123	0.002730 (1.71)[a]
V_{ne}	-925.577317	-922.593720	-2.983597
V_{ee}	270.222449	268.734095	1.488354
V_{nn}	191.281449	189.775879	1.505570
V	-464.073419	-464.083746	0.010327
T	231.921026	231.928623	-0.007597

[a]The value in parentheses is in kcal/mol.

Table 12. Atom Electron Populations of Methyl n-Propyl Ether Conformers

	aa-3	ga-3
Heavy Atoms		
C(1)	5.1516	5.1539
O(2)	9.3111	9.3086
C(3)	5.2241	5.2209
C(4)	5.7817	5.7973
C(5)	5.7809	5.7812
Hydrogens		
C(1)H$_3$	1.0438	1.0436
	1.0767	1.0781
	1.0767	1.0769
C(3)H$_2$	1.0912	1.0609
	1.0912	1.0896
C(4)H$_2$	1.0722	1.0743
	1.0722	1.0918
C(5)H$_3$	1.0706	1.0705
	1.0809	1.0789
	1.0809	1.0810
Total Electrons	42.0058[a]	42.0075[a]

[a]Group populations: for **aa-3**; CH$_3$(8.3488), O(9.3111), CH$_2$(7.4065), CH$_2$(7.9261), CH$_3$(9.0633): for **ga-3**; CH$_3$(8.3525), O(9.3086), CH$_2$(7.3714), CH$_2$(7.9634), CH$_3$(9.0116).

Analysis of Atom Electron Populations, Atom Energies and Bond Charge Densities. Atom electron populations are given in Table 12 along with the group populations. The oxygen loses charge and C(1) and C(4) gain charge in going from aa-3 to ga-3. As observed for ag-1, the hydrogen which becomes aligned antiperiplanar to a lp loses charge. Nevertheless, there is a small increase in the O-CH_2 and CH_2-CH_2 bond lengths in going from aa-3 to ga-3 as was found in the case of 1. The values of ρ, -$\nabla^2\rho$, and ϵ at the C-O (3,-1) bond cps are given in Table 13. The (3,-1) bond critical points of the CH_3-O bonds of aa-3 and ga-3 are found at nearly identical points between the nuclei as in the case of gg-1, ag-1, and aa-1. However, the -$\nabla^2\rho$ values are larger than the values for gg-1, ag-1, and aa-1 and the ellipticities are similarly small. But the O(2)-C(3) bonds of aa-3 and ga-3 exhibit much lower ellipticities (0.0263 and 0.0033) than the O(2)-C(3) bonds of aa-1 and gg-1 even though the bond cps are found virtually at identical points along the bond paths. This result is a further indication that the methylene groups of gg-1, ag-1, and aa-1 have π character. That the -$\nabla^2\rho$ values at the critical points of the C(3)-C(4) and C(4)-C(5)

Table 13. Properties of Bond Critical Points of Methyl Propyl Ether Conformers

Bond and Property	aa-3	ga-3
CH_3-O-		
ρ	0.2690 (0.4427, 0.9495)[a]	0.2677 (0.4436, 0.9501)
-$\nabla^2\rho$	0.1364	0.1572
ϵ	0.0145	0.0089
-O-CH_2-		
ρ	0.2661 (0.9531, 0.4429)	0.2634 (0.9551, 0.4494)
-$\nabla^2\rho$	0.1138	0.1322
ϵ	0.0263	0.0033
-CH_2-CH_2-		
ρ	0.2670 (0.7808, 0.7368)	0.2636 (0.7734, 0.7518)
-$\nabla^2\rho$	0.7458	0.7225
ϵ	0.0484	
-CH_2-CH_3		
ρ	0.2540 (0.7605, 0.7670)	0.2534 (0.7611, 0.7671)
-$\nabla^2\rho$	0.6691	0.6654
ϵ	0.0043	

[a]The values in parentheses correspond to the geometric distances to the appropriate atoms from the bond critical point.

bonds are larger than the values at the C-O bonds of aa-3 and ga-3 is in keeping with with the fact that the C-C bonds are less polar than C-O bonds.

Analysis of Nonbonded Charge Concentrations. Table 14 lists the -$\nabla^2\rho$ values at the (3-3) cps of the nbccs of aa-3 and ga-3. Surprizingly, the Laplacians of aa-3 are smaller (Δ = 0.0525) than the Laplacians of aa-1. This indicates that the lps of aa-1 are more localized than in aa-3. The data obtained with DOMAIN (Table 9) (V, L, and R are larger for the nbccs of aa-1 than for aa-3) and the ellipticities of the nbccs

(1.19 for **aa-1** and 1.28 for **aa-3**) are in accord with this analysis as is the angle between the cps which is slightly larger for aa-1 (139.9°) than for **aa-3** (138.6°). For **ga-3**, the Laplacian of the nbcc which is antiperiplanar to the C(3)-H bond was found

Table 14. Properties of the Laplacian at Critical Points of Nonbonded Charge Concentrations

Conformer	$-\nabla^2\rho$	distance/Å	$\in$	cp, cp angle/deg	cp, c-c dihedral angle/deg
aa-3	6.7697	0.3342	1.28	138.6	72.4
	6.7697	0.3342	1.28		
ga-3	6.6945	0.3343	1.49	137.1	26.4
	6.7941	0.3341	1.41		169.5

to be slightly smaller ($\Delta = 0.0996$) than $-\nabla^2\rho$ of the nbcc antiperiplanar to the C(3)–C(4) bond and the average is 6.7443. This suggests that the hyperconjugation between the lp and the antiperiplanar C(3)–H bond is greater than the interaction involving the lp and the C(3)–C(4) bond. For **ga-3** the ellipticities are smaller than the ellipticities of the nbccs of **gg-1**. That the ellipticity of the nbcc antiperiplanar to C(3)–H is larger than the ellipticity of the nbcc anti to the C(3)–C(4) bond correlates with the results obtained for **aa-3**.

CONCLUSIONS

The AIMPAC study of several conformations of methyl vinyl ether has established that the magnitude and the ellipticity of the Laplacian at critical points of nonbonded charge concentrations are sensitive probes of the differential delocalization/localization of geminal lone pairs on oxygen. On the basis of the analysis of the properties of the Laplacians, the nonbonded charge concentrations corresponding to the geminal lone pairs of the gauche-gauche conformer of dimethoxymethane, although virtually identical, are smaller than the nonbonded charge concentrations of the anti-anti conformation. This result is in accord with a greater localization of the lone pairs in the anti-anti conformer than in the gauche-gauche conformer. However, it is not in keeping with the view that the anomeric effect in dimethoxymethane derives from a differential interaction of a lone pair with a CH_2-O bond the antiperiplanar to it. If the delocalization of the lone pairs is the source of the stabilizing component of the anomeric effect, then the difference in energy between the anti-gauche conformer and the anti-anti conformer is due to a decrease in the localization of both geminal pairs of electrons.

CALCULATIONS

Optimized geometries and wavefunctions were computed for **1**, **2**, and **3**, with Gaussian 90 running on a Multiflow Trace 14/300 computer and Gaussian 92, and

Gamess running on IBM RS/6000 computers (Models 350 and 530 equipped with 64 Mb RAM and 2 Gb of disk storage) using standard techniques. The AIMPAC programs were run on FPS Stellar (Model 300) and IBM RS/6000 (Model 530) computers.

ACKNOWLEDGEMENT

We thank the Natural Sciences and Engineering Council of Canada for financial support. The author wishes to express his gratitude to Professor R.F.W. Bader for the hospitality afforded him during the course a six-month research leave spent in Professor Bader's laboratory. The author is also grateful to Dr. J. Cheeseman and Dr. T. Keith for assisting him in learning the operational aspects of the AIMPAC programs.

LITERATURE CITED

1. Deslongchamps, P. Stereoelectronic Effects in Organic Chemistry. Pergamon Press, Oxford, 1983.
2. Kirby, A.J. The Anomeric Effect and Related Stereoelectronic Effects at Oxygen. Springer Verlager, Berlin, 1983.
3. Gorenstein, D.G. *Chem. Rev.* **1987**, *87*, 1047.
4. Wiberg, K.B.; Murcko, M.A. *J. Am. Chem. Soc.* **1989**, *111*, 4821.
5. Astrip, E,E. *Acta. Chem. Scand.* **1973**, *27*, 3371.
6. Jorgensen, F.S. *J. Chem. Research (S).* **1981**, 212.
7. Grein, F; Deslongchamps, P. *Can. J. Chem.* **1992**, *70*, 1562.
8. Bader, R.F.W. Atoms in molecules - a quantum theory. Clarendon Press, Oxford, 1990.
9. Gaussian 90, Revision F. Frisch, M.J.; Head-Gordon, M.; Trucks, G.W.; Foresman, J.B.; Schlegel, H.B.; Raghavachari, K.; Robb, M.A.; Binkley, J.S.; Gonzalez, C.; Defrees, D.J.; Fox, D.J.; Whiteside, R.A.; Seeger, R.; Melius, C.F.; Baker, J.; Martin, R.L.; Khan, L.R.; Stewart, J.J.P.; Pople, J.A., Gaussian Inc., Pittsburgh PA, 1990.
10. Gaussian 92, Revision B. Frisch, M.J.; Trucks, G.W.; Head-Gordon, M.; Gill, P.M.W.; Wong, M.W.; Foresman, J.B.; Johnson, B.G.; Schlegel, H.B.; Robb, M.A.; Replogle, E.S.; Gomperts, M.; Andres, J.L.; Raghavachari, K.; Binkley, J.S.; Gonzalez, C.; Martin, R.L.; Fox, D.J.; Defrees, D.J.; Baker, J.; Stewart, J.J.P.; Pople, J.A., Gaussian Inc., Pittsburgh PA, 1992.
11. Gamess, Version April 4, 1991. Schmidt, M.W.; Baldridge, K.K.; Boatz, I.A.; Jensen, J.H.; Elbert, S.T. QCPE Bull. **1990**, *10*, 52.
12. DOMAIN, Version 1. Dr. I. Bytheway and Mr. R. Dewitte. McMaster University, 1992.
13. Dunitz, J.D.; Seiler, P. *J. Am. Chem. Soc.* **1983**, *105*, 7056.

RECEIVED August 17, 1993

Chapter 11

Anomeric and Reverse Anomeric Effect in Acetals and Related Functions

F. Grein

Department of Chemistry, University of New Brunswick, Bag Service 45222, Fredericton, New Brunswick E3B 6E2, Canada

Ab initio geometry optimizations were performed on various conformations of XH_m-CH_2-OH and XH_m-CH_2-NH_2 and corresponding protonated systems with XH_m = F, OH and NH_2. By combining Fourier analysis with energy decomposition methods developed in a previous paper, the electronic part of the anomeric stabilization energy, e, could be extracted. For neutral systems, the values of e_O are -2.0 for X = F, -1.0 for OH and -0.7 for NH_2. The values of e_N are -3.6 for F, -1.9 for OH and -1.0 for NH_2. e_O refers to stabilization due to O and e_N due to N. For protonated systems the results for e_O are -20.5* for FH^+, -6.2 for OH_2^+, and -3.5 for NH_3^+. e_N values are -30.6* for FH^+, -17.3* for OH_2^+, and -5.7 for NH_3^+. The units are kcal/mol. In the cases marked by asterisk, charge-dipole complexes are formed. Energy values for the reverse anomeric effect, operative in protonated systems, are v_O = -7.5 for OH_2^+, -8.0 for NH_3^+, and v_N = -7.5 for NH_3^+. In other cases, complexes are formed. The mechanism for electronic stabilization is explained by partial π-bonding in the CH_2-OH or CH_2-NH_2 fragment of the molecule.

Numerous theoretical studies on the anomeric effect have been published, with the first ab initio papers appearing in 1971 by Wolfe et al. (1) and Radom, Hehre and Pople (2,3). The energetic and structural changes, as observed experimentally and amply documented (4,5), are well reproduced by ab initio geometry optimizations, providing a basis for the more difficult task of interpreting the nature of this effect.

A breakdown of the anomeric stabilization energy of model acetals into components was attempted in a recent paper by Grein and Deslongchamps (6), to be called paper I. The main purpose was to extract the underlying electronic energy of stabilization, and to separate it from other forms of energy, such as steric and electrostatic. Overall, this study was able to rationalize energy lowerings for systems XH_m-CH_2-YH_n and their protonated counterparts XH_m^+-CH_2-YH_n, where X and Y are N and O. Paper I will be reviewed in the main part of the present

0097–6156/93/0539–0205$06.50/0

study under the heading "energy decomposition model". Results for systems with X = F will be added.

In a second article, to be referred to as paper II (7), more detailed studies were performed on systems XH_m-CH_2-OH, with X = N, O, F, C, where the OH hydrogen was rotated about the C-O bond. This is followed by a Fourier analysis of the energy as function of the rotation angle. Fourier analyses with the purpose of studying the anomeric effect were first performed by Radom et al. (8,9). Again, a decomposition of the energy into an electronic component (V_2 parameter), a steric component for the interaction of the rotating OH with the neighboring CH_2X group (V_3 parameter), and a steric/electrostatic component for the interaction of OH with XH_m (V_1 parameter) was performed. The electronic component of the anomeric energy stabilization, considered to be the fundamental driving force of the anomeric effect, was related to the ability of the CH_2-OH portion of the molecule to form a partial double bond provided that OH carries a partial positive charge. The findings of paper II will be summarized under the heading "OH rotation model for XH_m-CH_2-OH".

The present work is based on the two previous papers. Additional systems, such as XH_m-CH_2-NH_2, with NH_2 rotation, and protonated systems XH_m^+-CH_2-OH and XH_m^+-CH_2-NH_2, with OH rotation and NH_2 rotation, respectively, will be presented and discussed.

When applied to protonated systems, both the energy decomposition model and the OH or NH_2 rotation model show the need for an additional energy parameter which is not encountered in neutral systems. This new parameter will be associated with the "reverse anomeric effect", as introduced by Lemieux and Morgan (10,11).

Based on calculated energies and other properties of systems XH_m-CH_2-YH_n and their protonated counterparts, with X = F, O, N, (C), and Y = O and N, as function of the rotation angle of YH_n, the ideas presented in papers I and II will be combined, leading to much improved values for the electronic portion of the anomeric stabilization energy, as well as estimates of the reverse anomeric stabilization energy. The π-bonding model will be further developed.

Energy Analysis for Neutral Systems

On systems XH_m-CH_2-YH_n, with X,Y=F,O,N,C, ab initio SCF calculations were performed using the GAUSSIAN 86 program (12) and 6-31G** or 6-31+G* basis sets. For selected dihedral angles ϕ of one or two XH_m and YH_n hydrogens, geometry optimizations of all bond distances, bond angles, and other dihedral angles were carried out. Structure 1 shows dihedral angles ϕ of 0°. In general, ϕ values were chosen to be 0°, 60°, 120° and 180° (and 90°, not reported here). Assuming lone pairs (ℓp) of electrons to be located in sp^3 positions, and defining in a simple-minded way a lone pair in antiperiplanar position (app) (e.g. a lone pair on Y app to the C-X bond) to cause an anomeric effect (ae), then for structure 1 there is 0 ae, whereas structures 2, 3 and 4 have 1 ae(O), 1 ae(N), and 2 ae's, respectively, exemplified by NH_2-CH_2-OH.

Energy Decomposition Model.

In Table I, relative energies and optimized C-X and C-Y distances are summarized for the 0 ae, 1 ae and 2 ae conformers of OH-CH_2-OH, NH_2-CH_2-OH and NH_2-CH_2-NH_2 (paper I). It is seen that the 1 ae conformers are more stable than their 0 ae counterparts, and that in turn the 2 ae conformers are more stable than the 1 ae ones, as expected. However, the energy lowerings in the NH_2-CH_2-NH_2 system are marginal, and for NH_2-CH_2-OH, 2 ae is only slightly more stable than 1 ae. The only system where there is a distinct energy lowering from 0 ae to 1 ae to 2 ae is OH-CH_2-OH.

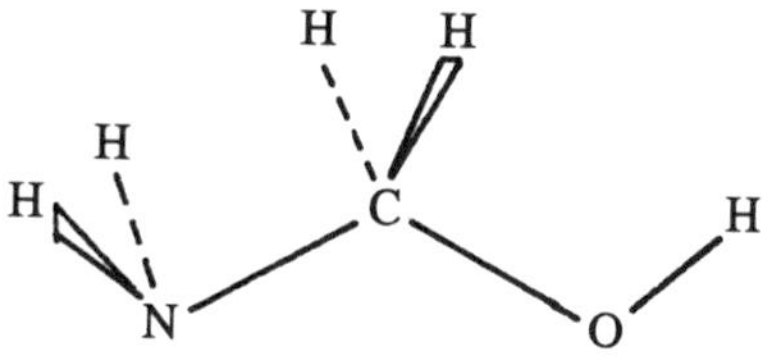

$\phi = 0°$

1

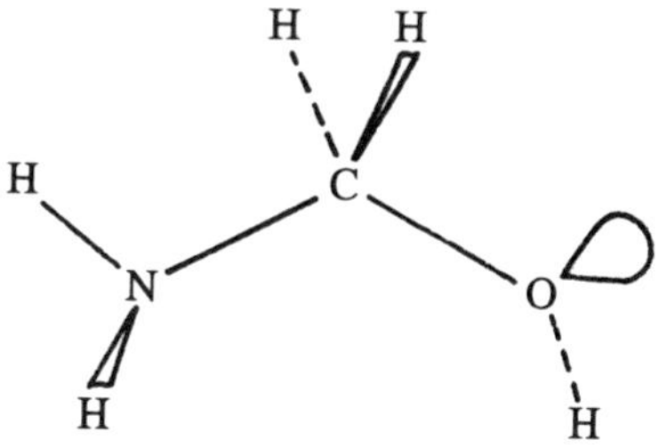

1 ae(O)

2

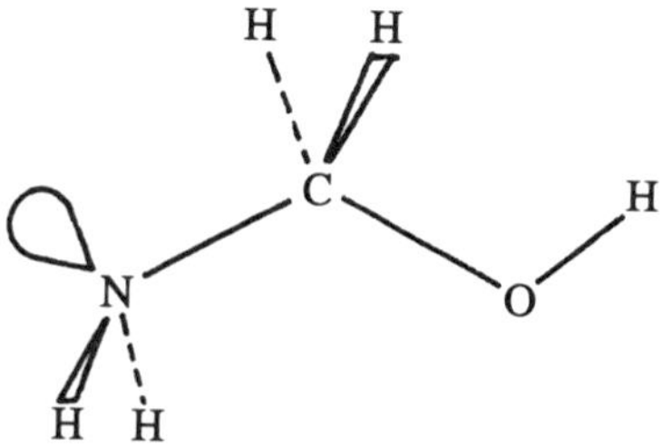

1 ae(N)

3

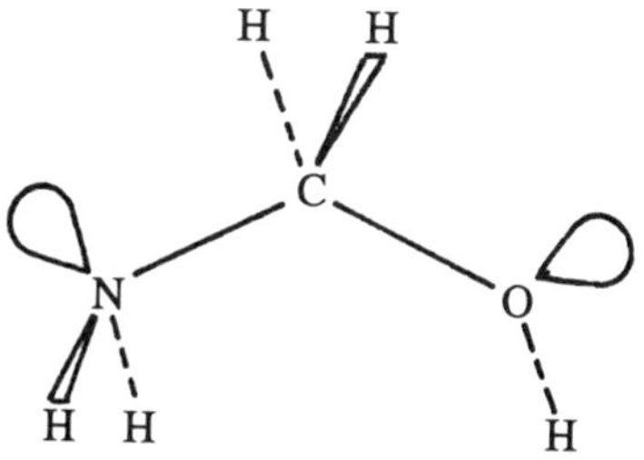

2 ae

4

In paper I, an energy analysis has been performed in order to explain qualitatively such different energetic behaviour. Energy differences between various conformers are described by an electronic component, labelled e; 1,3-diaxial H-H repulsions, labelled r; 1,3-diaxial ℓp-ℓp repulsions, called ℓ; and intramolecular "hydrogen bonding", labelled h, to be used for 1,3-diaxial H-ℓp interactions. Since in all conformations of Table I the 1,2-hydrogens and bonds are staggered, they need not be included in this analysis. Based on ab initio calculations, approximate values $r \cong \ell \cong$ -h $\cong 1$ kcal/mol were obtained (paper I). Applying this model to the systems given in Table I leads to $e_O \cong$ -2 kcal/mol, $e_N \cong$ -2.5 kcal/mol, where e_O (e_N) is the electronic energy due to an ae from O (N), when one of the ℓp's on O (N) is app to the C-X (or C-Y) bond. Surprisingly, the small energy differences between the conformers of NH_2-CH_2-NH_2, and between 1 ae and 2 ae of

NH_2-CH_2-OH, are explained by steric and electrostatic effects, while still allowing for a sizeable electronic component e_N caused by the anomeric effect. The trends in calculated bond distances agree with qualitative models and many previous calculations. An ae from X causes the C-X bond to shorten and the C-Y bond to lengthen.

Table I.   Relative energies ΔE in kcal/mol (first line) and optimized C-X/C-Y distances in Å (second line), obtained by 6-31G** geometry optimizations on systems XH_m-CH_2-YH_n

XH_m-CH_2-YH_n	0 ae	1 ae(X)	1 ae(Y)	2 ae
OH-CH_2-OH	8.35	3.86	3.86	0.0
	1.381	1.372/1.393	1.393/1.372	1.385
NH_2-CH_2-OH	5.06	0.17	0.95	0.0
	1.435/1.394	1.423/1.407	1.443/1.388	1.431/1.403
NH_2-CH_2-NH_2	0.82	0.64	0.64	0.0
	1.445	1.439/1.456	1.456/1.439	1.448

OH-rotation Model.   In Table II, relative energies, bond distances and Mulliken charges are listed for rotating the OH hydrogen of XH_m-CH_2-OH around the C-O bond, while leaving the XH_m hydrogens at fixed dihedral angles. XH_m was chosen to be F, a-OH, s-NH_2 and s-CH_3, with notations a (anti) and s (staggered) explained in structures 5 to 7. The dihedral angles of the XH_m hydrogens were held fixed at 0° for a-OH, $\pm$ 120° for s-NH_2, and 0, $\pm$ 120° for s-CH_3. In structures 5 to 7, the rotating OH is shown at a dihedral angle ϕ of 0°. CH_3-CH_2-OH was included as a reference system for which the anomeric effect is not operable. Results for ϕ values of 0°, 60°, 120° and 180° are given in Table II. Calculations were also performed for other angles, but they do not give additional insight, and have therefore been omitted. The dihedral angle ϕ = 0° corresponds to the 0 ae conformation in the case of the first two systems, and the 1 ae(N) conformation in

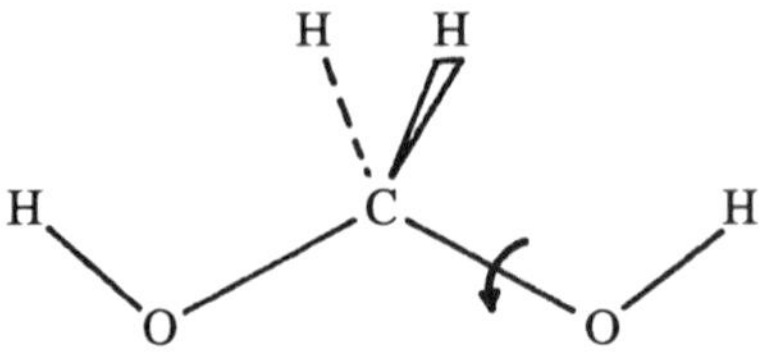

a-OH-CH_2-OH

5

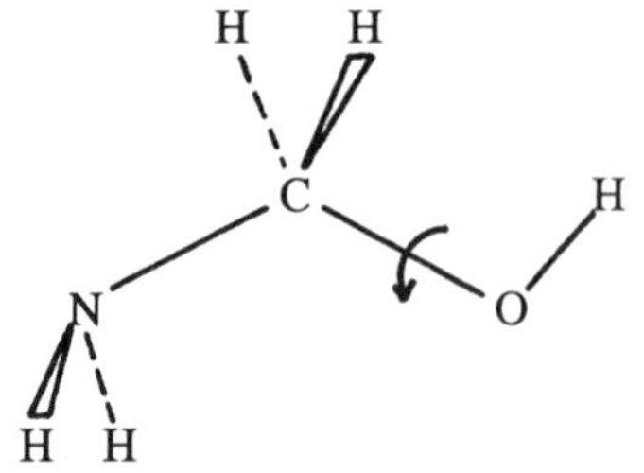

s-NH_2-CH_2-OH

6

Table II.   Relative energies ΔE in kcal/mol (first line), optimized C-O distances in Å (second line), and Mulliken charges on OH (third line), obtained by 6-31G** geometry optimizations on XH_m-CH_2-OH for OH dihedral angles of 0°, 60°, 120° and 180°

XH_m-CH_2-OH	0°	60°	120°	180°
F-CH_2-OH	0.00	-1.63	-4.79	-2.37
	1.3749	1.3689	1.3630	1.3742
	-0.279	-0.273	-0.269	-0.288
a-OH-CH_2-OH	0.00	-0.83	-4.49	-3.38
	1.3815	1.3787	1.3727	1.3802
	-0.280	-0.278	-0.279	-0.294
s-NH_2-CH_2-OH	0.00	0.91	0.06	2.28
	1.4082	1.4090	1.4029	1.4087
	-0.318	-0.318	-0.318	-0.332
s-CH_3-CH_2-OH	0.00	1.28	0.13	1.77
	1.4035	1.4074	1.4026	1.4053
	-0.304	-0.307	-0.309	-0.316

the case of NH_2-CH_2-OH. For $\phi = 120°$, one ae has to be added, leading to the 1 ae conformation of F-CH_2-OH and OH-CH_2-OH, and the 2 ae conformation of the NH_2-CH_2-OH.

In Table III, Fourier constants V_1 to V_3 are given for the systems XH_m-CH_2-OH, obtained by fitting the energy values of Table II to the formula

$$\Delta E(\phi) = (V_1/2)(1-\cos\phi) + (V_2/2)(1-\cos2\phi) + (V_3/2)(1-\cos3\phi) \quad (1)$$

It is seen that at $\phi = 0°$ and $\phi = 360°$ $\Delta E = 0$. According to equation (1), the V_1 term is periodic with $\phi = 360°$, the V_2 term with 180°, and the V_3 term with 120°. The V_3 term is easily identified with the threefold repulsion received when the rotating OH group is in eclipsed position relative to the neighboring CH_2X group. Since at $\phi = 0°$ OH is staggered relative to CH_2X, the maxima of the V_3 term occur at 60° and 180°. Therefore V_3 is positive, and close in value to the energy difference between the staggered and eclipsed positions of CH_3OH (1.29 kcal/mol in the same basis set).

The V_1 term has a maximum (or minimum) at 180° (and is zero at 0°). According to equation (1), $V_1 = \Delta E(180)$-V_3. For F-CH_2-OH and a-OH-CH_2-OH, $\Delta E(180)$ is the difference between intramolecular 1,3-hydrogen bonding at 180° (negative) and lone pair repulsions at 0° (positive), causing $\Delta E(180)$ and V_1 to be negative. Structure **8** illustrates this situation in the case of a-OH-CH_2-OH. For s-NH_2-CH_2-OH and s-CH_3-CH_2-OH, $\Delta E(180)$ is the difference between 1,3-H-H repulsions at 180° (positive) and 1,3-internal hydrogen bonding at 0° (negative),

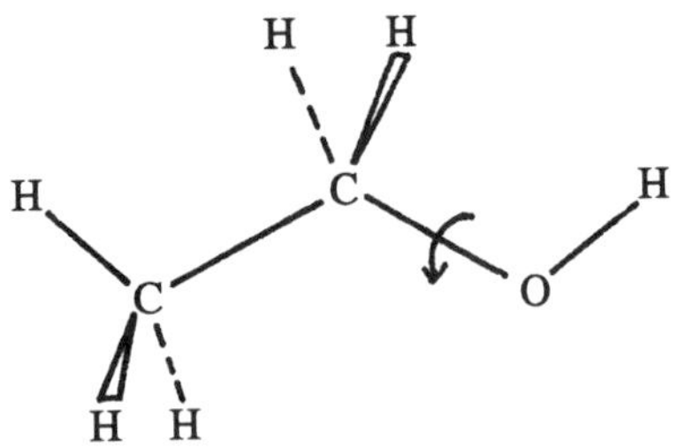

s-CH_3-CH_2-OH

7

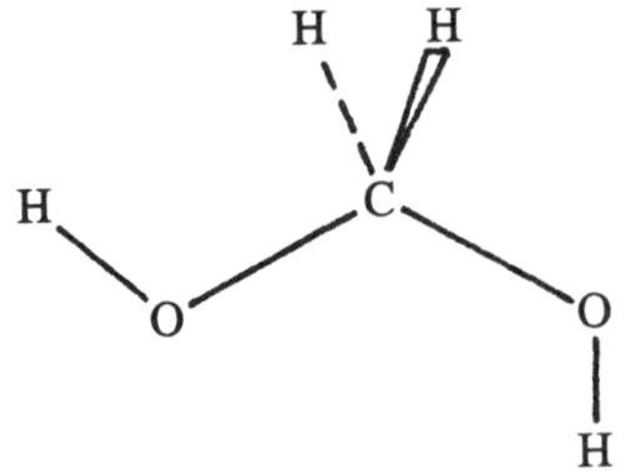

ϕ(OH) = 180°

8

Table III.   Fourier constants V_1 to V_3 in kcal/mol for OH rotation in systems XH_m-CH_2-OH, obtained from relative energies of Table II

XH_m-CH_2-OH	V_1	V_2	V_3
F-CH_2-OH	-3.69	-2.71	1.31
a-OH-CH_2-OH	-4.69	-1.29	1.31
s-NH_2-CH_2-OH	0.95	-0.87	1.33
s-CH_3-CH_2-OH	0.41	-0.24	1.36

causing ΔE(180) and V_1 (since ΔE(180) > V_3) to be positive (**9**). Obviously, no internal hydrogen bonding in the classical sense is expected for s-CH_3-CH_2-OH, as reflected in the small positive value of V_1.

To summarize, V_3 is explained by 1,2-steric interactions between the rotating OH and the CH_2X group, whereas V_1 results from 1,3-interactions, both steric (H-H) and electrostatic (ℓp-ℓp and ℓp-H). The components of V_1 correspond to the parameters r, ℓ and h used in the previous section. Due to the choice of different reference points, a simple direct relationship between V_1 and r, ℓ and h cannot be seen.

The remaining parameter V_2 results from an electronic effect that may loosely be called the electronic part of the anomeric effect. The origin of this electronic effect will be discussed later. The V_2 term of equation (1) has a maximum (or minimum) at 90°, and is 0 at 0° and 180°. Table III shows that V_2 is negative for all systems, however decreasing in magnitude, and essentially 0 for the s-CH_3 system.

We note from Table II that anomeric stabilization is achieved for the F and a-OH systems (E(120) lower than E (0), or ΔE(120) negative), but not for the s-NH_2 and s-CH_3 systems. Using the values of the Fourier constants combined with the interpretation given above, this is easily understood. According to equation (1), ΔE(120) = 0.75 (V_1+V_2). For the F and a-OH systems, both V_1 and V_2 are negative, meaning that ϕ = 180° is more stable than ϕ = 0 via the V_1 term (steric and electrostatic stabilization), and $\phi = 90°$ is more stable than $\phi = 0$ via V_2 (electronic stabilization). A combination of negative V_1 and negative V_2 gives rise to a large stabilization at $120°$. The anomeric stabilization in the first two systems is a combination of a favourable electronic effect, as well as favourable steric and electrostatic effects as one moves from $\phi = 0$ (0 ae) to $\phi = 120$ (1 ae).

For the s-NH$_2$ and s-CH$_3$ systems, V_1 is positive and V_2 negative, such that $V_1 + V_2$ is close to 0. In the case of s-NH$_2$, the electronic effect, V_2, is still about -1 kcal/mol. However, this is compensated for by the similarly large steric and electrostatic terms which favor the $0°$ conformation over the $180°$ conformation (**9**).

Applying the energy analysis of the previous section with $\ell = r = -h = 1$ kcal/mol to F-CH$_2$-OH leads to $e_O = -2.75$ kcal/mol, more negative than e_O for the OH-CH$_2$-OH and NH$_2$-CH$_2$-OH systems due to the larger electronegativity of F.

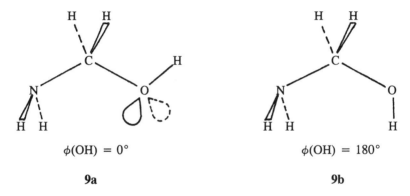

$\phi(\text{OH}) = 0°$ $\phi(\text{OH}) = 180°$

9a   **9b**

The OH rotation model, as discussed in the present chapter, associates the electronic component with V_2, and $\Delta E(120)$, the anomeric stabilization energy, is given by $\Delta E(120) = 0.75 \ (V_1 + V_2)$. Therefore, e_O can also be calculated as $0.75 \ V_2$, giving -2.0 for F-CH$_2$-OH, -1.0 for a-OH-CH$_2$-OH, and -0.7 for s-NH$_2$-CH$_2$-OH (in kcal/mol). All values are smaller than obtained earlier (-2.75 for F and -2.0 for O and N).

NH$_2$ Rotation Model. Similar to the OH rotation for systems XH$_m$-CH$_2$-OH, as described in the previous section, ab initio calculations and energy analyses were performed for NH$_2$ rotation in systems XH$_m$-CH$_2$-NH$_2$, where again XH$_m$ = F, a-OH, s-NH$_2$ and s-CH$_3$. For symmetry reasons, $\phi(\text{NH}_2) = 0$ was chosen such that the rotating NH$_2$ hydrogens have dihedral angles of $\pm 60°$ relative to the plane ZCN (**10a**), and dihedral angles of $\pm 120°$ (**10b**) for $\phi(\text{NH}_2) = 180°$. Except for the fixed dihedral angles of the NH$_2$ hydrogens and the XH$_m$ hydrogens (as described earlier), all other geometry parameters were allowed to optimize.

When comparing with the conformations of Table I, 0 ae of F-CH$_2$-NH$_2$ and a-OH-CH$_2$-NH$_2$ is accomplished at $\phi = 60°$, where one NH$_2$ hydrogen is app to the CX bond, and 1 ae occurs at $\phi = 180°$. For the s-NH$_2$ system, one moves from 1 ae at $60°$ to 2 ae at $180°$. The calculated energy differences, C-N distances and charges on NH$_2$ are given in Table IV. The energy difference between $60°$ and $180°$ corresponds to the anomeric stabilization energy. Accordingly, the

F-CH$_2$-NH$_2$ and a-OH-CH$_2$-NH$_2$ systems receive significant anomeric stabilization, whereas for the s-NH$_2$ system this quantity is very small.

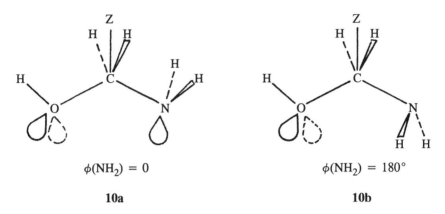

ϕ(NH$_2$) = 0 ϕ(NH$_2$) = 180°

10a   10b

Table IV.   **Relative energies ΔE in kcal/mol (first line), optimized C-N distances in Å (second line), and Mulliken charges on NH$_2$ (third line), obtained by 6-31+G* geometry optimizations on XH$_m$-CH$_2$-NH$_2$ for NH$_2$ dihedral angles of 0°, 60°, 120° and 180°**

XH$_m$-CH$_2$-NH$_2$	0°	60°	120°	180°
F-CH$_2$-NH$_2$	0.00	0.43	1.33	-5.44
	1.4233	1.4305	1.4360	1.4110
	-0.099	-0.122	-0.143	-0.127
a-OH-CH$_2$-NH$_2$	0.00	-1.50	-0.94	-6.26
	1.4356	1.4365	1.4422	1.4229
	-0.117	-0.127	-0.147	-0.148
s-NH$_2$-CH$_2$-NH$_2$	0.00	-1.27	1.73	-1.63
	1.4626	1.4578	1.4654	1.4509
	-0.127	-0.132	-0.152	-0.132
s-CH$_3$-CH$_2$-NH$_2$	0.00	-2.03	0.60	-1.84
	1.4643	1.4561	1.4635	1.4545
	-0.109	-0.138	-0.153	-0.127

The corresponding Fourier constants are shown in Table V. As before, V_3 is steric in nature, and results from the interaction of the rotating NH$_2$ group with the CH$_2$X moiety. At ϕ = 0° (and 120°) there is maximal steric interaction, whereas at 60° (and 180°) this is minimal. Therefore, V_3 is negative. Its value of about -2.4 kcal/mol resembles the energy difference between the staggered and eclipsed conformation of CH$_3$-NH$_2$ (-2.10 kcal/mol). V_1 results from steric and electrostatic effects. Using V_1 = ΔE(180) -V_3, one sees that ΔE(180), the energy difference between 180° and 0°, can be explained for F and a-OH by internal hydrogen bonding at 180° (**10b**) (negative), and lone pair and CH_2X-NH_2 (V_3) repulsion at 0° (**10a**) (positive), causing $\Delta E(180)$ and V_1 to be negative. For s-NH_2-CH_2-NH_2, there is 1,3-H,H repulsion at 180° (positive), and internal hydrogen bonding (negative) as well as CH_2X-NH_2 repulsion (positive) at $\phi = 0$, leaving the sign of $\Delta E(180)$ open. In Table IV, $\Delta E(180°)$ is seen to be about -1.6 kcal/mol, leading to a positive V_1.

Table V.   Fourier constants V_1 to V_3 in kcal/mol, for NH_2 rotation in systems XH_m-CH_2-NH_2, obtained from relative energies of Table IV

XH_m-CH_2-NH_2	V_1	V_2	V_3
F-CH_2-NH_2	-3.03	4.80	-2.41
a-OH-CH_2-NH_2	-3.80	2.55	-2.46
s-NH_2-CH_2-NH_2	0.91	1.39	-2.54
s-CH_3-CH_2-NH_2	0.53	0.27	-2.37

To summarize, for the systems studied under NH_2 rotation, V_3 is a steric term resulting from 1,2-interaction, whereas V_1 is steric and electrostatic, resulting from 1,3-interactions. V_1 contains the parameters r, ℓ and h as used in the original energy decomposition model.

The V_2 constant derives from electronic effects to be detailed later. It is seen to be positive for all systems, again of decreasing amount, and essentially 0 for the s-CH_3 system.

The anomeric stabilization energy, $\Delta E(180)$-$\Delta E(60) \equiv \Delta E_{ae}$, is explained in the following way. For the F and a-OH systems, V_1 is negative and V_2 positive. Since $\Delta E_{ae} = 0.75 (V_1$-$V_2)$, both terms stabilize these systems when going from 0 ae to 1 ae. Therefore, the observed energy lowering has both electronic and steric/electrostatic origins. On the other hand, for s-NH_2 and s-CH_3, both V_1 and V_2 are positive. The very small (essentially 0) stabilization energy obtained for s-NH_2-CH_2-NH_2 results from a combination of an unfavourable V_1 (steric/electrostatic) and a small but favourable V_2 (electronic). These ideas confirm the conclusions drawn in the first section where the electronic part of the stabilization energy due to nitrogen was estimated to be -2.5 kcal/mol, however unfavourable steric and electrostatic components prevented anomeric stabilization for NH_2-CH_2-NH_2 and NH_2-CH_2-OH (from 1 ae to 2 ae).

Recalculating e_N as -0.75 V_2 gives e_N = -3.6 for F-CH_2-NH_2, -1.9 for a-OH-CH_2-NH_2, and -1.0 for s-NH_2-CH_2-NH_2 (in kcal/mol). Again, these values are smaller than obtained in the energy decomposition model (-3.75 for F, -2.5 for O and N), but larger than the corresponding e_O values. This finding is in line with the larger electronegativity differences for nitrogen.

Energy Analysis for Protonated Systems

Energy Decomposition Model. Similar to their neutral counterparts, geometries were optimized for the protonated systems $XH_m{}^+$-CH_2-YH_n, where X and Y are N and O, in their 0 ae, 1 ae and 2 ae conformations. This means that one hydrogen each on X and on Y was restricted to a dihedral angle of $0°$ (0 ae) or $120°$ (1 ae). Relative energies and optimized CX and CY bond distances are given in Table VI.

Table VI.   Relative energies ΔE in kcal/mol (first line) and optimized C-X/ C-Y distances in Å (second line), obtained by 6-31G** geometry optimizations on systems $XH_m{}^+$-CH_2-YH_n

$XH_m{}^+$-CH_2-YH_n	0 ae	1 ae(X)	1 ae(Y)	2 ae
$OH_2{}^+$-CH_2-OH	4.73	1.18	0	1.81
	1.521/1.341	1.503/1.348	1.656/1.303	1.630/1.311
$OH_2{}^+$-CH_2-NH_2	13.45	15.04	0	1.10
	1.546/1.401	1.536/1.406	2.381/1.289	2.381/1.290
$NH_3{}^+$-CH_2-NH_2	0.12		0^a	
	1.513/1.424		1.561/1.401	
$NH_3{}^+$-CH_2-OH	0		3.80	
	1.501/1.365		1.534/1.350	

a.   The inverted structure (see text) has ΔE = -5.62 kcal/mol, R_{CX} = 1.513 Å and R_{CY} = 1.427 Å.

Contrary to the observations for neutral systems, 1 ae(Y) is in three of four cases the most stable conformation (not 2 ae). This will be explained later. Table VI shows that $OH_2{}^+$-CH_2-NH_2 optimizes for 1 ae(Y) and 2 ae to a charge-dipole complex of type $OH_2 \cdots CH_2 = NH_2{}^+$, as indicated by the large C-X distance, leading to significant stabilization over the 0 ae and 1 ae(X) conformers. For $NH_3{}^+$-CH_2-NH_2 and $NH_3{}^+$-CH_2-OH, 1 ae(X) and 2 ae conformers are not possible due to the lack of a lone pair of electrons on $NH_3{}^+$. For $NH_3{}^+$-CH_2-NH_2, the 1 ae conformation **11a** is unstable and converts to **11b**, which is 5.62 kcal/mol more stable than **11a**.

The energy analysis, performed in detail in paper I, necessitates the introduction of an additional energy parameter v, representing the "reverse anomeric effect" (rae). This new parameter is used when a lone pair on Y lies close to the (formal) positive charge on X, as for example in **11b**. It is negative due to Coulombic interactions between positive and negative charges.

In the energy analysis, $\ell \cong r \cong$ -h $\cong$ 1 kcal/mol is again used. For $OH_2{}^+$-CH_2-OH, e_O is found to be about -2 kcal/mol, and $e_O+ \cong 0$. e_O+ is the electronic part of the anomeric effect due to a lone pair on O^+.

11a   **11b**

As mentioned above, energy differences for OH_2^+-CH_2-NH_2 are very large due to the formation of a charge-dipole complex at 1 ae(Y) and 2 ae. For this system, e_N is fitted to be -15 kcal/mol, whereas e_O+ is again about zero.

The energetics between 0 ae and 1 ae(Y) of NH_3^+-CH_2-NH_2 are well explained with $e_N \cong$ -2.5 kcal/mol. The lower energy of the inverted structure **11b** leads to a parameter $v_N \cong$ -5 kcal/mol. For NH_3^+-CH_2-OH, 0 ae is much more stable than 1 ae. It is a clear example of the reverse anomeric effect, explained by the closeness of the lone pairs on O in the 0 ae conformer to the (formal) positive charge on NH_3^+. The calculated energies are fitted best by $v_O \cong$ -4 kcal/mol.

With the parameters as given here, all calculated energies for the protonated systems are modeled within 1 kcal/mol.

OH Rotation Model. In order to better understand the energetic behaviour of protonated acetals, a series of OH rotations was performed on the compounds XH_m^+-CH_2-OH, with XH_m = a-FH, s-OH_2 and s-NH_3, where the notations a (anti) and s (staggered) parallel those used earlier.

Energy differences, optimized CO/CX distances and Mulliken charges on OH/XH_m are given in Table VII. It is seen both from the large C-F distances and the close to zero charges on FH that a-FH^+-CH_2-OH forms a charge-dipole complex $FH\cdots CH_2=OH^+$ at all dihedral angles of the OH hydrogen, resulting in large energy differences. For the other systems, however, no such complexes are found, in agreement with our previous results. For the a-FH^+ system, the energy has a minimum at 120°, for s-OH_2^+ at 60°, and for s-NH_3^+ at 0°, providing for a variety of situations.

Fourier constants are shown in Table VIII. V_3 is due to the threefold repulsion of the rotating OH with the CH_2X group. It has the expected value of about 1.3 kcal/mol for the last two systems. V_3 is zero for the a-FH^+ system due to the formation of a complex. Similar to the corresponding XH_m-CH_2-OH series, V_1 results from steric and electrostatic interactions. Using $V_1 = \Delta E(180) - V_3$, one has for the a-$FH^+$ system lone pair repulsions at $\phi = 0°$ (positive), and hydrogen bonding at 180° (negative), causing $\Delta E(180)$ and V_1 to be negative. Due to the long-distance complex, such 1,3-interactions are weak, and V_1 is therefore small. For s-OH_2^+ and s-NH_3^+, there is H-bonding at 0° (negative), and 1,3-H-H repulsion at 180° (positive), leading to positive $\Delta E(180)$ as well as V_1. In addition, at 0° the reverse anomeric effect is operative, causing $\Delta E(180)$ and V_1 to be large (about 11 kcal/mol for $\Delta E(180)$ and 9.5 kcal/mol for V_1). V_2 results from the electronic part of the anomeric effect.   It is always negative (as for the corresponding neutral systems) and is seen to be very large for a-FH$^+$, due to the high stabilization received upon complex formation. Even for the other protonated systems, $|V_2|$ is much larger than for the corresponding neutral systems, indicating that electronic stabilization is more pronounced.

Table VII.   Relative energies ΔE in kcal/mol (first line), optimized C-O/C-X distances in Å (second line), and Mulliken charges on OH/XH$_m$ (third line), obtained by 6-31G** geometry optimizations on systems XH$_m$$^+$-CH$_2$-OH, for OH dihedral angles of 0°, 60°, 120° and 180°

XH$_m$$^+$-CH$_2$-OH	0° (0ae)	60°	120° (1 ae)	180°
a-FH$^+$-CH$_2$-OH	0.00	-21.17	-22.43	-2.51
	1.2373/2.3199	1.2334/2.5981	1.2322/2.5741	1.2467/2.2463
	0.028/0.055	0.097/0.033	0.103/0.031	0.013/0.064
s-OH$_2$$^+$-CH$_2$-OH	0.00	-2.29	0.63	10.66
	1.3479/1.5035	1.3241/1.5700	1.3111/1.6302	1.3359/1.5634
	-0.188/0.406	-0.149/0.352	-0.125/0.303	-0.178/0.343
s-NH$_3$$^+$-CH$_2$-OH	0.00	0.31	3.80	11.09
	1.3652/1.5013	1.3565/1.5210	1.3502/1.5340	1.3611/1.5224
	-0.207/0.584	-0.197/0.562	-0.193/0.535	-0.214/0.533

Table VIII.   Fourier   constants V_1 to V_3 in   kcal/mol   for   systems XH$_m$$^+$-CH$_2$-OH, obtained from relative energies of Table VII

XH$_m$$^+$-CH$_2$-OH	V_1	V_2	V_3
a-FH$^+$-CH$_2$-OH	-2.51	-27.39	0.0
s-OH$_2$$^+$-CH$_2$-OH	9.05	-8.21	1.61
s-NH$_3$$^+$-CH$_2$-OH	9.72	-4.65	1.37

Using the calculated values of V_1 to V_3, the anomeric stabilization energies of Table VII will be explained for the protonated systems. Again, use is made of the relationship $\Delta E(120) = 0.75(V_1+V_2)$.

The a-FH$^+$ system has a large anomeric stabilization due in most part to the large electronic effect V_2, caused by the formation of the complex. For s-OH$_2$$^+$, $\Delta E(120°) \cong 0.5$ kcal/mol since V_1 and V_2 are of similar magnitude. The sizeable electronic (anomeric) stabilization (V_2) is almost completely compensated for by the steric and electrostatic terms, and by a sizeable reverse ae. It should be noted that the rae was not included previously in the energy analysis of OH$_2$$^+$-CH$_2$-OH (Table VI).   Without the energy value at 180°, V_1 cannot be obtained, and therefore electronic contributions cannot be separated from contributions due to the reverse ae. From the new results obtained for OH_2^+-CH_2-OH, v_O is estimated to be -5 kcal/mol, and $e_O \cong$ -7 kcal/mol.

For the s-NH_3^+ system, V_i (mainly reverse ae) is much larger than -V_2 (electronic part of the ae), causing the system to be more stable at $\phi = 0°$ (0 ae) than at $\phi = 120°$ (1 ae) by about 4 kcal/mol. In the energy analysis of Table VI, the large size of the rae for NH_3^+-CH_2-OH was noted, and v_O was fitted to be -4 kcal/mol, while e_O was left at -2 kcal/mol. Using the new analysis based on Fourier constants, v_O is revised to be about -6 kcal/mol, and e_O therefore -4 kcal/mol. The energy decomposition model applied to a-FH^+-CH_2-OH gives e_O = -20.5 kcal/mol.

Recalculations of e_O from 0.75 V_2 gives -20.5 for a-FH^+, -6.2 for s-OH_2^+, and -3.5 for s-NH_3^+, in kcal/mol, in good to reasonable agreement with values listed above (-20.5 for a-FH^+, -7 for s-OH_2^+, and -5 for s-NH_3^+). In the energy decomposition model, v_O was calculated as the portion of the reverse anomeric effect corresponding to an OH-rotation from $\phi = 0°$ (0 ae) to $\phi = 120°$ (1 ae). It is better to redefine the v parameter for a 180° rotation, from $\phi = 0°$ (full rae) to $\phi = 180°$ (no rae). Using for the s-OH_2^+ and s-NH_3^+ systems E(0°) = 2h + v_O and E(180°) = r, gives ΔE(180°) = r - 2h - v_O. Equating with ΔE(180) of Table VII, and setting r = -h = 1 kcal/mol, gives v_O = -7.5 kcal/mol for the s-OH_2^+ system, and -8 kcal/mol for the s-NH_3^+ system. Due to the revised definition, these values are expected to be more negative than those obtained above for a 120° rotation.

NH_2 Rotation Model. NH_2 rotations were studied for the protonated systems XH_m^+-CH_2-NH_2, where XH_m = a-FH, s-OH_2 and s-NH_3. The notations a-FH, s-OH_2 and s-NH_3 are the same as used before. The dihedral angle $\phi(NH_2)$ for the NH_2 hydrogens is defined the same as for NH_2 rotations in neutral systems, as shown in **10a** and **10b** for a-OH-CH_2-NH_2.

Relative energies, optimized C-N and C-X distances, and Mulliken charges on NH_2 and XH_m are given in Table IX. Fourier constants are shown in Table X. The 0 ae conformer corresponds to $\phi = 60°$, and the 1 ae one to $\phi = 180°$, as for the corresponding neutral systems. The a-FH^+ system forms a charge-dipole complex FH···CH_2=NH_2^+ at all angles ϕ, with C-F distances ranging from 2.45 to 2.7 Å. The CFH angle had to be held fixed at 109.5°, otherwise the FH hydrogen would move to the syn position. For $\phi = 60°$ and 120°, the dihedral angles of the CH_2 hydrogens had to be fixed at ± 60°. Otherwise, due to the large CF distance, the CH_2 group rotates with NH_2 to achieve a more stable CH_2=NH_2^+ unit. Also, the FCH angles for the two CH_2 hydrogens were held the same in order to prevent one of them from becoming very small. Calculated energies, distances and charges are therefore dependent on these restrictions. V_3 and V_1 are quite small. V_1 results from internal hydrogen bonding at 180° and lone pair repulsion at 0°, and is therefore expected to be negative. Due to the large separation between NH_2 and FH, such interactions are small. For s-OH_2^+-CH_2-NH_2, complex formation is noted at 0° and 180°, causing ΔE(180) to be close to zero, since the 1,3-interactions used to explain this quantity are very small, and therefore V_1 = ΔE(180) - V_3 is dominated by V_3.

Table IX.   Relative energies ΔE in kcal/mol (first line), optimized C-N/C-X distances in Å (second line), and Mulliken charges on NH_2/XH_m (third line), obtained by 6-31+G* geometry optimizations on systems XH_m^+-CH_2-NH_2, for NH_2 dihedral angles of 0°, 60°, 120° and 180°

XH_m^+-CH_2-NH_2	0°	60° (0 ae)	120°	180° (1 ae)
a-FH$^+$-CH$_2$-NH$_2$[a]	0.00	30.23[b]	29.16[b]	-1.89
	1.2826/2.6960	1.3173/2.4907	1.3184/2.4582	1.2818/2.6325
	0.336/0.022	0.230/0.035	0.235/0.035	0.319/0.036
s-OH$_2$$^+$-CH$_2$-NH$_2$	0.00	14.09	19.91	-0.62
	1.2893/2.4783	1.4088/1.5356	1.4065/1.5764	1.2903/2.4497
	0.291/0.043	0.002/0.408	0.008/0.362	0.289/0.056
s-NH$_3$$^+$-CH$_2$-NH$_2$	0.00	4.88	11.33	4.83
	1.4177/1.5306	1.4267/1.5126	1.4312/1.5226	1.4044/1.5583
	-0.020/0.555	-0.024/0.533	-0.027/0.503	0.003/0.520

a.   The CFH angle was held at 109.5°.
b.   For 60° and 120°, the dihedral angles of the CH_2 hydrogens were held at ± 60°, and the FCH angles were forced to be the same.

Table X.   Fourier constants V_1 to V_3 in kcal/mol for systems XH_m^+-CH_2-NH_2, obtained from relative energies of Table IX

XH_m^+-CH_2-NH_2	V_1	V_2	V_3
a-FH$^+$-CH$_2$-NH$_2$	-1.97	40.86	-0.08
s-OH$_2$$^+$-CH$_2$-NH$_2$	3.46	23.08	-4.09
s-NH$_3$$^+$-CH$_2$-NH$_2$	7.52	7.59	-2.69

For the s-NH$_3$$^+$ system, no complex is formed. V_1 results from H-H repulsions at 180°, and H-bonding as well as the reverse anomeric effect at 0°, and is therefore positive and large. The electronic term V_2 is very large for a-FH$^+$ and s-OH$_2$$^+$ (complex), and still quite large for s-NH$_3$$^+$.

Using $\Delta E_{ae} = \Delta E(180) - \Delta E(60) = 0.75 \ (V_1 - V_2)$, one can explain the significant stabilization of the a-FH$^+$ and s-OH$_2$$^+$ systems by a very favourable electronic effect (complex), and small 1,3-steric and electrostatic effect (small due to large C-O distance). For the s-NH$_3$$^+$ system, no anomeric stabilization is observed. This is due to the cancellation of V_2 (favourable electronic effect) by V_1 (unfavourable 1,3 effects and reverse ae).

Calculating e_N as -0.75 V_2 leads to values (in kcal/mol) of -30.6 for a-FH$^+$, -17.3 for s-OH$_2$$^+$ and -5.7 for s-NH$_3$$^+$. Due to the long C-X distance in a-FH$^+$-CH$_2$-NH$_2$ and s-OH$_2$$^+$-CH$_2$-NH$_2$ (at 0 and 180°), the rae as defined earlier does not apply. For s-NH$_3$$^+$-CH$_2$-NH$_2$, v_N may be calculated from the energy difference between 0° (h + v_N) and 180° (2r + e_N) by the formula 2r + e_N - h - v_N = $\Delta E(180°)$, to give -7.5 kcal/mol assuming again r = -h = 1 kcal/mol.

Discussion

The π-bonding Model for Systems XH$_m$-CH$_2$-OH and XH$_m$$^+$-CH$_2$-OH. The energy analysis performed in the previous sections showed that the anomeric stabilization energy can be decomposed into steric (H-H repulsion), electrostatic (H-ℓp attraction, ℓp-ℓp repulsion) and electronic effects. For systems XH$_m$-CH$_2$-OH and XH$_m$$^+$-CH$_2$-OH, the electronic energy is represented by the second term in the Fourier expansion of the energy, associated with the Fourier constant V_2. The V_2 term has a minimum at ϕ(OH) = 90°, and is zero at ϕ(OH) = 0 and 180°. This situation may be compared with the π-bonding in CH$_2$=OH$^+$ which is optimal (minimum energy) at ϕ = 90° (planar) and absent at ϕ=0° and 180° (OH perpendicular to CH$_2$). The anomeric effect is usually explained by donation of electrons from O to the electronegative X, being most effective when the OH hydrogen has $\phi \cong 120°$, such that the lone pair on oxygen is app to the CX bond. This implies partial charges XH$_m$$^{(-)}$-CH$_2$-OH$^{(+)}$. The larger these partial charges, the closer is CH$_2$-OH$^{(+)}$ to the cation CH$_2$=OH$^+$. We shall refer later to the origin of the partial charges. The electronic part of the ae will now be explained as the tendency of the CH$_2$-OH$^{(+)}$ moiety of XH$_n$$^{(-)}$-CH$_2$-OH$^{(+)}$ to form a π-bond. For testing this hypotheses, calculated properties of XH$_m$-CH$_2$-OH and XH$_m$$^+$-CH$_2$-OH, such as energies, bond distances and charges, as given earlier, will be compared with corresponding quantities of CH$_2$=OH$^+$. In order to make this comparison more realistic, the CH$_2$ hydrogens of CH$_2$=OH$^+$ were held in pyramidal position, with dihedral angles of $\pm$ 60° (rather than $\pm$ 90°, planar) with respect to Z-C-O (12a, 12b). The results are given in Table XI. It is obvious that the minimum energy occurs around 90°, where π-bonding is optimal due to the cation being closest to planarity. At ϕ = 0 and 180°, OH is perpendicular to CH$_2$, and π-bonding is not possible. The C-O bond distance has a minimum around 90°, for which angle the OH charge, always very small, is most positive. V_3 is almost zero due to the absence of threefold repulsion. V_1 = -12.7 kcal/mol is determined by $\Delta E(180)$, and V_2 = -27.1 kcal/mol incorporates the electronic energy stabilization due to the formation of the π-bond.

ϕ(OH) = 0°

12a

ϕ(OH) = 90°

12b

First, the neutral systems XH_m-CH_2-OH will be studied by comparing the changes of ΔE, R_{CO} and Q(OH) as function of ϕ (Table II) with the changes of corresponding quantities of CH_2OH^+ given in Table XI.

For F-CH_2-OH and a-OH-CH_2-OH, all three properties show the same trend as for CH_2OH^+. For example, ΔE has a minimum at $\phi = 120°$. For the same dihedral angle, R_{CO} is smallest and Q(OH) least negative. (In Table XI, R_{CO} is slightly smaller at 60° than at 120°). Therefore, one may conclude that the electronic stabilization in the F and a-OH systems is governed by π-bonding. On the other hand, the NH_2 and CH_3 systems of Table II do not follow the trend of CH_2OH^+. For both systems, the energy has a minimum at 0°. R_{CO} for s-NH_2 is shortest at 120°, as for the F and a-OH systems, but R_{CO} is almost the same at 0°, 60° and 180°, contrary to the trend established by CH_2OH^+. The OH charges of s-NH_2-CH_2-OH behave quite different from those of CH_2OH^+. For the s-CH_3 system, no similarities to CH_2OH^+ remain.

Table XI.   6-31G** results for CH_2=OH^+ with OH rotating about C-O, and CH_2 held in pyramidal position. For definition of the OH dihedral angle see structures 12a and 12b[a]

	0°[b]	60°	90°	120°	180°
ΔE(kcal/mol)	0.00	-24.28	-32.62	-29.89	-13.48
$R_{C\text{-}O}$ (Å)	1.2508	1.2370	1.2350	1.2394	1.2425
Q(OH)	0.048	0.126	0.140	0.127	0.085

a.   For CH_2=OH^+, $V_1 = -12.7$, $V_2 = -27.1$, $V_3 = -0.8$ kcal/mol.
b.   For 0°, the COH angle was held fixed at 120°. Without setting this angle, OH would move from cis(0°) to trans (180°).
c.   The energy minimum occurs at 97.95°, for which $\Delta E = -33.0°$ kcal/mol.

Such behaviour can easily be rationalized by looking at partial charges. F is more, N and C less electronegative than O. Therefore, F-CH_2-OH should carry a partial positive charge on CH_2-OH, but the NH_2 and CH_3 systems are not expected to have a positively charged CH_2-OH fragment. The fact that a-OH-CH_2-OH also follows the pattern of CH_2OH^+ can be understood from the fixed position (dihedral angle) of the a-OH hydrogen which does not allow for π-bonding, whereas the rotating OH hydrogen can accomplish this at appropriate dihedral angles. Calculated Mulliken charges confirm the electronegativity arguments. XH_m is much more negative than OH for the F system (-0.410 vs -0.269), less so for the a-OH system (-0.306 vs. -0.279), and more positive for the other two systems (-0.170 vs -0.318 for s-NH_2 and -0.015 vs. -0.309 for s-CH_3). The charges given in parentheses apply to $\phi = 120°$ (OH charges from Table II, XH_m charges from Table V of paper II).

When comparing properties of the protonated systems XH_m^+-CH_2-OH (Table VII) with those of CH_2OH^+, it is noted that the ΔE's have the same trend only for a-FH$^+$-CH$_2$-OH, whereas R$_{CO}$ and Q(OH) follow the pattern of CH$_2$OH$^+$ for all three systems. Therefore, electronically all three systems behave like CH$_2$OH$^+$, as also seen by the negative V$_2$ constants. The energetics for s-OH$_2$$^+$ and s-NH$_3$$^+$ differ from the CH$_2$OH$^+$ model due to 1,3-steric and electrostatic effects, as expressed by their positive rather than negative V$_1$ (see earlier explanation). The C-O bond distances and charges of the a-FH$^+$ system resemble those of CH$_2$OH$^+$ quite closely, which is obvious from the large CF distance in this system.

In summary, the systems undergoing OH rotation are seen to receive electronic anomeric stabilization by their tendency to move charge from OH to XH$_m$, followed by maximal π-bond formation for dihedral angles ϕ between 90° and 120°. While this applies to all protonated systems, only F-CH$_2$-OH and a-OH-CH$_2$-OH of the neutral systems participate in π-bonding.

The π-bonding Model for Systems XH$_m$-CH$_2$-NH$_2$ and XH$_m$$^+$-CH$_2$-NH$_2$. For the molecules undergoing NH$_2$ rotation, the model system used for comparison is CH$_2$=NH$_2$$^+$. Energies, C-N distances and Mulliken charges for CH$_2$=NH$_2$$^+$, with CH$_2$ held in pyramidal position, are given in Table XII. The dihedral angles ϕ(NH$_2$) are defined in 13a and 13b. Maximal planarity, associated with best π-bonding, is accomplished at 0° and 180°, where CH$_2$NH$_2$$^+$ has the shortest C-N distance and the largest positive charge on NH$_2$, as expected, parallel to the situation seen for CH$_2$OH$^+$. The Fourier constants for the NH$_2$ rotation in CH$_2$NH$_2$$^+$ are V$_1$ = -10.9 and V$_2$ = 37.4 kcal/mol.

ϕ(NH$_2$) = 0°

13a

ϕ(NH$_2$) = 180°

13b

Table XII.   6-31+G* results for CH$_2$=NH$_2$$^+$, with NH$_2$ rotating about C-N, and CH$_2$ held in pyramidal position. For definition of NH$_2$ dihedral angles see structures 13a and 13b [a]

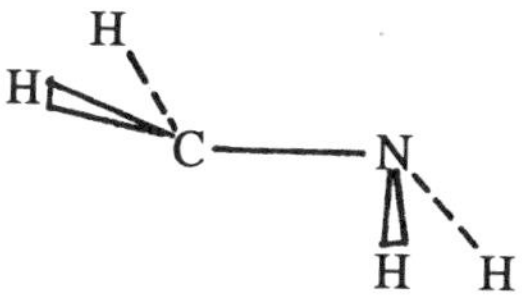

	0°	60°	90°	120°	180°
ΔE(kcal/mol)	0.00	26.60	45.83	19.82	-9.62
R$_{C-N}$(Å)	1.2801	1.3031	1.3592	1.3129	1.2882
Q(NH$_2$)	0.397	0.313	0.187	0.306	0.346

a.   For CH$_2$=NH$_2$$^+$, V$_1$= -10.93, V$_2$=37.36, V$_3$=1.31 kcal/mol.

Of the neutral systems XH_m-CH_2-NH_2, only F-CH_2-NH_2 and a-OH-CH_2-NH_2 follow the trend of $CH_2NH_2^+$, the latter system only in terms of R_{CN} and $Q(NH_2)$ (Table IV). For both systems, V_1 is negative and V_2 positive, as for $CH_2NH_2^+$. The electronic part of the anomeric stabilization is again explained by the tendency of the $CH_2NH_2^{(+)}$ moiety to form a π-bond at 180° (1 ae). The C-N distances of the s-NH_2 and s-CH_3 systems show an alternating long-short behaviour, in line with steric interaction of NH_2 singly bonded to CH_2X. The π-model is not operative, or only weakly in the case of s-NH_2.

All of the protonated systems XH_m^+-CH_2-NH_2 (Table IX) follow the trend of $CH_2NH_2^+$, and the electronic portion of their anomeric stabilization can be explained by the tendency of the $CH_2NH_2^{(+)}$ moiety of the molecule to form a π bond.

The Nature of the Anomeric Effect. For systems undergoing OH rotation, the dihedral angle $\phi(OH)$ of 120° corresponds to a lone pair on O being app to the C-X bond (in the sp^3 hybridization scheme). The conventional picture of the anomeric effect associates this position of the lone pair with the anomeric energy stabilization. The present study, however, sees the explanation in the formation of the π system (and indirectly in the location of the OH hydrogen), and not in the orientation of the lone pair. Between $\phi = 90°$ and 120°, the CH_2OH fragment of the molecule, when holding a partial positive charge, has optimal tendency to form a partial π bond, and thereby stabilize the system. The availability of a positive charge on CH_2OH depends on relative electronegativities between O and X. Therefore, electronic anomeric stabilization (due to OH) is given for F-CH_2-OH and less so for a-OH-CH_2-OH, but not for X being N and C. Steric and electrostatic factors may amplify or reduce the anomeric stabilization caused by the electronic (π-bonding) component.

The overall positive charge of the protonated systems XH_m^+-CH_2-OH ensures anomeric stabilization due to π-bonding in all cases studied. However, for X = O and N, steric and electrostatic effects counteract the electronic stabilization.

In the systems undergoing NH_2 rotation, the lone pair on NH_2 is in app position for $\phi(NH_2) = 180°$ (see earlier definition of $\phi(NH_2)$). This position coincides with the CH_2-$NH_2^{(+)}$ fragment of the molecule having optimal π bonding. CH_2-NH_2 carries a partial positive charge if X is more electronegative than N, as is the case for X = F, O in the neutral systems. In the protonated systems, the partial positive charge on CH_2NH_2 is always operative.

For the systems subjected to OH rotation, one has 0 ae at $\phi(OH) = 0°$, and 1 ae at $\phi(OH) = 120°$. The electronic part of the anomeric stabilization energy is 0.75 V_2. The V_2 term of the Fourier expansion has its maximum or minimum at 90°. It follows that $\phi(OH) = 60°$ receives the same amount of electronic stabilizations as $\phi(OH) = 120°$, as also seen from the relation $\Delta E(60) = 0.25\ V_1 + 0.75\ V_2 + V_3$. At $\phi = 60$, one ℓp of O is synperiplanar to C-X rather than antiperiplanar. The energetic preference of $\phi = 120$ over $\phi = 60$ is not a result of the position of the lone pair, but of steric and electrostatic effects expressed by the parameters V_1 and V_3.

A similar situation applies to the systems undergoing NH_2 rotation. Electronically, $\phi(NH_2) = 180°$ and 0° are equivalent. $\Delta E(180) = V_1 + V_3$ has no electronic component V_2. Again, a ℓp in synperiplanar position causes the same electronic stability as a ℓp in antiperiplanar position. A preference of 180° (over 0°) results from V_1 and V_3, steric and electrostatic terms. For all systems with NH_2 rotation, V_3 is negative, however V_1 may be positive or negative.

Finally, a word about the well studied effect that for systems XH_m-CH_2-YH_n which receive anomeric stabilization charges on Y are more positive when a lone pair on Y is app to C-X. For example, consider F-CH_2-OH. For ϕ(OH) = 120° (ℓp app to C-X), the Mulliken charge on OH is least negative (Table II). This effect is usually explained by electron donation from O into the σ^* orbital of C-F, caused by its suitable overlap with the ℓp on O. Associated with the depletion of charge from O is the accumulation of charge on F (actually, according to Table 5 of paper II, Q(F) is most negative at 180°).

Following the proposed π-bonding model for explaining the electronic part of the anomeric effect, one may look at the OH charges in CH_2OH^+, Table XI. They are most positive at 90° and less so at 120° and 60°, and much less again at 0° and 180°, similar to the OH charges in F-CH_2-OH. (This feature was used earlier in establishing those systems that follow the π-bonding model.) Therefore, conformations suitable for π-bonding cause OH to be more positive (or less negative), as seen in CH_2-OH^+, resulting from the flow of electrons from OH towards CH_2 upon double bond formation, eliminating the need of electron donation from O via overlap of its lone pair with the σ^* orbital of C-X.

For systems undergoing NH_2 rotation, Q(NH_2) is no longer most positive (or least negative) for ϕ = 180° (1 ae, ℓp on N app to C-X). Table IV shows that for F-CH_2-NH_2 Q(NH_2) is least negative at 0°, followed by 60° and 180°. For CH_2=NH_2^+ (Table XII), NH_2 is most positive at 0°, followed by 180° and 60°. Maximal electron donation from the lone pair on N, when app to C-X, is not borne out by the Mulliken charges. The spp position of the lone pair appears to be more suitable for charge donation. Again, it is better to explain charges as being governed by the ability of the CH_2-NH_2 fragment to form a π-bond, as simulated by $CH_2NH_2^+$.

Summary and Conclusion

In Table XIII, the energy parameters e_O, e_N, v_O and v_N are summarized. The parameters e_O (e_N) are defined as the electronic component of the anomeric stabilization energy when OH (NH_2) of XH_m-CH_2-OH (XH_m-CH_2-NH_2) is rotated from ϕ(OH) = 0° to ϕ(OH) = 120° (ϕ(NH_2) = 60° to ϕ(NH_2) = 180°) such that the anomeric effect from O (or N) becomes operative. All values e_O and e_N were calculated as 0.75 V_2 (or -0.75 V_2).

For neutral systems the ordering depends on the difference in electronegativity between X and O for e_O, and X and N for e_N. The parameter e_N is always more negative than the corresponding e_O. In the energy decomposition model, the same steric and electrostatic parameters r, h and ℓ were used for *all* systems. While the choice of parameters r = ℓ = -h = 1 kcal/mol was based on ab initio calculations done on specific systems, these parameters cannot be expected to be equally good in all cases. Therefore, e_O and e_N values obtained by the energy decomposition model are only approximate, and differ from those obtained by the rotation model by 0.15 to 1.5 kcal/mol, usually being more negative. Such deviations are quite understandable, and it should be emphasized that the energy decomposition model formed the basis for obtaining the more refined parameters of the rotation model.

Table XIII.   Energy parameters e_O, e_N for electronic component of anomeric effect due to O, N, and v_O, v_N for reverse anomeric effect due to O, N, in kcal/mol [a]

XH_m	e_O	e_N	v_O	v_N
F	-2.0	-3.6	0	0
a-OH	-1.0	-1.9	0	0
s-NH_2	-0.7	-1.0	0	0
a-FH^+	-20.5	-30.6	C	C
s-OH_2^+	-6.2	-17.3	-7.5	C
s-NH_3^+	-3.5	-5.7	-8.0	-7.5

a.   C stands for complex formation, which prevents the reverse anomeric effect from taking place.

Previous estimates of the anomeric stabilization based on experimental evidence are a minimum of 2.1 kcal/mol (13) (for O-alkyl systems), a minimum of 1.4 kcal/mol (14), and from theoretical studies a value of 2.8 kcal/mol (15). It is not clear to what extent the above estimates correspond to the definition of electronic anomeric stabilization energy as used in the present paper.

For protonated systems, e_O and e_N are much larger than for the corresponding neutral systems, due to the larger electronegativity of the protonated atoms. The formation of charge-dipole complexes in the case of a-FH^+-CH_2-OH, a-FH^+-CH_2-NH_2, and s-OH_2^+-CH_2-NH_2 is reflected in very large (negative) e_O and e_N values.

Schleifer et al., in a recent paper (16), conclude that structural parameters rather than energy characterize the anomeric effect. As shown here in detail, electronic stabilization may be opposed by steric and electrostatic effects, still allowing for bond distance and other structural changes to follow the π-electron model.

The energy parameters v_O and v_N were designed to quantify the stabilization due to the reverse anomeric effect. Such effect arises when the ℓp's on O or the ℓp on N are synperiplanar to the C-$X^{(+)}$ bond. The v parameters are defined as the energy difference between $\phi = 0$ (ℓp or ℓp's spp to C-X) and $\phi = 180°$ (ℓp or ℓp's app to C-X), corrected for steric and electrostatic effects. They only apply to protonated systems that do not form charge-dipole complexes at $\phi = 0°$ and $180°$. The numbers listed in Table XIII are derived from the Fourier constant V_1, corrected for steric and electrostatic effects, where in lack of a better description the assumption $r = \ell = -h = 1$ kcal/mol was made. It is seen from Table XIII that in all 3 cases where the rae is operative, the v parameters have about the same value, from -7.5 to -8 kcal/mol, irrespective of differences in electronegativities. In paper I, v_N was estimated to be -5, and v_O -4 kcal/mol for the s-NH_3^+ systems (based on a different definition, with $\Delta\phi = 120°$). As mentioned before, the e and v components cannot be separated without using energies at additional dihedral angles, and the calculation of v was based on the values of e_O and e_N as given in paper I.

The driving force for the anomeric energy stabilization is seen in the tendency of the CH_2-OH or CH_2-NH_2 fragment of XH_m-CH_2-OH or XH_m-CH_2-NH_2 to acquire a conformation suitable for π-bonding, which means as close to planarity as possible. The ability to form a π-bond requires a partial positive charge on CH_2OH or CH_2NH_2, which is assured if OH or NH_2 is less electronegative than X. For the neutral OH systems XH_m-CH_2-OH, X = F satisfies this requirement, but to a lesser degree also a-OH, where the hydrogen is held in fixed conformation. The small negative value of e_O for s-NH_2 lies probably within the uncertainty of the method or results from other smaller effects not considered here. It was noted earlier that C-O distances and OH charges of s-NH_2-CH_2-OH do not follow the pattern given by CH_2OH^+.

Of the neutral systems undergoing NH_2 rotation, F-CH_2-NH_2 and a-OH-CH_2-NH_2 qualify for having CH_2NH_2 partially positive (at $\phi = 120°$), in agreement with negative e_N values. The CH_2-NH_2 portion of these molecules follows the pattern of CH_2=NH_2^+ as far as energies, C-N distances and NH_2 charges are concerned. s-NH_2-CH_2-NH_2, although having $e_N = -1.0$ kcal/mol, does not follow the trend of $CH_2NH_2^+$.

The π-bonding model is easily verified and understood for all protonated systems by the high electronegativity of protonated atoms.

It was pointed out above that electronically, lone pairs in synperiplanar position give rise to as much stabilization as in antiperiplanar position (relative to C-X). This follows from considering the 60° vs 120° conformers of CH_2=OH^+, or the 0 vs 180° conformers of CH_2=NH_2^+, which are electronically equivalent. The same can also be seen from Fourier analysis, since the expressions for $\Delta E(60)$ and $\Delta E(120)$ both contain 0.75 V_2, whereas $\Delta E(0)$ as well as $\Delta E(180)$ both have no contribution from V_2. However, the possibility that there are smaller electronic effects that could give the app position a slight electronic preference over the spp position cannot be ruled out.

The theoretical study covering the anomeric effect in model acetals and protonated acetals leads to electronic energy components that are seen to be reasonable, and also assigns energy values to the reverse anomeric effect. The mechanism leading to anomeric stabilization can be explained by the CH_2-YH_n portion of the molecule to form a partial π-bond.

Acknowledgments. Sincere thanks are expressed to Dr. Pierre Deslongchamps for suggesting this research, and for keeping an intense interest in its execution. I am grateful to Camilla Scott for performing geometry optimizations on several systems and for evaluating some of the results. Financial support by NSERC was essential to this project.

Literature Cited
1. Wolfe, S.; Rauk, A.; Tel, L.M.; Csizmadia, I.G. J. Chem. Soc. B, **1971** 136.
2. Radom, L.; Hehre, W.J.; Pople, J.A. J. Amer. Chem. Soc. **1971**, 93, 289.

3. Radom, L.; Hehre, W.J.; Pople, J.A. J. Chem. Soc. A, **1971**, 2299.

4. Deslongchamps, P. *Stereoelectronic effects in organic chemistry*; Pergamon Press, Oxford, 1983.

5. Kirby, A.J. *The anomeric effect and related stereoelectronic effects at oxygen*; Springer Verlag, Berlin, 1983.

6. Grein, F.; Deslongchamps, P. Can. J. Chem. **1992**, 70, 1562.

7. Grein, F.; Deslongchamps, P. Can. J. Chem. **1992**, 70, 604.

8. Radom, L.; Hehre, W.J., Pople, J.A. J. Am. Chem. Soc. **1972**, 94, 2371.

9. Jeffrey, G.A.; Pople, J.A.; Radom, L. Carbohydr. Res. **1972**, 25, 117.

10. Lemieux, R.U.; Morgan, A.R. Can. J. Chem. **1965**, 43, 2205.

11. Lemieux, R.U. Pure Appl. Chem. **1971**, 25, 527.

12. Frisch, M.J.; Binkley, J.S.; Schlegel, H.B. et al. Gaussian 86, Carnegie-Mellon Quantum Chemistry Publishing Unit, Pittsburgh, PA, 1984.

13. Frank, R.W. Tetrahedron **1983**, 39, 3251.

14. Deslongchamps, P.; Rowan, D.P.; Pothier, N.; Sauve, T.; Saunders, J.K. Can. J. Chem. **1981**,59, 1105.

15. Wolfe, S.; Whangbo, M.H.; Mitchell, D.J. Carbohydr. Res. **1979**, 69, 1.

16. Schleifer, L.; Senderowitz, H.; Aped, P.; Tartakovsky, E.; Fuchs, B. Carbohydr. Res. **1990**, 206, 21.

RECEIVED May 12, 1993

Chapter 12

Glycosylmanganese Complexes and Anomeric Anomalies

The Next Generation?

Philip DeShong[1], Thomas A. Lessen[1], Thuy X. Le[1], Gary Anderson[1], D. Rick Sidler[1], Greg A. Slough[2], Wolfgang von Philipsborn[3], Markus Vöhler[3], and Oliver Zerbe[3]

[1]Department of Chemistry and Biochemistry, University of Maryland, College Park, MD 20742
[2]Department of Chemistry, Pennsylvania State University, University Park, PA 16802
[3]Organisch-chemisches Institut, University of Zurich, Zurich, Switzerland

Sequential and migratory insertion processes involving pyranosyl and furanosylmanganese pentacarbonyl complexes have been performed to afford C-glycosyl derivatives. The rates of these insertion processes depends upon the configuration of the metal at the anomeric center. Kinetic analysis of the insertion reaction and determination of the solution conformation of the respective glycosyl complexes from 1D and 2D-NMR analysis and molecular mechanics calculations demonstrates that the orientation of the carbon-metal bond with regard to the lone pairs of electrons on oxygen has an influence on the rates of the insertion process.

In recent years, we have demonstrated that alkylmanganese pentacarbonyl complexes can be utilized for the synthesis of carbonyl derivatives *via* the intermediacy of manganacycles **1** and **2**.(*1*) These processes are summarized in Scheme 1. In these sequential insertion reactions, a molecule of carbon monoxide and an alkene (or an alkyne) is incorporated into the manganese complex with the concomitant formation of at least two carbon-carbon bonds. Subsequent demetalation of manganacycle **1** or **2** results in the formation of a variety of carbonyl derivatives as indicated.

The generality of the sequential insertion methodology with highly function-alized alkylmanganese complexes has been demonstrated(*1*) and the approach has been extended to the preparation of biologically relevant substances. One appealing feature of this methodology would involve the synthesis of C-glycosyl derivatives as outlined in Scheme 2. As originally conceived, glycosyl halide **3** and sodium

0097–6156/93/0539–0227$06.00/0
© 1993 American Chemical Society

manganate pentacarbonyl (**4**) react to afford glycosylmanganese complex **5**. Following the protocols established in the earlier studies, glycosyl complex **5** was expected to serve as the precursor for either ester **6** or ketone **7**, respectively. The overall process results in formation of a carbon-carbon bond to the anomeric center, a reaction which is traditionally difficult to achieve with high stereoselectivity.

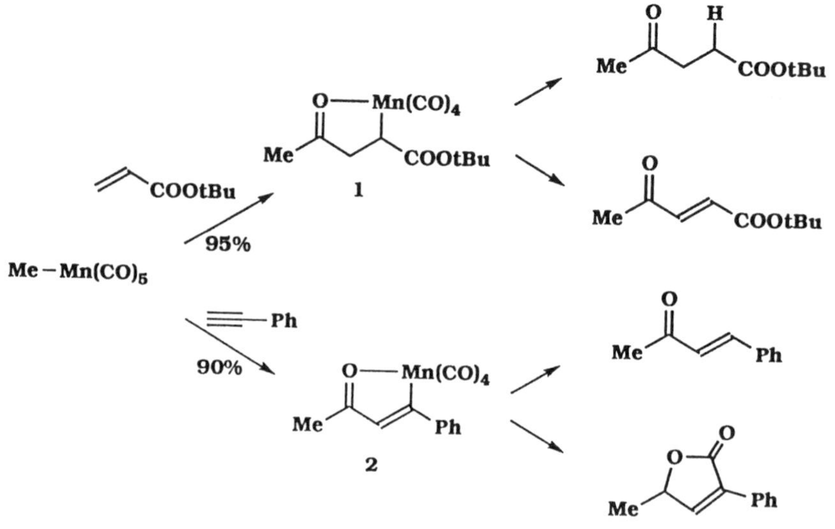

Scheme 1

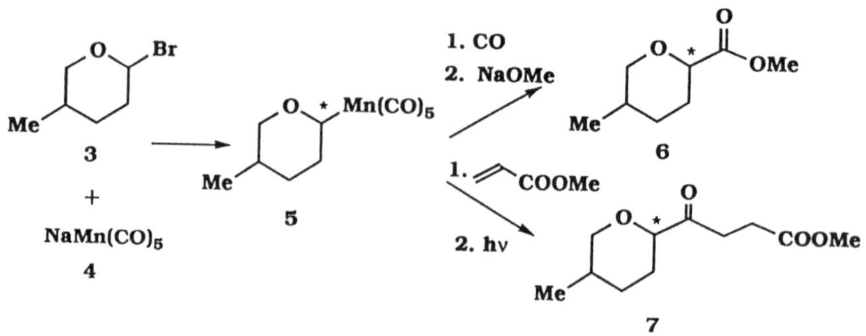

Scheme 2

One aspect of this process which deserved particular attention from the outset was the condensation of manganate anion **4** with halide **3** which could provide, in principle, a diastereomeric mixture of complexes **5** (at *). Since subsequent migratory insertion of carbon monoxide, the initial step in formation of **6** and **7**, must occur with retention of configuration at the anomeric center, the configuration of the carbon-metal bond in complex **5** would be retained in the sequential insertion product. We were able to demonstrate subsequently that the anomeric configuration of glycosylmanganese complexes can be controlled in the condensation reaction. For example, condensation of potassium manganate (**9**) and glucopyranosyl bromide **8** in THF at low temperature leads *exclusively* to β-complex **10** in high yield. On the other hand, if the condensation of **8** and **9** is performed at higher temperatures and in the presence of tetrabutylammonium bromide, a mixture of α-anomer **11** and β-anomer **10** is obtained with the α-anomer predominating.(*2,3*) This protocol has been applied to the preparation of a variety of pyranosyl- and furanosylmanganese complexes. In this paper, we shall focus on four families of glycosyl derivatives: permethylated glucopyranosyl (**12**), mannopyranosyl (**13**), arabinofuranosyl (**14**), and ribofuranosyl (**15**) complexes. In each glycosyl family, both the α- and β-anomers can be prepared in anomerically pure form.(*2,3*)   See Scheme 3.

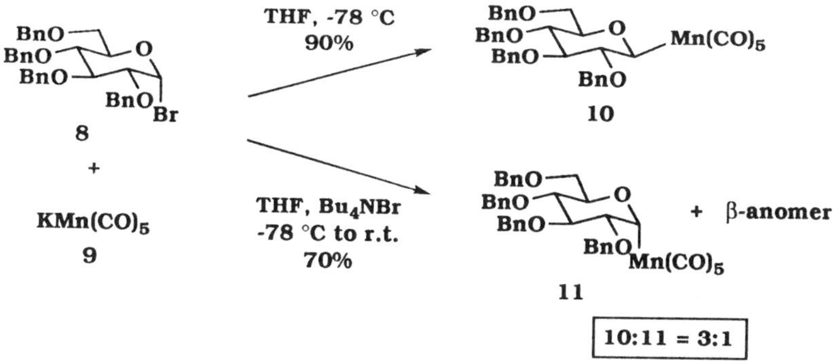

Scheme 3

As the sequential insertion processes were studied for each of these compounds, **it was observed that the rate at which the respective anomers underwent sequential insertion was very different**. For instance, sequential insertion of methyl acrylate with the α-anomer of arabinosylmanganese pentacarbonyl (α-**14**) occurred readily and in high yield to afford manganacycle **16** (see Scheme 4). However, anomer β-**14** was inert to manganacycle formation under identical conditions. Other glycosyl derivatives displayed qualitatively analogous

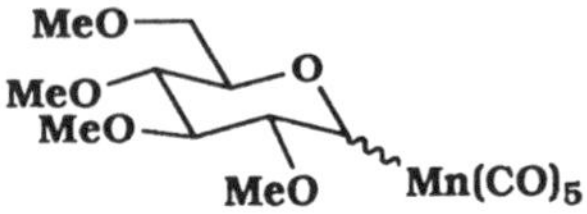

Glucopyranosyl (12)

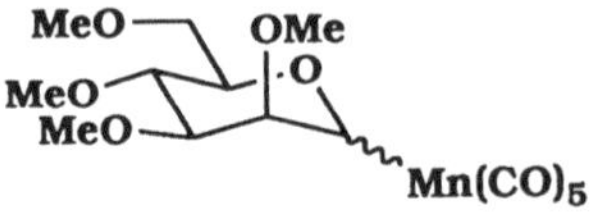

Mannopyranosyl (13)

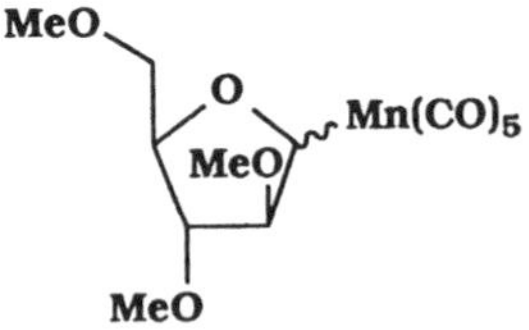

Arabinofuranosyl (14)

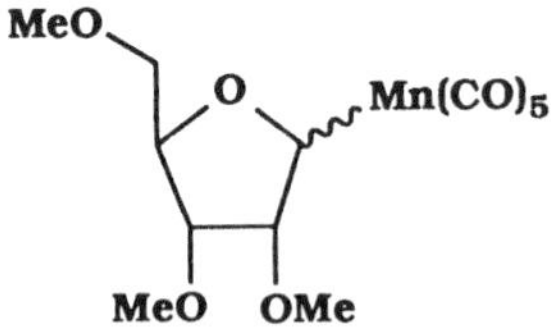

Ribofuranosyl (15)

behavior. Employing control experiments which cannot be discussed at this time,(3) we were able to demonstrate that it was the migratory insertion of carbon monoxide into the anomeric carbon-metal bond of complex **12**, and not the alkene insertion step, which was responsible for the differential rates of manganacycle production. Accordingly, a preliminary kinetic study of the migratory insertion of carbon monoxide into glycosyl complexes **12-15** was perfrmed to investigate this phenomenon. The results of this study are the subject of this presentation.

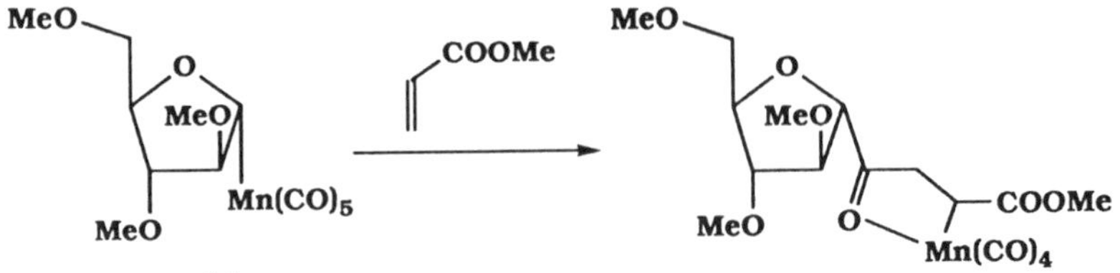

α-14   16

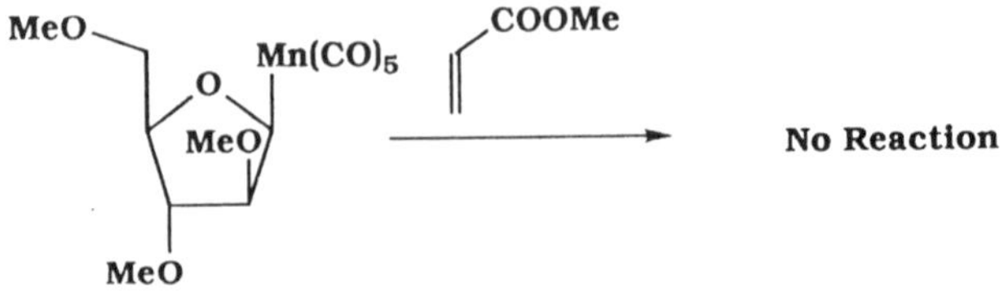

β-14

Scheme 4

The preliminary results of this kinetic investigation are summarized in the Table.(4) In each case, the relative rate of migratory insertion promoted by carbon monoxide to afford the acyl derivative (i.e., **17**) was determined. The kinetics of this process were complicated by the reversibility of the insertion reaction (**17** slowly reverts to **16**); however, under pseudo first-order conditions with excess carbon monoxide, the rates of the forward reaction could be measured. Several trends are apparent. First, the relative rates of migratory insertion of the anomers are remarkably different in each glycosyl family. For example, the α-anomers of the glucopyranosyl (**12**) and ribofuranosylmanganese (**15**) complexes are approximately an order of magnitude more reactive than the corresponding β-anomers. The difference in the relative rates of mannopyranosyl (**13**) and arabinofuranosyl (**14**) derivatives is even greater (see Table I). This extreme difference in insertion proclivity suggests to us that electronic features are influencing the rates of migratory insertion (*vide infra*).

Table I. Relative Rates of Migratory Insertion

	k_α / k_β
12	≈10
13	≈100
14	≈1000
15	≈10

A second trend was noted for the pyranosyl derivatives - the glucosyl (12) and mannosyl (13) series, respectively. While in each family the α-anomer was more reactive toward migratory insertion of carbon monoxide than the β-anomer (see Table I), the relative difference was ca. 10 in the glucosyl derivatives, but ca. 100 in the mannosyl complexes. We attribute the "hyper"-reactivity of the α-mannosyl complex to anchimeric stabilization of the transition state by the C-2 alkoxy substituent in the migratory insertion reaction (see Figure1). An analogous anchimeric assistance has been proposed to explain the preference for α-anomer formation in glycosidic coupling of mannosyl halides. (5)

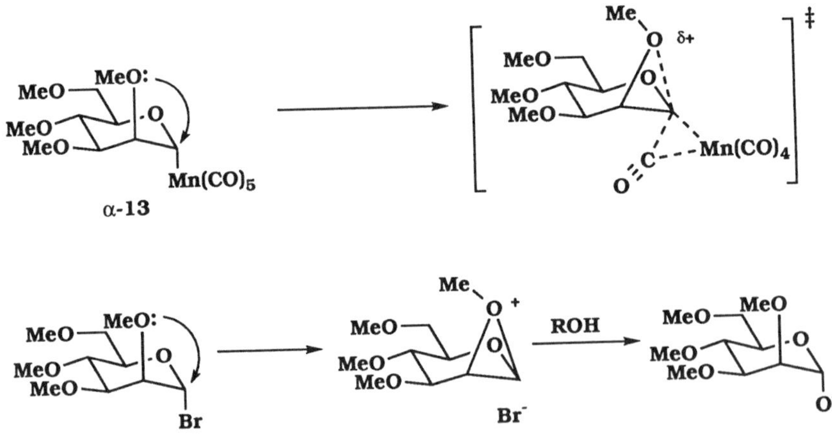

Figure 1

Based upon our preliminary kinetic data we hypothesized that the orientation of the lone pairs of the ring (or glycosidic) oxygen and the carbon-metal bond dramatically influenced the rate of migratory insertion of the complexes. The hypothesis is summarized in Scheme 5 for the glucopyranosyl complexes. For example, glucopyranosyl complex β-12 should exist in the chair conformation with the carbon-metal bond *gauche* to the pyranosidic lone pairs on oxygen. In this conformation, migratory insertion is slow. On the other hand, for the α-anomer in the chair conformation, the carbon-metal bond lies antiperiplanar to one lone pair of the pyranosidic oxygen and undergoes migratory insertion more rapidly. A similar situation can be envisioned for the other glycosyl derivatives studied.

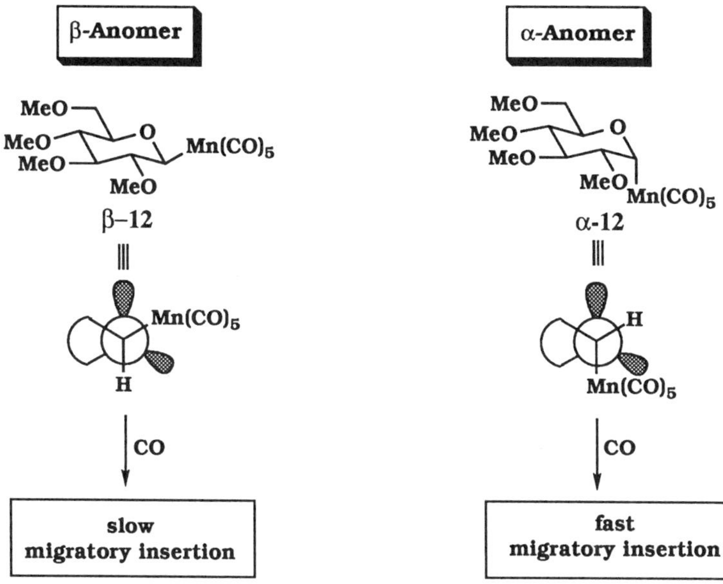

Scheme 5

One assumption implicit in this hypothesis is that the pyranosyl derivatives exist in chair conformations with the orientation of lone pairs of electrons and bonds as indicated in Scheme 5. To provide support for this hypothesis, we were, accordingly, compelled to unambiguously determine the preferred solution conformations of the respective glycosyl derivatives. As indicated in Schemes 6 and 7, the solution conformations of the anomers of the glucopyranosyl (**12**) and mannopyranosyl (**13**) complexes were established by 1D TOCSY and 2D NOESY NMR experiments.(6) These experiments conclusively demonstrated that as anticipated the β-anomers in each series adopted the chair conformation of the pyranosidic ring. As such, the carbon-manganese bond in question was *gauche* to the lone pairs on the pyranosidic oxygen in accord with the hypothesis outlined in Scheme 5. However, the α-anomers of the glucopyranosyl (α-**12**, Scheme 6) and mannopyranosyl derivatives (α-**13**, Scheme 7) did not adopt the anticipated chair conformations in solution. Analysis of the NMR data for gluco-complex α-**12** was consistent with an approximately 1:1 mixture of boat conformer **18** and chair conformer **19**. Similarly, the α-anomer of the mannosyl complex (α-**13**) exists in conformation **20**.

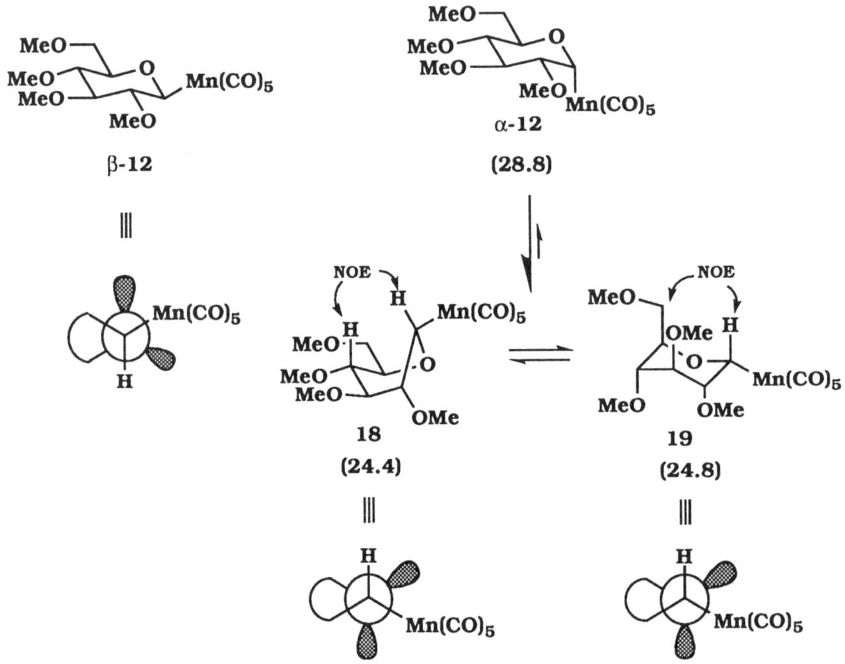

Scheme 6

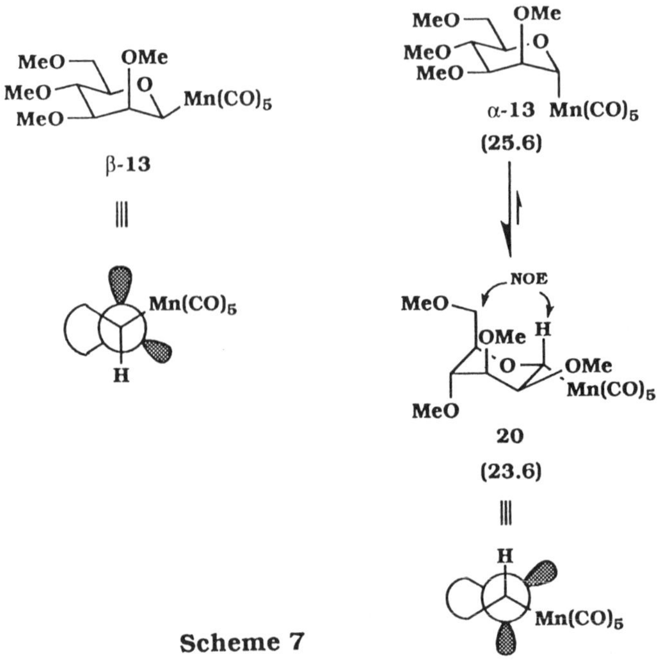

Scheme 7

Molecular mechanics calculations on these derivatives employing the augmented force field parameters of the CAChe molecular modeling system (7) were consistent with the results from the NMR experiments. The calculated energies of the respective global and local minima of the glycosyl conformers are indicated in Schemes 6 and 7 in parantheses.

Once the solution conformations of the pyranosyl complexes had been determined, it was clear that the original hypothesis for the differences in rate of migratory insertion was not tenable. As originally proposed, the β-anomers of **12** and **13** adopt the *gauche* orientation of lone pairs and carbon-manganese bond. However, the α-anomer of gluco-complex **12** exists as conformers **18** and **19** in which *the orientation of lone pairs and carbon-manganese bonds are also gauche* (see Scheme 6)! Therefore, by the original hypothesis (Scheme 5), migratory insertion of carbon monoxide in this anomer should not be increased in comparison to the β-anomer. An analogous situation exists for the mannosyl anomers (see Scheme 7). A revision of the hypothesis was required to accomodate the new data.

Careful analysis of the conformational situation has allowed us to retain the basic tenet of the hypothesis developed in Scheme 5: if the carbon-manganese bond is *gauche* to the lone pairs, the migratory insertion reaction is not enhanced. However, when the carbon-metal bond is either antiperiplanar or synperiplanar to a lone pair, the migratory aptitude in increased. Application of this hypothesis to the results in the gluco and manno series are summarized in Schemes 8 and 9.

For the β-anomer of glucopyanosylmanganese pentacarbonyl, the ground state conformation (β-**12**) is unreactive according to the hypothesis. One can envision that conformational mobility of the pyranosidic ring would provide either conformers **21** or **22**, both of which should display "enhanced" reactivity. However, molecular mechanics calculations (7) indicate that neither **21** nor **22** is accessible energetically ($\Delta\Delta E$ vs β-**12** > 6 kcal/mole in each instance). Accordingly, the insertion reaction is not accelerated, and anomer β-**12** is assigned a relative insertion rate of 1. The situation for the α-anomer is that conformers **18** and **19** (Scheme 6) are the predominant solution conformers and by the hypothesis are "unenhanced" due to the *gauche* orientation of lone pairs and bonds. However, conformer α-**12**, an "enhanced" orientation, lies only 4.4 kcal/mole above **18** and **19** and is conformationally accessible. It is the small amount of conformer α-**12** in the equilibrium (Curtin-Hammett principle) that is responsible for the increased reactivity of the α-anomer. Analogous reasoning can be applied to the situation in the mannosyl series.

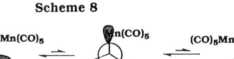

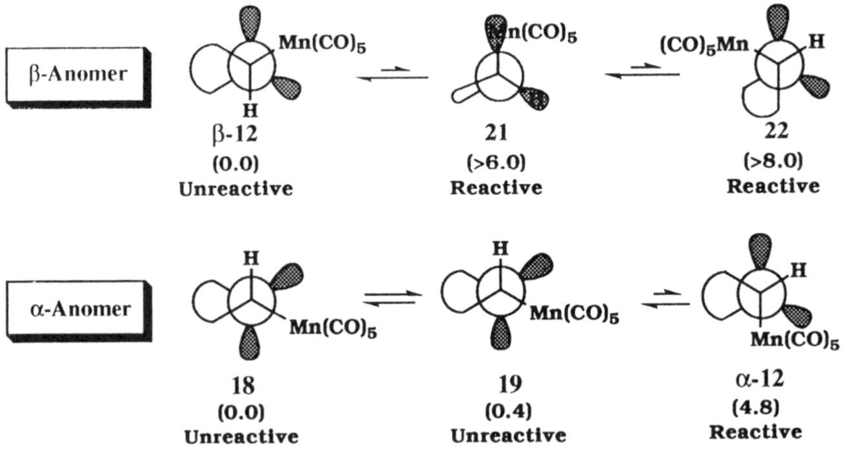
β-Anomer
Mn(CO)$_5$
H
β-12
(0.0)
Unreactive
Mn(CO)$_5$
21
(>6.0)
Reactive
(CO)$_5$Mn
H
22
(>8.0)
Reactive
α-Anomer
H
Mn(CO)$_5$
18
(0.0)
Unreactive
H
Mn(CO)$_5$
19
(0.4)
Unreactive
H
Mn(CO)$_5$
α-12
(4.8)
Reactive

Conclusion: α-anomer undergoes migratory insertion of CO with slight preference
Experimental Result: k_α/k_β = *ca.* 10

Scheme 8

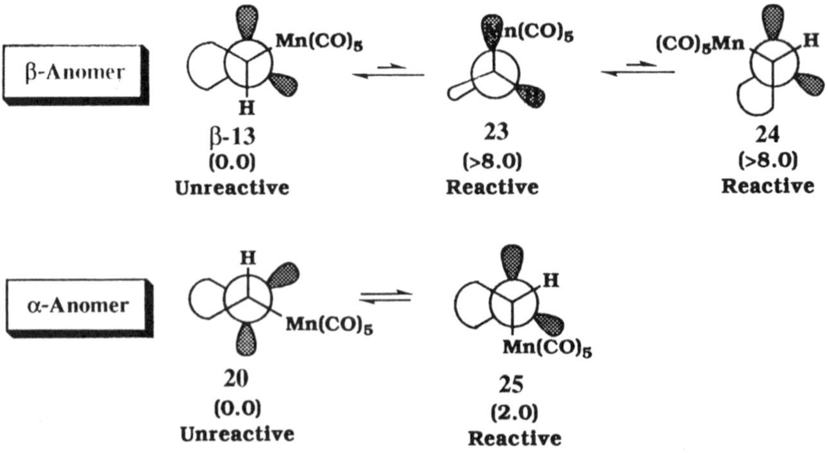

β-Anomer
Mn(CO)$_5$
H
β-13
(0.0)
Unreactive
Mn(CO)$_5$
23
(>8.0)
Reactive
(CO)$_5$Mn
H
24
(>8.0)
Reactive
α-Anomer
H
Mn(CO)$_5$
20
(0.0)
Unreactive
H
Mn(CO)$_5$
25
(2.0)
Reactive

Conclusion: α-anomer undergoes migratory insertion of CO with strong preference
Experimental Result: k_α/k_β = *ca.* 100

Scheme 9

This analysis also allows us to rationalize why the difference in rates in the gluco complexes (k_α/k_β = *ca.* 10) is less than in the manno derivatives (k_α/k_β = *ca.* 100). The energy difference between the ground state conformers of the α-gluco anomer (**18** and **19**) and the "reactive" conformers (α-**12**) is *ca.* 4.4 kcal/mole; while the $\Delta\Delta E$ between the "unreactive" conformer **20** and the "reactive" conformer **25** in the mannose family is only 2.0 kcal/mole. Conformer **25** assumes a more dominant role in the conformation equilibrium of the mannosyl series than conformer α-**12** has in the glucose system. Accordingly, the difference in rates (k_α/k_β) for the mannopyranosyl derivatives is greater than in the glucose derivatives.

The important question that is inherent in the hypothesis presented above is why is it that antiperiplanar and synperiplanar orientations lead to "enhanced" rates of migratory insertion? We propose that the rate enhancements are the result of electronic interactions between the lone pairs on oxygen and the carbon-metal bond. In the antiperiplanar arrangement, the predominant orbital interaction is between the filled n-orbital of oxygen and the empty σ*-orbital of the carbon-manganese bond. This interaction results in population of the σ*-orbital and weakening of the carbon-manganese bond. Since this bond is broken during migratory insertion, the net result is a lowered energy in the transition state for CO insertion.

The situation is altered for the synperiplanar orientation. In this case, interaction between the filled n-orbital on oxygen and the filled, high-lying σ-carbon-metal orbital results in destabilization of the system. In the transition state for migratory insertion in which the carbon-manganese bond is being broken, this destabilizing interaction with the synperiplanar lone pair on oxygen is diminished. Accordingly, the overall activation energy for migratory insertion is reduced.(8)

The kinetic and computational results presented in this forum are preliminary and require refinement. Migratory insertion reactions in these arnd related derivatives **26-29** under conditions in which precise kinetic data can be obtained are underway and shall be reported when complete. To date, the results from these experiments are qualitatively identical to those presented above and are consistent with the hypothesis outlined in Scheme 5.

Acknowledgments. We thank the National Institutes of Health (P. D.) and the Swiss National Science Foundation (W. v. P.) for generous financial support of this program. We also acknowledge numerous discussions with our colleagues, especially Bryan Eichhorn and Rinaldo Poli. Finally, P. D. wishes to thank CAChe Scientific, Inc. for the generous donation of software, hardware, advice, and intellectual stimulation during the course of this investigation.

References and Notes

1. (a) DeShong, P.; Slough, G.A. *Organometallics* **1984**, *3*, 636. (b) DeShong, P.; Slough, G. A.; Rheingold, A. L. *Tetrahedron Lett.* **1987**, *28*, 2229. (c) DeShong, P.; Slough, G.A.; Sidler, D. R. *Tetrahedron Lett.* **1987**, *28*, 2233. (d) DeShong, P.; Sidler, D.R.; Rybczynski, P. J.; Slough, G. A.; Rheingold, A. L. *J. Am. Chem. Soc.* **1988**, *110*, 2575. (e) DeShong, P.; Sidler, D. R. *J. Org. Chem.* **1988**, *53*, 4892. (f) DeShong, P.; Slough, G. A.; Sidler, D. R.; Rybczynski, P. J.; von Philipsborn, W.; Kunz, R.; Bursten, B. E.; Clayton, T. W. Jr. *Organometallics* **1989**, *8*, 1381.

2. (a) DeShong, P.; Slough, G. A.; Trainor, G. *J. Am. Chem. Soc.* **1985**, *107*, 7788. (b) DeShong, P.; Slough, G.A.; Elango, V. Carbohydr. *Res.* **1987**, *171*, 342.

3. Unpublished results of T.L. Lessen and T. X. Le.

4. Unpublished results of G. Anderson and T.X. Le.

5. Bochkov, A. F.; Zaikov, G. E. "Chemistry of the O-Glycosidic Bond". New York: Pergamon Press, 1979, pp. 5-80; and references cited therein.

6. Unpublished results of M. Vöhler, O. Zerbe, and W. von Philipsborn.

7. CAChe Scientific, Inc.; Beaverton, Oregon 97077.

8. Preliminary extended Hückel and ZNDO calculations (7) on model systems of related manganese complexes have provided results consistent with this interpretation. Unpublished results by P. DeShong.

RECEIVED June 25, 1993

Chapter 13

Do Stereoelectronic Effects Control the Structure and Reactivity of Trigonal-Bipyramidal Phosphoesters?

Philip Tole and Carmay Lim

Departments of Molecular and Medical Genetics, Chemistry, and Biochemistry, University of Toronto, 1 King's College Circle, Toronto, Ontario M5S 1A8, Canada

The stationary points for trigonal–bipyramidal transition states and intermediates of phosphate and phosphoamidate esters have been calculated using *ab initio* methods. The fully optimized geometries have been examined to establish whether stereoelectronic control through antiperiplanar lone pair effects can influence the structure and reactivity of these phospho–esters. The results show that although stereoelectronic effects cannot completely be ruled out, a more appropriate concept that is consistent with the results is conformational rearrangement to optimize electrostatic interactions. This is especially important in the long–range transition states corresponding to the rate–limiting hydroxyl ion attack of **methyl ethylene phosphate** and its acyclic analog, **trimethyl phosphate**. The results also show that stereoelectronic effects do not appear to play a significant role in enhancing the reactivity of five–membered cyclic esters relative to their six–membered or acyclic analogs.

It is very useful to identify general principles that govern chemical structure and reactivity which would allow us to construct predictive models. One such principle is stereoelectronic control of X–C–Y systems, where overlap between the antiperiplanar (app) lone pair on Y (n_Y) and the antibonding orbital of the C–X bond (σ^*_{C-X}) is thought to strengthen the C–Y bond and weaken the C–X bond; for reviews see (1–4). This concept has been extended to X–P–Y systems, originally by Lehn and Wipff (5) to account for the preferred conformations of phosphoric acid, and subsequently by Gorenstein and co–workers (1,6) to explain part of the enhanced reactivity of five–membered cyclic phospho–esters relative to their acyclic counterparts.

0097–6156/93/0539–0240$06.00/0

Based on semi–empirical CNDO/2 and Hartree–Fock/STO–3G calculations on acyclic trigonal-bipyramidal (TBP) neutral and dianionic dimethoxy phosphoranes, Gorenstein and co–workers (*6*) proposed that the lowest energy reaction path for acyclic and cyclic phospho–ester hydrolyses should have a TBP transition state with the equatorial oxygen lone pair(s) app (or anticlinal) to the apical leaving group. This $n-\sigma^*$ interaction would then cause the axial bond to weaken and facilitate its cleavage. It was stated that the activation entropy for acyclic phospho–ester hydrolysis, which required two rotational degrees of freedom to be fixed about the acyclic P–O(Me) bonds, should be more negative than that for hydrolysis of its cyclic counterpart, where the equatorial oxygen lone pairs were already partially app to the axial P–O ring bond due to the constraint of the ring. The authors (*6*) suggested that such stereoelectronic effects in the TBP transition states were consistent with the large entropic contribution (18 cal/mol/K) to the rate acceleration in ethyl propyl phosphonate observed by Aksnes and Bergesen (*7*).

However, Kluger and Taylor (*8*) measured an activation entropy for ethyl propyl phosphonate of –27 cal/mol/K. They found that the activation entropy differences between methyl esters of ethylene phosphate or propyl phosphonate and their acyclic analogs was less than 1 cal/mol/K. On the other hand the differences for the ethyl esters were about 8 cal/mol/K. The conclusion was that activation entropy differences stemming from stereoelectronic effects were not the source of the rate enhancement. Furthermore, the rate constant ratio of methyl ethylene phosphate (MEP)/phosphonate compared with the acyclic analogs is similar to the rate constant ratio of methyl propyl phosphonate/phosphinate to their acyclic counterparts (*8*). This is inconsistent with stereoelectronic theory which would predict that methyl ethylene phosphonate should have a larger cyclic/acyclic rate constant ratio than methyl propyl phosphinate whose TBP transition state has no equatorial app lone pairs to stereoelectronically aid in ring opening.

Doubts about the stereoelectronic control of TBP phosphorane transition states have also been raised due to the fact that CNDO/2 calculations do not reproduce the anomeric effect at the carbon centre (*9*) and the STO–3G basis does not predict the relative stability of TBP phosphorane structures (*10,11*). When Gorenstein and co–workers (*6*) suggested that stereoelectronic effects were responsible for the extra stabilization of five–membered cyclic versus acyclic phosphate diester transition states, ring opening was implicitly assumed to be the rate–determining step. However, calculations on the alkaline hydrolysis of ethylene phosphate (*10*), MEP (*12, 13*), dimethyl phosphate (*11*) and trimethyl phosphate (TMP) (Tole, P.; Lim, C. *manuscript in prep.*) established that hydroxyl ion attack is rate–limiting. Thus, stereoelectronic effects originating from one app lone pair or two anticlinal lone pairs on the equatorial ring atom in the TBP transition state cannot control the observed rate acceleration.

In previous work we have employed quantum mechanical and continuum dielectric calculations to map out the detailed gas–phase and solution reaction profiles for the alkaline hydrolysis of MEP (*12, 13*) and its acyclic analog, TMP (Tole, P.; Lim, C. *manuscript in prep.*). The geometries of TBP transition states and/or intermediates formed from hydrolysis of five–membered cyclic MEP, methyl aminoethylene phosphonate MNP (MEP with one of the ring oxygens replaced by NH) and acyclic TMP have been fully optimized using *ab initio* methods. The next section briefly outlines these methods. In the results and discussion section the TBP structures are examined to determine the factors, in particular stereoelectronic effects arising from $n-\sigma^*$ interactions, that affect their conformation and stability.

Method

Details of the method have been described in (*14*) and only a brief summary is given here. The *ab initio* calculations were performed using the program Gaussian 90 (*15*). All geometries were fully optimized at the Hartree–Fock (HF) level using the 3–21+G* basis set, which has 1s orbitals on hydrogens, 1s, 2s, $2p_x$, $2p_y$, $2p_z$ orbitals on carbon and oxygen atoms, 1s, 2s, $2p_x$, $2p_y$, $2p_z$, 3s, $3p_x$, $3p_y$, $3p_z$, $3d_{xx}$, $3d_{yy}$, $3d_{zz}$, $3d_{xy}$, $3d_{yz}$, $3d_{zz}$ orbitals on phosphorus as well as diffuse s, p_x, p_y, p_z functions on all atoms except hydrogens. The geometries of cyclic and acyclic tetrahedral phospho–esters have been found to be in good agreement with experiment (*13*). The electronic correlation energy was estimated by second–order Møller–Plesset perturbation theory (MP2) with the 6–31+G* basis for single points corresponding to the 3–21+G* optimized geometries. The electrostatic potential was calculated from the HF/3–21+G* wavefunctions and fitted to an atomic point charge model using the CHELP program (*16*). The stereoelectronic effect was analyzed in terms of $n-\sigma^*$ interactions where the lone pairs are either sp³ or sp² hybrids, as indicated by the X–O–P (X= C, H) angle.

Results and Discussion

The following nomenclature is adopted in this paper. The first letter of each structure defines the starting material with M, A and T representing MEP, methyl aminoethylene phosphonate and acyclic TMP respectively. The second letter denotes either a TBP intermediate (I) or transition state (TS) with a superscript a/e indicating an apical/equatorial hydroxyl group. The third letter describes the nature of the complex: L denoting a long–range ion–dipole interaction where the P–O(H) bond is not fully formed (a superscript N denotes an apical ring nitrogen), P for pseudorotation, H and Me for hydroxyl and methoxy group rotation about P–O(H/Me), N and X for endo- and exo–cyclic cleavage. Table I contains the HF and MP2 energies and the thermodynamic parameters. The structural information is listed in Tables II–IV where the 2' and 3' denote an oxygen or nitrogen in the axial 2' or equatorial 3' position. The Mulliken and CHELP atomic charges are given in Tables V and VI, respectively.

The results will be discussed in the following order: (i) the long–range transition state complexes between the reactants that determine the rate–limiting

Table I. Energies and Thermodynamic Parameters

	$E_{HF}{}^a$	$E_{MP2}{}^b$	$E_{TRV}{}^c$	$S_{TRV}{}^d$
	atomic units (a.u.)		kcal/mol	cal/mol/K
MEP	-754.13769	-759.15445	82.196	92.391
MNP	-734.40294	-739.30189	90.499	90.957
TMP	-755.31038	-760.32676	96.903	101.807
MTS^{a}L	-829.15439	-834.76724	89.774	97.420
ATS^{a}L	-809.44313	-814.94414	98.392	95.836
ATS^{a}L^N	-809.40901	-814.90689	97.935	98.653
TTS^{a}L	-830.33481	-835.94541	104.632	108.273
MTS^{a}P	-829.20178	-834.81786	90.763	91.328
MI^{e}N	-829.21132	-834.82843	91.835	92.857
MTS^{e}H	-829.19343	-834.81163	90.389	93.157
MI^{e}X	-829.21107	-834.82866	91.846	92.429
AI^a	-809.46421	-814.96539	99.882	93.687
AI^{e}N	-809.47454	-814.97331	100.107	92.900
ATS^{e}H	-809.45551	-814.95770	98.751	93.060
AI^{e}X	-809.46705	-814.97009	99.927	93.922

a) HF energies for fully optimized 3-21+G* geometries.
b) Single point MP2/6-31+G* energy calculation at 3-21+G* geometry.
c) $E_{TRV} = E_{trans} + E_{rot} + E_{vib}$ in kcal/mole.
d) $S_{TRV} = S_{trans} + S_{rot} + S_{vib}$ in cal/mol/K.

Table II. HF/3–21+G* Bond lengths (Å)

	P O1	P O2	P 2'	P 3'	P O5'	C2' 2'	C3' 3'	C(Me) O	C2' C3'
MEP	1.46	1.57	1.60	1.60	–	1.47	1.47	1.47	1.55
MNP	1.47	1.58	1.64	1.60	–	1.48	1.47	1.47	1.55
TMP	1.47	1.57	1.59	1.57	–	1.46	1.47	1.47	–
MTS^{a}L	1.46	1.56	1.65	1.61	2.92	1.45	1.45	1.48	1.55
ATS^{a}L	1.48	1.59	1.65	1.65	2.53	1.45	1.46	1.46	1.54
ATS^{a}L^N	1.47	1.57	1.69	1.61	2.90	1.47	1.46	1.48	1.54
TTS^{a}L	1.48	1.59	1.58	1.62	2.59	1.44	1.46	1.46	–
MTS^{a}P	1.51	1.63	1.79	1.66	1.69	1.42	1.44	1.44	1.54
MI^{e}N	1.50	1.63	1.82	1.67	1.67	1.42	1.44	1.44	1.54
MTS^{e}H	1.51	1.65	1.76	1.68	1.69	1.43	1.44	1.44	1.54
MI^{e}X	1.50	1.64	1.73	1.68	1.73	1.43	1.44	1.43	1.54
AI^a	1.52	1.63	1.74	1.69	1.78	1.43	1.46	1.44	1.54
AI^{e}N	1.51	1.65	1.81	1.69	1.70	1.42	1.46	1.43	1.54
ATS^{e}H	1.51	1.65	1.75	1.68	1.76	1.43	1.46	1.43	1.54
AI^{e}X	1.51	1.63	1.73	1.70	1.80	1.43	1.46	1.43	1.54

Table III. HF/3-21+G* Bond Angles (Degrees)

| A | O1 | O2 | 3' | O1 | O2 | 2' | 3' | O1 | O1 | O2 | P | P | P | P | 2' | 3' |
| B | P | P | P | P | P | P | P | P | P | P | O | 2' | 3' | O | C2' | C3' |
C	2'	2'	2'	O5'	O5'	O5'	O5'	O2	3'	3'	C(Me)	C2'	C3'	H	C3'	C2'
MEP	116.9	105.5	95.14	–	–	–	–	114.2	118.1	104.6	129.0	113.6	114.7	–	103.6	104.0
MNP	118.8	107.4	94.29	–	–	–	–	111.9	117.9	104.6	126.7	114.1	114.9	–	102.7	105.3
TMP	114.8	102.5	105.8	–	–	–	–	116.6	113.1	102.5	126.5	128.0	125.7	–	–	–
MTS^{a}L	111.6	97.87	91.98	89.70	73.07	158.4	73.84	120.2	118.5	110.8	130.9	115.1	112.2	110.0	104.4	103.5
ATS^{a}L	109.7	97.03	91.58	89.96	82.37	158.3	69.16	114.5	123.8	113.4	126.7	114.9	115.9	115.2	105.5	102.4
ATS^{a}L^N	111.5	100.6	91.22	87.84	74.88	159.6	72.49	116.0	120.8	110.7	129.9	113.5	116.4	105.1	101.2	105.9
TTS^{a}L	106.3	95.72	95.54	82.32	79.09	171.4	79.56	120.8	121.8	109.3	134.1	124.3	135.5	101.4	66.66	15.31
MTS^{a}P	96.20	85.98	84.32	98.38	88.59	165.3	86.68	120.0	119.3	120.6	128.4	111.4	119.9	110.6	101.9	104.5
MI^{e}N	96.82	82.26	83.83	102.2	87.15	160.9	88.77	122.5	113.9	123.0	121.5	111.0	119.8	110.8	102.6	105.2
MTS^{e}H	90.90	84.82	83.70	102.5	86.78	161.6	85.34	117.3	114.1	128.5	122.0	111.8	119.0	119.1	102.6	104.5
MI^{e}X	99.41	85.22	86.10	99.17	84.31	161.4	86.65	123.2	114.3	122.5	122.1	111.3	117.9	110.4	102.6	104.4
AI^a	98.73	86.66	85.59	93.82	89.75	167.2	85.10	118.1	124.9	117.0	125.3	114.1	120.3	109.5	104.2	101.6
AI^{e}N	98.41	86.99	83.95	97.76	89.66	163.3	83.44	115.3	125.7	119.0	121.5	112.6	121.0	117.6	103.7	102.0
ATS^{e}H	98.27	87.42	85.00	96.66	90.37	164.5	82.69	113.4	127.6	119.0	122.5	114.0	121.1	114.6	104.0	102.0
AI^{e}X	99.58	87.29	86.18	96.24	84.14	164.2	86.18	120.8	120.7	118.4	122.0	112.6	119.3	110.2	104.2	101.8

The rows A, B, C give the atom connectivities for the respective bond angles in degrees.

Table IV. HF/3-21+G* Dihedral angles (Degrees)

A	O1	O1	O2	O2	2'	2'	3'	P	P	O1	O2	2'	3'	O5'	O5'	O1
B	P	P	P	P	P	C2'	2'	2'	3'	P	P	P	P	P	P	P
C	2'	3'	2'	3'	3'	C3'	C2'	C2'	C3'	O	O	O	O	2'	3'	O
D	C2'	C3'	C2'	C3'	3'	3'	3'	C3'	C2'	C(Me)	C(Me)	C(Me)	C(Me)	C2'	C3'	H
MEP	-141.2	120.1	90.67	-111.6	-4.19	-29.09	-15.97	29.67	20.95	3.78	–	133.5	-126.8	–	–	–
MNP	-139.2	119.2	92.60	-115.2	-6.07	-29.14	-14.08	27.57	22.56	28.95	–	161.0	-99.79	–	–	–
TMP	170.4	36.13	42.92	162.6	-90.35	–	-64.17	–	–	48.66	–	175.0	-75.44	–	–	–
MTS^{a}L	-134.4	146.8	98.66	-68.33	30.86	26.64	-12.61	-7.43	-38.44	37.74	–	158.4	-106.4	35.37	-132.6	71.75
ATS^{a}L	119.7	-129.3	-121.1	84.14	-14.21	-30.87	-7.34	24.45	28.83	-38.96	–	-154.4	110.8	-34.11	155.5	-6.19
ATS^{a}L^N	-146.7	116.7	88.64	-101.5	0.614	-31.78	-22.65	34.96	19.41	41.84	–	162.7	-101.9	13.70	-166.9	80.19
TTS^{a}L	-179.8	-50.62	55.69	98.24	-163.7	-19.36	-54.40	35.94	-168.2	45.86	–	158.7	-103.4	3.02	23.68	-1.82
MTS^{a}P	-82.76	78.99	157.5	-96.87	-14.96	31.55	36.18	-45.17	-7.28	-73.02	–	21.83	102.8	88.83	176.7	-11.53
MI^{e}N	-75.96	75.93	162.0	-95.41	-18.81	28.39	37.43	-43.92	-2.07	47.61	170.3	-133.5	-66.56	105.2	178.8	-85.11
MTS^{e}H	-78.77	80.69	164.3	-95.72	-16.21	29.67	36.33	-44.66	-5.44	56.00	173.3	-120.7	-59.32	97.98	-179.0	1.21
MI^{e}X	-77.31	82.05	159.8	-98.33	-16.61	30.03	36.74	-44.93	-5.58	50.20	172.9	-131.1	-99.60	104.0	-179.4	88.38
AI^a	98.54	-94.87	-143.6	86.67	2.70	-33.58	-26.08	40.41	18.05	-56.50	–	-154.6	122.1	-69.61	173.9	-2.90
AI^{e}N	86.64	-84.00	-155.6	89.86	11.30	-31.96	32.53	43.56	10.08	48.92	-67.68	-133.2	170.3	-89.22	176.2	84.94
ATS^{e}H	100.2	-92.32	-146.5	88.54	4.28	-32.89	-27.17	40.72	16.46	38.92	–	-156.4	166.1	-64.49	174.9	-16.61
AI^{e}X	90.06	-91.14	-149.1	92.58	7.734	-33.45	-30.44	43.42	14.07	-28.01	-148.5	153.8	92.48	-91.74	173.9	-84.05

The rows A, B, C, D give the atom connectivity associated with each dihedral angle in degrees.

Table V. Mulliken Atomic Charges

	P	O1	O2	2′	3′	O5′
MEP	2.13	-0.76	-0.52	-0.47	-0.49	–
MNP	2.19	-0.79	-0.49	-0.51	-0.86	–
TMP	2.20	-0.81	-0.52	-0.57	-0.49	–
MTS^{a}L	2.12	-0.76	-0.50	-0.51	-0.42	-1.35
ATS^{a}L	2.18	-0.84	-0.46	-0.52	-0.74	-1.29
ATS^{a}L^N	2.14	-0.77	-0.47	-0.87	-0.40	-1.33
TTS^{a}L	2.21	-0.82	-0.56	-0.45	-0.55	-1.28
MTS^{a}P	2.11	-1.02	-0.45	-0.50	-0.45	-0.95
MI^{e}N	2.21	-0.99	-0.90	-0.63	-0.53	-0.46
MTS^{e}H	2.11	-1.03	-0.92	-0.51	-0.48	-0.41
MI^{e}X	2.22	-0.99	-0.90	-0.52	-0.52	-0.57
AI^a	2.36	-1.02	-0.46	-0.46	-0.79	-1.04
AI^{e}N	2.43	-0.98	-0.98	-0.58	-0.78	-0.48
ATS^{e}H	2.38	-1.00	-0.98	-0.45	-0.79	-0.47
AI^{e}X	2.35	-0.95	-0.92	-0.51	-0.86	-0.63

Table VI. CHELP Charges

	P	O1	O2	2′	3′	O5′
MEP	2.19	-0.97	-0.86	-0.79	-0.82	–
MNP	2.14	-1.00	-0.89	-0.73	-1.12	–
TMP	2.26	-1.01	-0.81	-0.79	-0.88	–
MTS^{a}L	2.37	-1.04	-0.89	-0.88	-0.74	-1.20
ATS^{a}L^N	2.46	-1.10	-0.91	-1.25	-0.67	-1.15
ATS^{a}L	2.29	-1.09	-0.89	-0.93	-0.91	-1.10
TTS^{a}L	2.27	-1.03	-0.79	-0.79	-0.84	-1.00
MTS^{a}P	2.48	-1.15	-0.95	-0.79	-0.77	-1.09
MI^{e}N	2.50	-1.12	-1.05	-1.01	-0.77	-0.98
MTS^{e}H	2.56	-1.17	-1.02	-0.95	-0.70	-1.05
MI^{e}X	2.42	-1.11	-0.97	-0.83	-0.77	-0.97
AI^a	2.47	-1.17	-0.93	-0.85	-1.23	-1.09
AI^{e}N	2.28	-1.10	-0.87	-0.80	-1.09	-1.05
ATS^{e}H	2.57	-1.20	-1.08	-0.91	-1.25	-1.04
AI^{e}X	2.43	-1.16	-1.04	-0.80	-1.09	-0.99

step; (ii) the TBP phosphorane complexes formed from nucleophilic addition of hydroxyl ion to MEP and (iii) the TBP complexes from (ii) with the basal oxygen replaced by NH. Since Mulliken population analysis, in particular overlap populations and dipole moments are very sensitive to the system and basis set, it has not been relied upon in the analysis of the results. In particular, the Mulliken (Table V) and CHELP (Table VI) atomic charges do not always show the same qualitative trends (see below); thus, care in general must be exercised in employing such charges to arrive at conclusions.

Long–range Transition States.   Figure 1 depicts four transition states formed by rate–limiting hydroxyl ion attack on MEP (MTS^{a}L), methyl aminoethylene phosphonate (ATS^{a}L, ATS^{a}L^N) and TMP (TTS^{a}L). In order to accommodate the incoming hydroxyl ion, the phosphorus centre changes from a tetrahedral ligand geometry in the reactants (average O–P–O angle of 109°) to a distorted TBP geometry in the long–range transition states (Figure 1) with average O^e–P–2', O^e–P–O5' and O^e–P–O^e angles of 99–100°, 78–81° and 116–117°, respectively, where O^e denotes an equatorial oxygen atom (see Table III). The geometry changes, *e.g.;* elongation of the P–O/N bond length, angle deformation, rotation of the basal methoxy groups towards the nucleophile, are consistent with minimization of the electrostatic repulsion between the negatively charged hydroxyl oxygen and the basal atoms directly bonded to the phosphorus centre. The location of the ion–dipole transition state complex varies among the four species. The MTS^{a}L and ATS^{a}L^N have the longest reactant separation since the two negatively charged basal oxygens (O1 and O3'), whose movements are restricted by the ring, repel the approach of the hydroxyl ion. Without the constraint of a ring in TTS^{a}L, both basal methoxy groups can rotate about the P–O(Me) bond towards the hydroxyl ion; this serves two purposes: (i) it places the lone pairs of the basal oxygens away from the nucleophile and (ii) the proximity of *both* methyl groups reduces the negative charge density on the hydroxyl oxygen (Tables V and VI) and the P–O(H) distance relative to its cyclic MTS^{a}L analog. In ATS^{a}L the nucleophile is attracted to a positively charged hydrogen of the basal amide so that the P–O(H) distance is reduced relative to MTS^{a}L.

In the long–range transition states (Figure 1), the basal methoxy group, whose O–C(Me) bond is gauche to the P–O(H) bond, has a lone pair that is app to the P–O(H) bond. This apparent stereoelectronic effect is similar to that found in the acyclic transition states resulting from hydroxyl ion attack of dimethyl phosphate (*11,17,18*). In the acyclic TTS^{a}L transition state, the second basal methoxy group has a lone pair anticlinal to the P–O(H) bond and the apical methoxy oxygen has a lone pair app to the P–O3' bond. For optimal stereoelectronic control one would have expected the O3'–C bond to be oriented *gauche* instead of *cis* to the P–O(H) bond so that the O3' lone pair is app instead of anticlinal to the P–O(H) bond. Furthermore, the relative axial P–O(ring) and P–O(H) bond lengths in the long–range transition states (Figure 1) do not appear to be consistent with stereoelectronic effects. For example, a lone pair

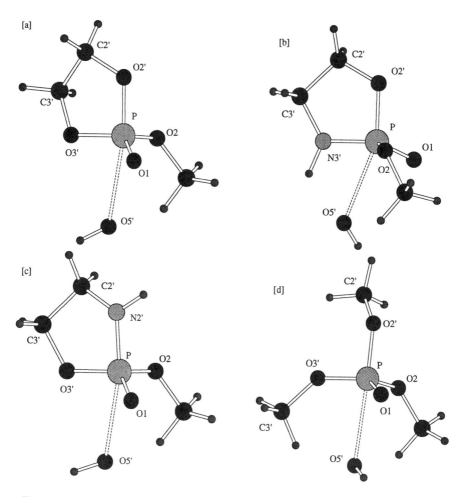

Figure 1. Long–range transition states formed by rate–limiting hydroxyl ion attack on **methyl ethylene phosphate** (MTS^{a}L) [a], **methyl aminoethylene phosphonate** opposite the ring oxygen (ATS^{a}L) [b], and opposite the ring nitrogen (ATS^{a}L^N) [c], and **trimethyl phosphate** (TTS^{a}L) [d].

on the basal ring atom is app to the axial P–O(ring) bond in MTS^{a}L but not in ATS^{a}L; yet, the axial P–O(ring) bond lengths in MTS^{a}L and ATS^{a}L are similar (1.65 Å). The stereoelectronic arguments of Gorenstein and co–workers (1,6) do not seem to contribute significantly to the observed enhanced reactivity of five–membered cyclic phospho–esters relative to their acyclic analogs. There appears to be no entropic advantage stemming from stereoelectronic effects in the cyclic long–range transition states since both MTS^{a}L and TTS^{a}L require a rotational degree of freedom about the P–O(Me) bond to be frozen so that a lone pair on the methoxy oxygen is app the P–O(H) bond. This is supported by the negligible activation entropy difference at 298 K (0.4 kcal/mole) between MEP and TMP alkaline hydrolysis (Tole, P.; Lim, C. *manuscript in prep.*), which is in accord with experiment (8).

TBP Phosphorane Complexes from MEP alkaline hydrolysis. The observed product distribution in the hydrolysis of MEP as a function of pH (19,20) was originally based upon a TBP phosphorane with the nucleophile in an apical position, which could undergo endo–cyclic cleavage directly or exo–cyclic cleavage via a pseudorotation process. However, *ab initio* calculations (12, 13) found a TBP phosphorane MTS^{a}P with an apical hydroxyl group (Figure 2a) to be unstable since it could spontaneously pseudorotate to yield a lower energy phosphorane with the hydroxyl group equatorial. The MTS^{a}P structure in Figure 2a was found to be a transition state characteristic of the pseudorotation process (12). Calculations of the gas–phase and solution activation free energy profiles for the alkaline hydrolysis of MEP showed that the observed products do not depend on the relative rates of ring–opening and pseudorotation of a TBP phosphorane with an apical hydroxyl group. Instead, the product distribution depends on the endo–cyclic cleavage rate of the pseudorotated phosphorane MI^{e}N (Figure 2b) relative to its isomerization rate via MTS^{e}H (Figure 2c) to MI^{e}X (Figure 2d) and the exo–cyclic cleavage rate of MI^{e}X (13). The calculations (13) showed that formation of MI^{e}N was rate–limiting and its ring opening was faster than exo–cyclic cleavage of MI^{e}X, consistent with the observed absence of exo–cyclic cleavage products (19).

The factors that affect P–O(H) rotation, endo–cyclic and exo–cyclic cleavage can be rationalized in terms of electrostatic effects. The MI^{e}N or MI^{e}X intermediates are connected by two isoenergetic MTS^{e}H transition states with P–O(H) *cis* (Figure 2c) or *trans* to P–O1. When the O–H bond is *cis* to the P–O bond to be cleaved, the latter becomes weaker (longer axial P–O bond length) in MI^{e}N and MI^{e}X relative to MTS^{e}H. The proximity of the hydroxyl hydrogen to the departing oxygen facilitates intramolecular hydroxyl hydrogen transfer to the departing oxygen. This will incur less entropic loss compared to intermolecular proton transfer from water molecules. It also stabilizes the endo–cyclic and exo–cyclic cleavage transition states via favourable electrostatic interactions between the hydroxyl hydrogen and the developing charge on the departing oxygen in the transition state. Ring opening of MI^{e}N is faster than exo–cyclic cleavage of MI^{e}X since the axial P–O bond in the ring is longer and

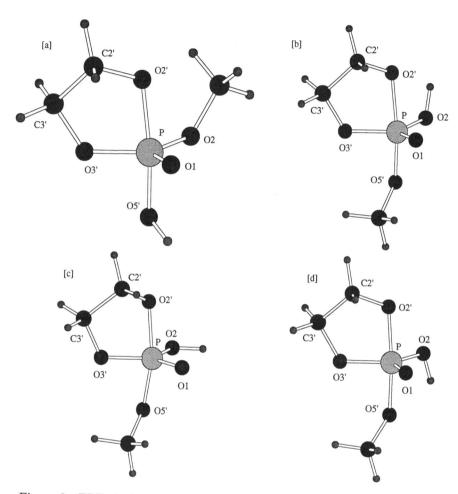

Figure 2. TBP stationary points along the reaction pathway for the alkaline hydrolysis of **methyl ethylene phosphate**: the pseudorotation transition state (MTS^{a}P) [a], the endo–cyclic cleavage intermediate (MI^{e}N) [b], transition state for OH rotation (MTS^{e}H) [c], and the exo–cyclic cleavage intermediate (MI^{e}X) [d].

weaker than that outside the ring due to ring strain in MI^{e}N (13). The latter is evidenced by the smaller O2$'$–P–O3$'$ (84°) and C2$'$–O2$'$–P (111°) angles in MI^{e}N compared to the corresponding angles (89° and 123°) in the acyclic analog.

The relative P–O bond lengths in MI^{e}N (Figure 2b), MTS^{e}H (Figure 2c) and MI^{e}X (Figure 2d) correlate with the number of app lone pairs, consistent with stereoelectronic theory. In MI^{e}N (Figure 2b) there are two lone pairs (one on the basal sp^2–hybridized ring oxygen and one on the sp^3–hybridized hydroxyl oxygen) anticlinal to the axial P–O(ring) bond and no lone pairs app or anticlinal to the axial P–O(Me) bond. Thus, the P–O2$'$ bond (1.82 Å) is longer than the P–O(Me) bond (1.67 Å) and the charge on O2$'$ is greater than that on O3$'$ or O(Me) (Tables V and VI). MTS^{e}H (Figure 2c) differs from MI^{e}N (Figure 2b) in that the hydroxyl oxygen is sp^2–hybridized with a lone pair in the basal plane; thus, the axial P–O(ring) bond (1.76 Å) is shorter and the charge on O2$'$ is less than that in MI^{e}N (Tables V and VI). In MI^{e}X (Figure 2d) a lone pair on the basal sp^2–hybridized ring oxygen is anticlinal to the P–O2$'$ bond and a lone pair on the sp^3–hybridized hydroxyl oxygen is anticlinal to the P–O(Me) bond. Thus, the stereoelectronic effects cancel as manifested by the equal axial P–O bond lengths (1.73 Å). The P–O(Me) bond length in MI^{e}X (1.73 Å) is longer than that in MTS^{e}H (1.69 Å) and MI^{e}N (1.67 Å) where there are no lone pairs app to P–O(Me). Although the Mulliken charge on the methoxy oxygen in MI^{e}X (–0.57e) is greater than that in MI^{e}N (–0.46e) and MTS^{e}H (–0.41e), as expected from stereoelectronic theory, the O(Me) CHELP charge in MI^{e}X (–0.97e) is in contrast less than that in MI^{e}N (–0.98e) and MTS^{e}H (–1.06e). Stereoelectronic effects stabilize the transition states for the endo–cyclic and exo–cyclic cleavage and appear consistent with their relative rates; the axial P–O(ring) bond, which is app to two lone pairs, is weaker and thus requires less activation energy than the P–O(Me) bond which is app to only one lone pair.

TBP Phosphorane Complexes with a basal NH group. To separate the stereoelectronic effects originating from the basal ring and hydroxyl oxygen atoms, the structures in Figures 2a–2d were re–optimized by replacing the basal ring oxygen with an NH group which introduces minimal steric perturbations. The corresponding structures are shown in Figures 3a–3d. With a more electropositive NH group in the basal ring position the incoming OH nucleophile can be stabilized in an axial position to form a metastable AI^a intermediate. The basal nitrogen is sp^2–hybridized with no lone pair app to the axial P–O(ring) bond. Stereoelectronic effects stemming from the lone pair on the O(H) correlate with the relative axial P–O bond lengths in AI^{e}N (Figure 3b), ATS^{e}H (Figure 3c) and AI^{e}X (Figure 3d). In ATS^{e}H there are no lone pairs app or anticlinal to the axial P–O bonds which have almost equal bond lengths (1.75–1.76 Å). In AI^{e}N (Figure 3b) and AI^{e}X (Figure 3d) a lone pair on the sp^3–hybridized hydroxyl oxygen is anticlinal to the axial P–O(ring) bond and P–O(Me) bond respectively, which are 0.06–0.04 Å longer and weaker than the corresponding bond in ATS^{e}H.

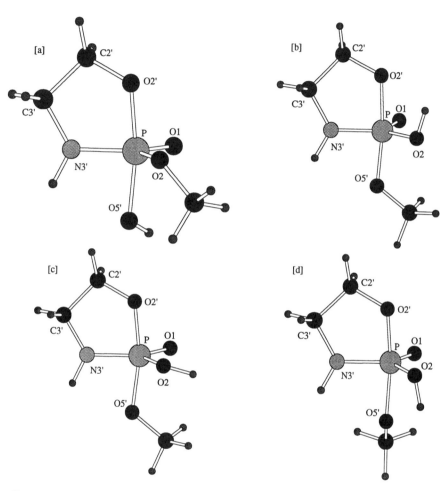

Figure 3. TBP stationary points along the reaction pathway for the alkaline hydrolysis of **methyl aminoethylene phosphonate**: the metastable intermediate (AI[a]) [a], the endo–cyclic cleavage intermediate (AI[e]N) [b], transition state for OH rotation (ATS[e]H) [c], and the exo–cyclic cleavage intermediate (AI[e]X) [d].

However, the axial P–O bond lengths in Figures 2a–2c are similar to the corresponding distances in Figures 3a–3c although the axial P–O(ring) bond is anticlinal to a basal ring oxygen lone pair in Figures 2a–2c and to no lone pairs in Figures 3a–3c. This is analogous to the situation in MTS^{a}L (Figure 1a) and ATS^{a}L (Figure 1b), as discussed in the section on long–range transition states. Thus, stereoelectronic effects arising from the basal ring oxygen do not seem to be significant. Although app effects originating from O(H) correlate with the lowest energy pathway for endo–cyclic and exo–cyclic cleavage, they do not control the observed rate enhancement of MEP relative to its acyclic analog. Otherwise, a similar rate enhancement might be expected in the six–membered ring analog methyl propylene phosphate.

Concluding Remarks

There are 3 key points that can be derived from the results described.

1. Stereoelectronic effects do not control the reactivity of five–membered cyclic and acyclic phospho–esters since there is no entropic advantage originating from app lone pairs in the rate-determining cyclic long–range transition states. The structure of the long–range transition states does not seem to be governed by stereoelectronic effects alone but appears consistent with the minimization of the electrostatic repulsion between the incoming nucleophile and the reactant.

2. Stereoelectronic effects, however, cannot be ruled out completely. The lowest energy pathway for endo–cyclic and exo–cyclic cleavage requires the O–H bond to be cis to the departing P–O bond. This is consistent with stereoelectronic effects stemming from an anticlinal lone pair on the basal hydroxyl oxygen rather than the basal ring atom.

3. The magnitude of the app contributions is very dependent on the substituent and is likely to be altered by other interactions, *e.g.*; steric, solvent. A more valid concept that appears to be consistent with the results would be conformational rearrangements that maximize favourable electrostatic interaction and minimize unfavourable electrostatic repulsion. However, any electrostatic and/or app lone pair interactions will be attenuated in solution by intermolecular water interactions and solvent effects could contribute significantly to the relative stability of the long–range transition states (*12, 13*).

Acknowledgments

This work was supported by the Protein Engineering Network Centre of Excellence of Canada.

Literature Cited

(*1*) Gorenstein, D. G. *Chem. Rev.* **1987**, *87*, 1047.
(*2*) Juaristi, E.; Cuevas, G. *Tetrahedron* **1992**, *48*, 5019.
(*3*) Kirby, A. J. *Concepts in Organic Chem.*; Springer Verlag: Berlin, 1983, Vol.15.

(*4*) Deslongchamps, P. *Stereoelectronic effects in Organic Chemistry*; Pergamon: New York, NY, 1983.

(*5*) Lehn, J. M.; Wipff, G. H. *J. Chem. Soc. Chem. Commun.* **1975**, *99*, 800.

(*6*) Gorenstein, D. G.; Luxon, A. L.; Findley, J. B.; Momii, R. *J. Am. Chem. Soc.* **1977**, *99*, 4170.

(*7*) Aksnes, G.; Bergesen, K. *Acta Chem. Scand.* **1966**, *30*, 2508.

(*8*) Kluger, R.; Taylor, S. *J. Am. Chem. Soc.* **1990**, *112*, 6669.

(*9*) Sinnott, M. L. *Adv. Phys. Org. Chem.* **1988**, *24*, 113.

(*10*) Lim, C.; Karplus, M. *J. Am. Chem. Soc.* **1990**, *112*, 5872.

(*11*) Dejaegere, A.; Lim, C.; Karplus, M. *J. Am. Chem. Soc.* **1991**, *113*, 4355.

(*12*) Lim, C.; Tole, P. *J. Phys. Chem.* **1992**, *96*, 5217.

(*13*) Tole, P.; Lim, C. *J. Phys. Chem.* **1993**, to be published.

(*14*) Lim, C.; Tole, P. *J. Am. Chem. Soc.* **1992**, *114*, 7245.

(*15*) Frish, M. J.; Head–Gordon, M.; Trucks, G. W.; Foresman, J. B.; Schlegel, H. B.; Raghavachari, K.; Robb, M.; Binkley, J. S.; Gonzalez, C.; Defrees, D. J.; Fox, D. J.; Whiteside, R. A.; Seeger, R.; Melius, C. F.; Baker, J.; Martin, R. L.; Kahn, L. R.; Stewart, J. J. P.; Topiol, S.; Pople, J. A. Gaussian Inc.; Pittsburgh, PA 15213, USA.

(*16*) Chirlian, L. E.; Francl, M. M. *J. Comp. Chem.* **1987**, *8*, 894.

(*17*) Uchimaru, T.; Tanabe, K.; Nishikawa, S.; Taira, K. M. *J. Am. Chem. Soc.* **1991**, *113*, 4351.

(*18*) Gorenstein, D. G.; Luxon, B. A.; Findlay, J. B. *J. Am. Chem. Soc.* **1979**, *101*, 5869.

(*19*) Kluger, R.; Covitz, F.; Dennis, E.; Williams, L. D.; Westheimer, F. H. *J. Am. Chem. Soc.* **1969**, *91*, 6066.

(*20*) Westheimer, F. H. *Acc. Chem. Res.* **1968**, *1*, 70.

RECEIVED August 17, 1993

Chapter 14

Stereoelectronic Effects in Pentaoxysulfuranes

Putative Intermediates in Sulfuryl-Group Transfer

Dale R. Cameron and Gregory R. J. Thatcher

Department of Chemistry, Queen's University, Kingston, Ontario K7L 3N6, Canada

Ab initio calcluations have been performed on rotamers of the trigonal bipyramidal species $(HO)_3SO_2^-$ and $CH_3O(HO)_2SO_2^-$ at the HF/3-21+G(*) and HF/6-31+G*//3-21+G(*) levels. Significant stereoelectronic effects on energy and bond length are observed, correlated with rotation about the equatorial S-O(H) bond in the non-methylated rotamers, *in simile* with those observed in phosphoranes. Three approaches are taken to decompose individual contributions to these effects: (1) rigid rotor conformational energy surface scans with Mulliken population analysis; (2) comparison of torsional profile periodicity with that calculated for $HOSO_3^+$; (3) full geometry optimization and natural bond order (NBO) analysis of charge transfer (CT) interactions for each rotamer. NBO analysis indicates that the combination of (a) internal hydrogen bonding and (b) the consequent increase in geminal $\sigma{\rightarrow}\sigma^*$ CT interactions, contributes significantly to changes in geometry and stabilization of the rotamer with the equatorial OH bond in the apical plane. This analysis is confirmed by observation of the loss of stabilization in the methylated congener. In neither system does any evidence exist for the $n{\rightarrow}\sigma^*$ interaction proposed to dominate in phosphoranes.

Sulfuryl group transfer is a reaction of some biological significance (*1*). However, mechanisms of nucleophilic substitution at S(VI) are poorly studied relative to the related reactions at phosphorus (*2*). As part of a study of the mechanism of sulfuryl group transfer, we have initiated computational studies on putative reaction intermediates. An associative process, depicted in Figure 1, results in a pentacoordinate trigonal bipyramidal (TBP) intermediate, for which $H_3SO_5^-$ (R=H) is a model .

An investigation of stereoelectronic effects in the $H_3SO_5^-$ series of pentaoxasulfuranes was undertaken to compare with similar pentaoxaphosphorane systems ($H_3PO_5^{2-}$) in which specific stereoelectronic effects have been postulated (*3,4*).

0097–6156/93/0539–0256$06.25/0

In the phosphoranes, the general trend is for a much longer apical bond as the equatorial OH is rotated from 90° to 0° (Figure 2). A corresponding bond shortening is also observed for the equatorial P-O bond. This has been rationalized as an n→σ* interaction between one of the non-bonding lone pairs (n) of the equatorial oxygen and the anti-bonding (σ*) orbital of the apical P-O bond. Figure 2 depicts this proposed stabilization. This argument has been proposed as the possible driving force for endocyclic cleavage in the hydrolysis of methyl ethylene phosphate and a manifestation of the anti-periplanar lone pair hypothesis (ALPH) (*3*).

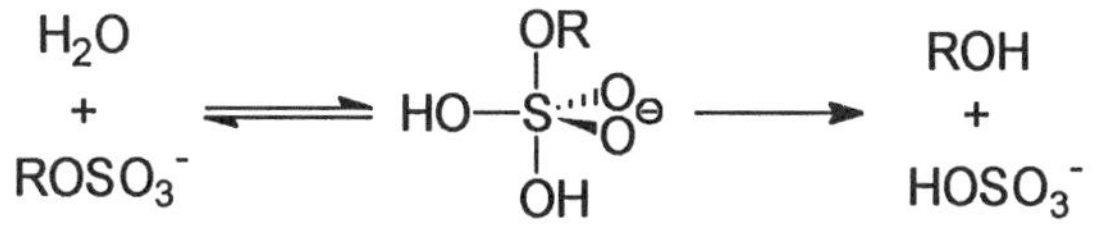

Figure 1. Reaction Scheme of Sulfate Esters with Putative Intermediate

The interaction suggests an in-phase relationship with the lobe of the anti-bonding orbital, increasing the strength of the P-O equatorial bond. The donation of electron density from the lone pair into the anti-bonding orbital would lead to a weaker P-O apical bond, with longer bond length.

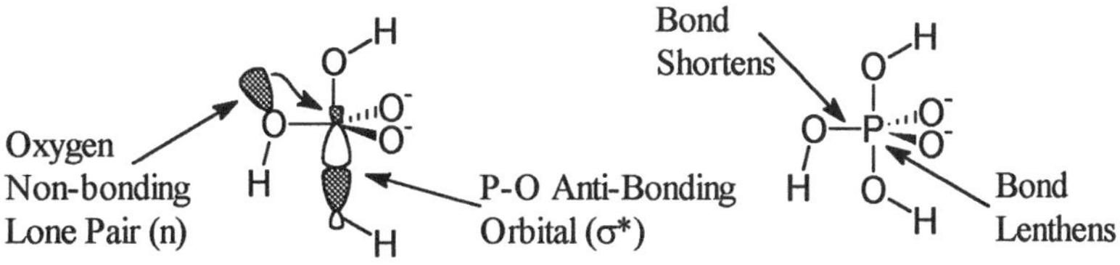

Figure 2. Pentaoxaphosphorane System Showing n→σ* Delocalization

Three approaches are employed to define the cause of the observed stereoelectronic effects. Firstly, the conformational energy surface is examined using rigid rotor calculations, accompanied by Mulliken population analysis. Secondly, the torsional profile obtained by full optimization at each constrained torsion angle is compared with $HOSO_3^+$. Finally, NBO analysis and comparison with $CH_3SO_5H_2^-$ allows definition of the dominant orbital mixing effects contributing to the observed stereoelectronic effect.

Method

The structural and electronic nature of penatoxasulfuranes was investigated by looking at various structures using RHF/3-21+G(*) level (5-8) geometry optimization.. The starting point for the rigid rotation of $H_3SO_5^-$ was obtained by full optimization using HONDO-8 (*9*) on an IBM RISC-6000/320E workstation. The subsequent rotations were performed as single point calculations and bond orders were calculated (*10*). The surfaces obtained were generated with the Surfer (*11*) series of programs. The optimized rotational points were calculated using Spartan 2.1 (*12*) on a Silicon Graphics Indigo workstation at the RHF/3-21+G(*) level. Natural Bond Orbital

(NBO) populations were also calculated using Spartan (*13-16*). Each of these points was recalculated with Gaussian 92 at the RHF/6-31+G*//3-21+G(*) level using the standard basis set supplied by Gaussian (*17*). The Gaussian 92 calculations were performed on an IBM RISC 6000/320E workstation. Pre-orthogonal NBO overlap matrices, second order perturbational analyses and full NBO analyses were performed with Gaussian 92. Both the eqec and apec structures were also optimized at the 6-31G* and 6-31+G* levels and NBO analysis was performed to investigate basis set trends.

The sulfurane series $H_3SO_5^-$ is depicted in Figure 3. The structures located were distorted trigonal bipyramids (TBPs). TBP structures are characterized by the positioning of three ligands, around the central sulfur atom, in a plane with angles close to 120°. This is called the equatorial plane. There are two more ligands, one above and one below the equatorial plane, each bonded to sulfur. These are called the apical ligands. We have further named the two rotamers in Figure 3 depending on the position of the equatorial OH (O1-H7). When O1-H7 is parallel to one of the apical bonds (S-O5) and thus, has a torsion angle (O5-S-O1-H7) of 0°, it is eclipsing the apical bond and is called apec. Similarly, when the O1-H7 is eclipsing the S-O4 equatorial bond, it has a torsion (O5-S-O1-H7) of 90° and is called eqec. Optimizations were performed along the rotational pathway from eqec to apec at 0°, 4.8°, 20°, 50°, 70°, 75°, 80°, 85° and 90° by freezing the O5-S-O1-H7 dihedral angle and optimizing the rest of the internal coordinates. Bond lengths and angles, as well as charges, derived from the electrostatic potential (CHELP) (*18*) were recorded for each of these points. Full NBO analysis was not calculated for the 75°, 80° or 85° points.

Two other points were calculated. The equatorial proton of both apec and eqec was replaced my a methyl substituent and the geometry of these was optimized to compare with the proton species. Both structures (apecMe and eqecMe) were located by full optimization with the Spartan package and the NBO analysis was performed using Gaussian 92. The fully optimized eqecMe species is not located at a torsion angle of 90° but at 75.74°.

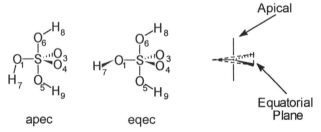

Figure 3. Pentaoxasulfurane Trigonal Bipyramidal (TBP) Structures

Results and Discussion

1. Rigid Rotation. The optimized structure for the eqec species was located and the apical (O1-S-O5-H9) and equatorial (O5-S-O1-H7) dihedral angles were rotated in 30° increments generating 144 rigid structures from which, energies and bond orders were obtained. The important surfaces generated are shown in Figures 4 and 5.

Energy. There are three main features of the energy surface to note (Figure 4a). First, there is a large energy maximum when both the apical and equatorial dihedral angles are $0°$. This is due to steric repulsion of the two hydrogens. This barrier is overestimated as the geometries were not optimized. The next feature to note concerns the location of minima on the surface. It is clear that there is a minimum located when the equatorial dihedral angle is at $0°$ and the apical dihedral angle is $\sim 180°$. There is clearly a much stronger dependence of energy on equatorial rotation than on apical rotation. This relative insensitivity to apical rotation has also been noted in the phosphorane case (2).

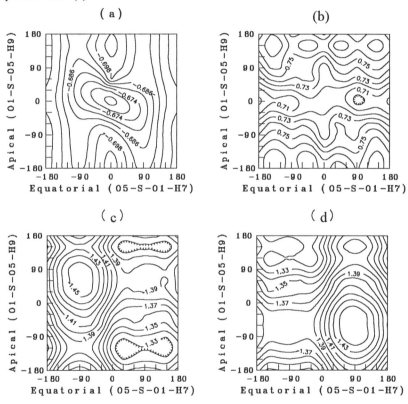

Figure 4. (a) Energy of rigid rotors in a.u. ($+769$, contour interval (c.i.) $= 0.006$); (b) Equatorial S-O1 Bond order (c.i. $= 0.01$); (c) Sulfuryl S-O3 Bond Order (c.i. $= 0.01$); (d) Sulfuryl S-O4 Bond Order (c.i. $= 0.01$).; All of the rigid rotor data was performed at the RHF/3-21+G(*) level.

Bond Orders. In the case of a single bond, ideally the bond order would be 1; similarly for a double bond, the bond order would be 2 (10). In highly delocalized, hypervalent, high energy intermediates, the bond orders rarely approach the ideal values.

The bond order surface for the S-O5 apical bond (Figure 5a) shows an interesting trend. First, there are minimum and maximum bond orders when the equatorial dihedral is at $0°$ and $180°$ respectively. This corresponds to a weakening of the bond at the apec structure. This also suggests a strengthening of the S-O5 bond when the O1-H7 is eclipsing the S-O6 apical bond. This is in keeping with the effects noted for phosphoranes (4). The same effect occurs for the opposite apical bond, S-O6 (Figure 5b). There is, for the S-O5 bond order, a slight effect from apical bond rotation, suggesting a minor stabilizing effect when the apical O5-H9 is either synperiplanar (sp) or antiperiplanar (ap) to the equatorial S-O1. This may be due to the delocalization of electron density from the π-type non-bonding lone pair on O5 into the two sulfuryl anti-bonding orbitals (S-O3σ* and S-O4σ*). It is a small effect, the nature of which will not be investigated further.

(a) (b)

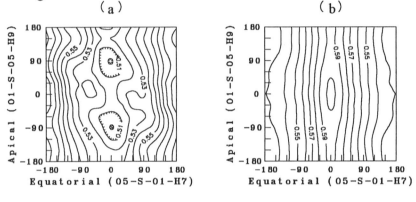

(c)

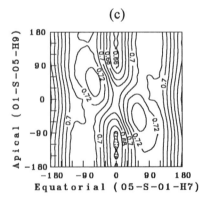

Figure 5. (a) Apical S-O5 Bond Order (contour interval (c.i.) = 0.01); (b) Apical S-O6 Bond Order (c.i. = 0.01); (c) Equatorial O1-H7 Bond Order (c.i.) = 0.01; All rigid rotor data was calculated at the RHF/3-21+G(*) level.

The bond order for the equatorial S-O1 bond also exhibits some interesting features (Figure 4b). First, it is clear that the S-O1 bond strength is at a maximum when the apical O5-H9 is eclipsing the S-O1 bond and a minimum when it is ap to it. This would suggest a strong dependence on apical rotation. This may be due to a n→σ* delocalization from the apical O5 lone pair to the S-O1 anti-bonding orbital. Also, there is a slight dependence on equatorial rotation with maxima when the equatorial dihedral angle is ± 150° or ± 30°. This would suggest that there is a stabilizing effect when the equatorial O1-H7 eclipses either an apical S-O bond or an equatorial S-O bond. This may be due to charge transfer as described by the ALPH but this analysis does not allow that conclusion. NBO analyses on the equatorial rotation may decipher this interaction.

The two sulfuryl S-O bonds also show a marked dependence on both apical and equatorial rotation. The S-O3 bond order (Figure 4c) is at a maximum when the apical O5-H9 is ap to the S-O3 bond and when the equatorial O1-H7 bond is ap to the S-O3 bond. This would suggest that any charge transfer into the S-O3 anti-bonding orbital is at a minimum in this conformation. This may be due to an n→σ* type interaction between the non-bonding lone pairs of both the equatorial O1 and the apical O5 and the S-O3 anti-bonding orbital (σ*). When the lone pair is ap to the antibonding orbital, maximum overlap occurs and the donation of electron density is accentuated. This would lead to a bond weakening when the O1-H7 and the O5-H9 bonds are sp to the S-O3 bond. This is indeed the case. The minimum for S-O3 bond order occurs when apical O5-H9 is sp to the S-O3 bond and when the O1-H7 bond is approx. sp to the   S-O3 bond. The same trend can be seen for the S-O4 bond order (Figure 4d)

There is only one more important surface to discuss: O1-H7 bond order dependence on equatorial and apical S-O bond rotation (Figure 5c). There is a very strong dependence on equatorial rotation with minimum bond orders occurring when the equatorial dihedral is either 0° or 180°. The dependence on apical rotation is minimal. Except for the large increase in bond order at the energy maximum (both dihedral angles = 0°) the apec structure seems to lead to sharp decrease in O1-H7 bond order. This effect is not important until the equatorial dihedral angle is between -60° and 60°. This suggests a distance dependant effect possibly indicating internal hydrogen bonding between H7 an O5 or O6.

2. Investigating Equatorial Rotation by Partially Optimized Torsional Scan. The rigid rotation data above may be interpreted in terms of bond strengthening and weakening coupled strongly to equatorial O1-H7 rotation. In an attempt to investigate this, a partially optimized torsional scan from the apec structure to the eqec structure was performed and the geometric parameters recorded (torsion angle frozen, all other parameters optimized).

Energy. The energy of the system as a function of equatorial rotation can be seen in Figure 6. The three sets of data represent an increasing level of treatment and are summed in Table I. It is clear that even at the highest level used, there is a substantial barrier to rotation from the apec species to the eqec species. RHF/6-31+G*//3-21+G(*) data suggests this barrier to be 10.91 kcal/mol. There must be strongly stabilizing interactions in the intermediate apec structure that either do not exist or are minimal in the transition state structure, eqec. The transition state normal mode of vibration (-556.11 cm^{-1}) consists almost exclusively of equatorial rotation with a small component of equatorial plane distortion and apical bond length changes, as expected.

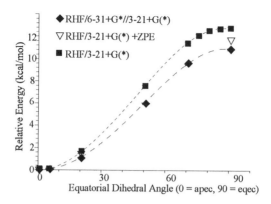

Figure 6. Relative Energy (kcal/mol) Vs Dihedral Angle

Table I. Energy Vs Equatorial Rotation[a]

Dihedral Angle O5-S-O1-H7	3-21+G(*) (a.u.)	6-31+G* // 3-21+G(*) (a.u.)	3-21+G(*) relative (kcal/mol)	3-21+G(*)+ ZPE (rel.) (kcal/mol)	6-31+G* // 3-21+G(*) (kcal/mol)
0.00	-769.711324	-773.466907	0.00	0.00	0.00
4.80	-769.711195	-773.466830	0.08	-	0.05
20.00	-769.708839	-773.465147	1.56	1.46	1.10
50.00	-769.699340	-773.457346	7.50	-	6.00
70.00	-769.693152	-773.451714	11.40	-	9.53
75.00	-769.692057	-	12.09	-	-
80.00	-769.691363	-	12.53	-	-
85.00	-769.691058	-	12.72	-	-
90.00	-769.690964	-773.449525	12.78	11.65	10.91

a) All energies measured in kcal/mol are relative.

Comparison with $HOSO_2^+$. Grein and Deslongchamps (*19*) have attempted to decompose the anomeric effect observed in dihydroxymethane using a Fourier analysis. The V_2 term in their analysis is attributed to an electronic contribution from π-bonding, compatible with the $n_p \rightarrow \sigma^*$ rationale for the anomeric effect. Quantification is garnered from the energy profile for rotation about the C-O(H) bond of $HOCH_2^+$ which shows the same periodicity as dihydroxymethane (Figure 7). In a similar fashion the energy

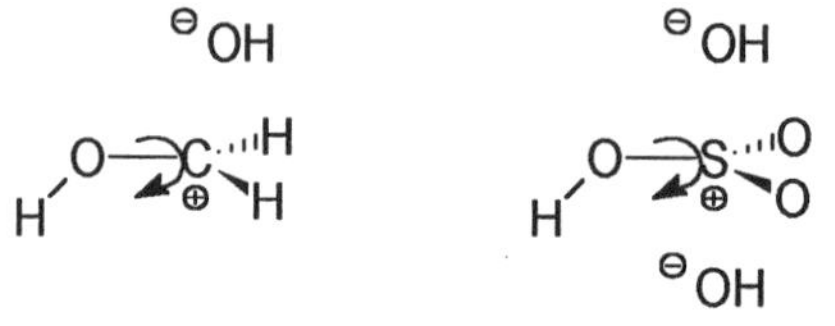

Figure 7. Dihydroxymethane Comparison with Pentaoxasulfurane Monoanion

profile for rotation about the S-O(H) bond of $HOSO_2^+$ was examined (Figure 7). However, π-bonding in this system would stabilize the coplanar rotamer corresponding to eqec, over the orthogonal (apec), contrary to the relative energies observed for the $H_3SO_5^-$ rotamers. This is confirmed by the relative stabilization energy of the coplanar rotamer of $HOSO_3^+$ calculated at the HF/6-31G* level to be 8.89 kcal/mol.

Bond Length. The apical S-O5 bond length is strongly correlated with equatorial bond rotation (Table II). The S-O5 bond increases from 1.743 Å in the eqec structure to 1.833 Å in the apec structure. This correlates with the rigid rotation surface above. The S-O6 bond length undergoes a decrease from 1.743 Å to 1.691 Å, a change of 0.052 Å, indicating an increase in bond strength going from eqec to apec. The S-O1 also undergoes a decrease from 1.626 Å to 1.610 Å, again indicating an increase in bond strength from eqec to apec. Notably, the O1-H7 bond length increases from 0.9691Å at apec to 0.9713 Å at eqec, indicating a bond weakening effect, which again may be attributed to an internal hydrogen bonding interaction.

Table II. Bond Lengths Vs Equatorial Rotation[a]

Dihedral Angle	S-O1	S-O3	S-O4	S-O5	S-O6	O1-H7
0.00	1.6099	1.4745	1.4745	1.8331	1.6914	0.9713
4.80	1.6103	1.4770	1.4723	1.8321	1.6916	0.9712
20.00	1.6131	1.4795	1.4720	1.8173	1.6951	0.9701
50.00	1.6236	1.4858	1.4723	1.7811	1.7078	0.9693
70.00	1.6250	1.4779	1.4807	1.7598	1.7255	0.9690
75.00	1.6246	1.4742	1.4831	1.7556	1.7315	0.9689
80.00	1.6253	1.4742	1.4839	1.7512	1.7360	0.9689
85.00	1.6254	1.4718	1.4851	1.7472	1.7403	0.9690
90.00	1.6259	1.4757	1.4825	1.7435	1.7433	0.9691

a) RHF/3-21+G(*) Level; Dihedral Angle = O5-S-O1-H7; Bond lengths measured in angstroms (Å); see figure 1 for atom labels.

There are less noticeable changes in both sulfuryl bonds (S-O3 and S-O4). The S-O3 bond has maximum length at 50° (1.486 Å) and two minima at apec (1.474 Å) and at eqec (1.475 Å). This suggests there is maximum donation to the S-O3 σ^* orbital between the two stationary points (eqec and apec). Minimal overlap to this bond must be occurring at the stationary points leading to shorter bond lengths. The opposite effect occurs for the S-O4 bond. These interactions will be investigated further with NBO analyses

Bond Angle. There are significant bond angle changes to note (Table III). First, the O1-S-O5 bond angle decreases from 83.36° to 79.61° while the O1-S-O6 bond angle increases slightly from 83.35° to 83.79° (eqec to apec) with a maximum at 84.39° (at the dihedral angle of 50°). The nature of these changes in bond angle is not easily elucidated. The changes in angle may be related to orbital overlap and changes in the polarization of the bonding and anti-bonding orbitals. This will be investigated in the NBO analyses. The angle between the two sulfuryl oxygens decreases from 122.73° to 119.79° (eqec to apec). This may be simply be due to a change in effective

Table III. Selected Bond Angles Vs Equatorial Dihedral Angle[a]

Bond Angle	Angle 0.00	Angle 4.80	Angle 20.0	Angle 50.0	Angle 70.0	Angle 90.0
S-O1-H7	107.8085	107.8506	108.3889	109.9875	111.6235	112.4621
O1-S-O3	119.9201	120.8395	122.0300	122.9659	120.7260	117.9589
O1-S-O4	119.9277	118.8968	117.4761	114.8142	116.9356	119.3070
O1-S-O5	79.6076	79.6427	80.1368	81.6250	82.6270	83.3584
O1-S-O6	83.7852	83.8685	83.7079	84.3850	83.3079	83.3456
O3-S-O4	119.7862	119.9002	120.2173	122.1059	122.3119	122.7341
O3-S-O5	92.1915	91.8792	91.5226	91.0826	92.5152	93.3449
O3-S-O6	96.1056	95.3595	94.5402	92.3771	93.1513	93.3272
O4-S-O5	92.1920	92.4844	93.1137	93.8272	93.0056	93.0214
O4-S-O6	96.0984	96.7414	96.9698	96.5347	94.4307	93.0241
O5-S-O6	163.3928	163.4493	163.5736	165.1555	166.3563	166.7025

a) RHF/3-21+G(*) level; Dihedral Angle = O5-S-O1-H7; see figure 1 for atom labels

charge on the two oxygens leading to less electrostatic destabilization in the apec species and a smaller angle. The final angular change of importance involves the S-O1-H7 angle. This angle drops greatly from 112.46° in eqec to 107.81° in apec. This angle change coupled with the large change in O1-H7 bond length when the hydrogen is close to the O5 oxygen, is compatible with the proposed internal hydrogen bonding interaction.

Charges. Charges (derived from the electrostatic potential) indicate a large change in charge on sulfur from +2.56 at eqec to +2.44 at apec (Table IV). There is also an overall change in the charge of the oxygens making up the equatorial plane from -2.85 (eqec) to -2.77 (apec). This change in equatorial charge makes up for 66% of the increase in negative character of the sulfur and it should be noted that the drop in charge occurs evenly among all of the oxygens in the equatorial plane, O1 changing from -0.87 to -0.86 , O3 changing from -0.963 to -0.959 and O4 changing from -1.01 to -0.96. The most profound change in atomic charge occurs on the O5 oxygen, an increase from -1.06 to -1.14. This and the change in charge on H7 are compatible with internal hydrogen bonding.

Natural Bond Orbital Analysis (NBO). Weinhold's NBO analysis provides a method for deriving localized orbitals from *ab initio* wavefunctions. This is desirable as in the traditional Lewis bonding picture, each bond has two electrons in a localized bonding orbital (σ). Any deviation from normal Lewis bonding schemes, such as hyperconjugative (or negative hyperconjugative) and charge transfer (CT) interactions, can be represented by population of non-Lewis, anti-bonding and rydberg orbitals, collectively refered to as σ^* orbitals. The populations of these σ^* orbitals is usually much smaller then their bonding counterparts (usually around 1%, higher for highly delocalized structures). The total energy of the system can then be represented as the sum of the energies of all bonding and anti-bonding type orbitals. It should be noted that the CT to a σ^* orbital is stabilizing. A more detailed description of NBO analysis can be found elsewhere (13-16).

Table IV. Charges Derived From Electrostatic Potentials (CHELP)[a]

Atom	Angle 0.0	4.8	20.0	50.0	70.0	90.0
S	2.4432	2.4420	2.4811	2.5150	2.5325	2.5646
O1	-0.8626	-0.8631	-0.8673	-0.8711	-0.8637	-0.8719
O3	-0.9589	-0.9571	-0.9649	-0.9715	-0.9591	-0.9627
O4	-0.9583	-0.9598	-0.9740	-0.9851	-1.0047	-1.0099
O5	-1.1368	-1.1454	-1.1265	-1.0885	-1.0697	-1.0561
O6	-0.9925	-0.9930	-0.9998	-1.0274	-1.0388	-1.0548
H7	0.5270	0.5311	0.5195	0.5021	0.4958	0.4959
H8	0.4612	0.4623	0.4596	0.4668	0.4542	0.4472
H9	0.4776	0.4830	0.4723	0.4597	0.4534	0.4477

a) RHF/3-21+G(*) level; Dihedral Angle = O5-S-O1-H7; see figure 1 for atom labels

In addition to NBO populations and orbital polarizations, the pre-orthogonalized overlap matrix can yield a measure of the strength of a CT interaction, and the second-order perturbation analysis gives a measure of the stabilization energy for a particular CT interaction.

NBO Populations. Deviations away from the ideal Lewis bonding picture indicate charge transfer interactions. These stabilizing interactions are characterized by an increase in the population of the anti-bonding σ^* orbitals. Since both bonding orbitals (σ) and non-bonding lone pairs (n) can donate into these anti-bonding orbitals, studying the change in populations of each of these groups is mandatory.

Table V summarizes the populations of the bonding orbitals for each of the points calculated. The S-O5 σ orbital drops in population from apec to eqec indicating an increase in its donation into other anti-bonding orbitals. Conversely, σ(S-O6) increases in population from apec to eqec indicating a loss of donation from the S-O6 σ orbital to adjacent anti-bonding orbitals. The change in population of σ(S-O5) would suggest the S-O5 bond is strengthening as the equatorial O1-H7 bond is rotated toward apec. This is the opposite to that suggested by the rigid rotation data and by the bond lengthening effects. Similarly, the S-O6 bond's strength increases upon rotation to the apec structure but the population of the S-O6 σ orbital decreases. These data suggest the bonding orbitals do not control the bond strengthening or lengthening of the apical bonds. Thus, the populations of the anti-bonding orbitals must be the controlling feature.

Table VI summarizes the populations of the anti-bonding orbitals. It is immediately apparent that there is a great deal of delocalization into the equatorial σ^*(S-O1) as well as the two apical σ^* (S-O5 and S-O6) orbitals. The patterns involved with these delocalizations follows the trends expected by examining both bond length changes and the bond order analysis from rigid rotor examinations. The equatorial S-O1σ^* and apical S-O6 σ^* orbitals decrease in population with rotation towards the apec structure indicating bond strengthening, as well as bond shortening. Conversely, the other apicalbond S-O5 lengthens in the apec structure indicating bond weakening, as predicted by the increased anti-bonding orbital population. Similarly, the S-O4 sulfuryl bond becomes weaker from apec to eqec as σ^*(S-O3) decreases in population. The nature of the delocalizations into the sulfuryl anti-bonding orbitals is complex and is not important in the stereoelectronic effect and thus, will not be investigated here.

Table V. Population of Important Bonding NBOs Vs Dihedral Angle[a]

Equatorial Dihedral	S-O1 σ population	S-O3 σ population	S-O4 σ population	S-O5 σ population	S-O6 σ population	O1-H7 σ population
0.00	1.93513	1.95009	1.95007	1.91401	1.90740	1.98910
4.80	1.93520	1.94778	1.95226	1.91389	1.90733	1.98910
20.00	1.93539	1.94505	1.95459	1.91255	1.90739	1.98893
50.00	1.93678	1.93908	1.95765	1.90963	1.90710	1.98783
70.00	1.93805	1.94298	1.95265	1.90873	1.90761	1.98695
90.00	1.93860	1.94423	1.95071	1.90807	1.90808	1.98670

a) RHF/6-31+G*//3-21+G(*) Level, Dihedral Angle = O5-S-O1-H7

Table VI. Population of Important Anti-Bonding NBOs Vs Dihedral Angle[a]

Equatorial Dihedral	S-O1σ* population	S-O3σ* population	S-O4σ* population	S-O5σ* population	S-O6σ* population	O1-H7σ* population
0.00	0.18723	0.11812	0.11812	0.21262	0.16719	0.01654
4.80	0.18750	0.11780	0.11835	0.21233	0.16712	0.01636
20.00	0.18903	0.11628	0.11940	0.20805	0.16815	0.01329
50.00	0.19560	0.11294	0.11947	0.19655	0.17127	0.00725
70.00	0.19778	0.10993	0.12032	0.18904	0.17716	0.00500
90.00	0.19915	0.10914	0.11997	0.18324	0.18318	0.00462

a) RHF/6-31+G*//3-21+G(*) Level, Dihedral Angle = O5-S-O1-H7

The nature of the donation into both apical anti-bonding orbitals and into the equatorial S-O1 anti-bonding orbitals is of prime importance as this is the source of the stereoelectronic effect.   Since the stereoelectronic effect in phosphoranes is proposed as donation from a lone pair to an anti-bonding orbital (3), investigating the effect of bond rotation on lone pair population is important.   Table VII outlines the population changes in lone pair orbitals on the important oxygens as well as changes in p-orbital character.   The population changes in the lone pairs on the equatorial O1 are of distinct importance as these would be the donor orbitals in any stereoelectronic effect similar to that proposed for phosphoranes.

There are two lone pairs on the equatorial O1, each of a different nature.   In the apec structure there is a π-type orbital (p-orbital) labelled n_π, which in to the equatorial plane and a σ-type orbital (52.4% p-orbital content) labelled n_σ, which is ap to the S-O5 bond.   There are two major points arising from the population data.   First, there is only a small change in the n_σ lone pair population coupled with rotation away from the apec structure.   This would suggest the lone pair is not contributing greatly to the change in population in the S-O5 σ* orbital as is required by the stereoelectronic effect suggested by Gorenstein (3).   There is a larger change in the n_π lone pair but this delocalization isinto the sulfuryl S-O3 and S-O4 bonds.

The second feature to note is the change in hybridization at the equatorial O1 through rotation to the eqec structure.   At 70°, the n_π orbital has gained in s-character by 9.4%. This changes at 90° where the n_π is 50% in p character, becoming the n_σ orbital while the other lone pair becomes a pure p-type, n_π, orbital.   This change in hybridization

Table VII. Oxygen Lone Pair Populations Vs. Equatorial Rotation[a]

	Angle 0.00	Angle 4.80	Angle 20.00	Angle 50.00	Angle 70.00	Angle 90.00
O1 n_σ pop	1.98294	1.98306	1.98355	1.98416	1.98110	1.97483
%p (n_σ)	52.40	52.34	52.13	53.68	59.48	100.00
O1 n_π pop	1.94682	1.94696	1.94957	1.96067	1.97809	1.97720
%p (n_π)	100.00	100.00	99.58	96.56	90.61	49.99
O3 n_σ pop	1.97782	1.97782	1.97781	1.97772	1.97809	1.97816
%p (n_σ)	26.36	26.27	26.09	25.62	25.89	25.98
O3 n_π pop	1.90983	1.91305	1.91754	1.92768	1.92275	1.92078
O3 n_π pop	1.84437	1.84448	1.84159	1.83506	1.82787	1.82734
O4 n_σ pop	1.97782	1.97780	1.97789	1.97807	1.97819	1.97813
%p (n_σ)	26.36	26.44	26.42	26.27	25.67	25.50
O4 n_π pop	1.90986	1.90710	1.90671	1.90938	1.91939	1.92249
O4 n_π pop	1.84438	1.84409	1.84539	1.84339	1.84253	1.83960
O5 n_σ pop	1.98310	1.98335	1.98581	1.99208	1.99225	1.99180
%p (n_σ)	47.59	48.01	59.81	48.72	47.89	48.02
O5 n_π pop	1.99057	1.99053	1.99025	1.98696	1.98600	1.98458
O6 n_σ pop	1.98894	1.98914	1.98970	1.99115	1.99131	1.99180
%p (n_σ)	49.58	50.02	50.39	50.33	48.65	47.97
O6 n_π pop	1.98029	1.97985	1.97912	1.97722	1.98211	1.98457

a) RHF/6-31+G*//3-21+G(*); %p character indicates hybridization (50%p = sp, 75%p = sp^2); Both O3 and O4 have two π-type lone pairs, the first is in the equatorial plane, perpendicular to the S-O1 bond and the second is also in the equatorial plane and is parallel to the S-O1 bond.

does not effect the donation into the apical plane and simply allows an energy level shift of the two lone pairs.

The large change in σ*(S-O5) population is not matched in magnitude by the change in O1 n_σ population thus, the source of the major stabilizing interaction compatible with the observed S-O1, S-O5 and S-O6 bond length changes is not an n_σ(O1)→σ*(S-O5) interaction. A strong internal hydrogen bond has been indicated by bond order and bond length changes. The large increase in σ* (O1-H7) population with rotation to the apec structure again represents a stabilizing hydrogen bonding effect.

The source of the stabilization can be deciphered by examining the second order perturbation energies for the important CT interactions for the apec species vs the eqec species. These NBO stabilization energies, although crude, give an estimate of the strength of an interaction.

Second Order Perturbational Analysis. There are three different types of CT interactions to consider in this type of analysis. Geminal interactions are between a donor and acceptor orbital which share one or more atoms. Vicinal interactions involve

CT between a donor and acceptor separated by one bond and remote interactions occur between donors and acceptor separated by two or more bonds. Examining each of these interactions will help to decipher the geometry controlling CT interactions.

Geminal Interactions. Geminal interactions usually involve bonding orbitals donating directly into adjacent anti-bonding orbitals. Due to the large amount of overlap, these interactions are usually quite large. Unlike vicinal and remote interactions, they are not sensitive to rotation. They are, however, much more sensitive to orbital polarization and bond angle.

Table VIII summarizes the important aspects of the geminal, σ (S-O1)$\rightarrow$ σ^* (S-O5) interaction. An overall stabilization of the apec structure over the eqec structure of 6 kcal/mol is observed simply by rotation of the equatorial OH. This stabilization is mainly due to an increase in orbital overlap, reflected in the increase in the overlap matrix, leading to larger CT. This change in overlap is caused, in part, by the decrease in bond angle and also, by the increase in polarization of both orbitals toward sulfur. The bond angle decrease may be due to the suspected intramolecular hydrogen-bonding interaction, investigated further as a remote interaction. This one geminal interaction may be the controlling CT interaction involved in S-O5 bond lengthening.

In total, the important geminal interactions (Tables IX and X) lead to destabilization of apec over eqec by 15.27 kcal/mol even though the S-O1 to S-O5 interaction is

Table VIII. S-O1$\sigma \rightarrow$S-O5σ^* Geminal Interaction Summary[a]

	Angle 0.00	Angle 4.80	Angle 20.00	Angle 50.00	Angle 70.00	Angle 90.00
Second Order Energy	44.07	44.02	43.52	41.40	39.44	38.09
Overlap Matrix	0.2213	0.2213	0.2200	0.2122	0.2095	0.2045
% Polarization at Sulfur (S-O1 σ)	25.95	25.94	25.80	25.62	25.75	25.79
% Polarization at Sulfur (S-O5 σ^*)	85.00	84.96	84.40	82.88	82.04	81.37

a) RHF/6-31+G*//3-21+G(*) level; Energies are in kcal/mol; overlap matrix values taken from the pre-orthogonalized NBO matrix

Table IX. Geminal Interactions in the Equatorial Plane[a]

σ^* Orbital	σ Orbital	Apec Energy	ΔE(eqec-apec)	Eqec Energy
S-O1	S-O3	0.00		0.64
	S-O5	30.33		30.92
	S-O6	31.06		30.92
	Total	61.39	1.09 (destab.)	62.48
S-O3	S-O5	19.93		25.00
	S-O6	21.38		24.99
	Total	41.31	8.68 (destab.)	49.99
S-O4	S-O5	19.94		21.46
	S-O6	21.39		21.45
	Total	41.33	1.58 (destab.)	42.91

a) RHF/6-31+G*//3-21+G(*) level; Energies in kcal/mol

Table X. Geminal Interactions Involving the Apical Bonds[a]

σ* Orbital	σ Orbital	Energy Apec	ΔE(eqec-apec)	Energy Eqec
S-O5	S-O1	44.07		38.09
	S-O3	36.77		39.17
	S-O4	36.78		32.85
	S-O6	56.20		51.25
	Total	173.82	-12.46 (stab.)	161.36
S-O6	S-O1	36.92		38.10
	S-O3	31.11		39.18
	S-O4	31.13		32.83
	S-O5	45.82		51.25
	Total	144.98	16.38 (destab.)	161.36

a) RHF/6-31+G*//3-21+G(*) level; Energies in kcal/mol

strongly stabilizing. Part of this destabilization is due to the decrease in bond angle between the S-O5 and sulfuryl bonds leading to an increased CT interaction in the eqec structure. The main destabilizing feature of geminal interactions is CT to the S-O6 antibonding orbital in the eqec structure. This increase in energy is mainly due to the stronger delocalization into the apical bond. It is probable that geminal interactions have a dominant influence on structure, but not on conformational energy, since the apec structure is ~10 kcal/mol more stable than the eqec. structure. Thus, either the vicinal and/or remote interactions must stabilize the apec conformer.

Vicinal Interactions. Vicinal (hyper-conjugative) CT interactions are frequently suggested as the cause of stereoelectronic interactions such as the anomeric effect (*20*) and the stereoelectronic effects in phosphoranes (*3*). The most important interaction we must consider is the $n_\sigma \rightarrow \sigma^*$ from the equatorial O1 lone pair to the anti-bonding orbital of the apical S-O5 bond. This is the main effect suggested by Gorenstein for similar pentaoxaphosphoranes (*3*).

Table XI outlines the major vicinal CT interactions important in the equatorial plane. There are two strong interactions with each of the equatorial S-O anti-bonding orbitals. For σ^*(S-O1), the major donor interactions arise from one of the lone pairs on each of the sulfuryl oxygens, but overall, donation to the S-O1 anti-bonding orbital from vicinal orbitals is destabilizing by 4 kcal/mol. The most remarkable change in energy is due to charge transfer into σ^*(S-O3). This is due mainly to the loss of the stabilizing n_π(O1)$\rightarrow \sigma^*$(S-O3) interaction which is not replaced when the O1n_σ orbital moves into the plane. This leads to a stabilization of apec over eqec of 8.75 kcal/mol. There is a similar loss of stabilization in eqec, from the n_π(O1)$\rightarrow \sigma^*$(S-O4) interaction in the S-O4 bond however, rotation of the O1n_σ into the plane does recover most of this lost energy by stabilizing the eqec structure. This leads to only a modest stabilization of 1.91 kcal/mol.

Table XI. Vicinal Interactions Involving the Equatorial Plane[a]

σ^* Orbital	σ Orbital	Energy Apec	ΔE(eqec-apec)	Energy Eqec
S-O1	O3 n_σ	3.45		4.30
	O3 n_π (2)	29.85		30.85
	O4 n_σ	3.46		2.40
	O4 n_π (2)	29.85		32.78
	O5 n_σ	0.00		2.84
	O5 n_π	1.99		0.00
	O6 n_σ	3.41		2.84
	Total	72.01	4.00 (destab.)	76.01
S-O3	O1 n_σ	1.44		0.00
	O1 n_π	10.60		0.00
	O4 n_σ	3.72		4.56
	O4 n_π (2)	21.12		19.67
	O5 n_σ	2.21		0.00
	O5n_π	0.00		3.03
	O6 n_π	4.38		3.00
	O1-H7σ	0.71		5.17
	Total	44.18	-8.75 (stab.)	35.43
S-O4	O1 n_σ	1.44		7.70
	O1 n_π	10.60		0.00
	O3 n_σ	3.72		4.33
	O3 n_π (2)	21.12		22.21
	O5 n_σ	2.21		0.00
	O5 n_π	0.00		3.64
	O6 n_π	4.37		3.67
	Total	43.46	-1.91 (stab.)	41.55

a) RHF/6-31+G*//3-21+G(*) level; Energies in kcal/mol

The vicinal interactions with the two apical anti-bonding orbitals are of principle importance as suggested by other similar systems. These interactions are summarized in Table XII and show some startling results. From the population analysis, it was speculated that the $O1n_\sigma \rightarrow$S-$O5\sigma^*$ CT interaction was not important although it has been previously suggested as the dominant interaction for phosphoranes. It is clear from the second order perturbation analysis that there is in fact no such interaction. In its place, there are three stabilizing interactions, from one of the lone pairs on each of the sulfuryl oxygens and from the equatorial O1-H7 bonding orbital (σ) totalling about 6.41 kcal/mol stabilization. This is offset by the destabilizing effect of the n_π (O1) donating into the S-O5 anti-bonding orbital (4.07 kcal/mol).

The absence of the proposed hyperconjugative interaction in sulfuranes, is of great importance as it implies that the same interaction suggested to be dominant for

Table XII. Vicinal Interactions Involving the Apical Bonds[a]

σ* Orbital	σ Orbital	Energy Apec	ΔE(eqec-apec)	Energy Eqec
S-O5	O1 n_π	0.00		4.07
	O3 n_π (1)	17.02		15.01
	O4 n_π (1)	17.01		14.67
	O6 n_σ	6.07		5.58
	O1-H7σ	2.07		0.00
	Total	42.17	-2.84 (stab.)	39.33
S-O6	O1 n_σ	2.66		0.00
	O1 n_π	0.00		4.11
	O3 n_π (1)	15.59		15.02
	O4 n_π (1)	15.59		14.66
	O5 n_σ	0.00		5.57
	O5 n_π	4.56		0.00
	Total	38.40	0.96 (destab.)	39.36

a) RHF/6-31+G*//3-21+G(*) level; Energies all in kcal/mol.

phosphoranes, is incorrect.   Instead, the stabilization of the apec over the eqec conformer is the result of remote interactions.

Dominant (Remote) Interactions.   Table XIII summarizes the main remote interactions, the largest two being the internal hydrogen-bonding interactions between the n_σ orbital of O5 and the O1-H7σ* orbital and the O1-H7 bonding orbital interaction with the S-O5σ* orbital.   These two interactions lead to a stabilization of the apec structure over the eqec structure by 10.99 kcal/mol.   This strong interaction, besides being important in controlling the energy, also influences the O1-S-O5 bond angle.   As we have seen above, the decrease in this bond angle leads to a large increase in the overlap between the geminal pair of orbitals σ(S-O1) and σ*(S-O5).   It is likely that a combination of these two interactions is responsible for the changes in S-O5 bond length and the stabilization of the apec strucure relative to the eqec structure.

Table XIII. Remote Interactions[a]

σ* Orbital	σ Orbital	Energy Apec	ΔE(eqec-apec)	Energy Eqec
O5-H9	O1-H7	2.09		0.00
O6-H8	O1 n_σ	1.13		0.00
O6-H8	O1 n_π	0.00		1.62
O5-H9	O1 n_π	0.00		1.61
O1-H7	O5 n_σ	8.90		0.00
O1-H7	O6 n_σ	0.77		0.00
	Total	12.89	-9.66 (stab.)	3.23

a) RHF/6-31+G*//3-21+G(*) level; Energies in kcal/mol.

Furthermore, the trends in the NBO analysis were tested for basis set dependence by full optimization of the apec and eqec structures at both the 6-31G* and 6-31+G* levels followed by full NBO analysis. The data supports the conclusions made at the 6-31+G*//3-21+G(*) level. The absence of the $n_\sigma \rightarrow \sigma^*$ negative hyperconjugative interaction in sulfuranes remains and is not a consequence of the highly delocalized 3-21+G(*) basis set. This hypothesis was examined further by geometry optimization and NBO analysis of the equatorial methyl substituted sulfuranes (apecMe and eqecMe) in which no comparable hydrogen bond is possible.

Effects of Equatorial Substitution of H by CH₃ on CT Interactions and Energy Stabilization. The substitution of a methyl group for the proton on the equatorial oxygen O1 should have a major effect both on stabilization energy and geometry as the hydrogen bonding interaction observed in apec will not be possible. A related $n_\sigma(O5) \rightarrow \sigma^*(C-O1)$ interaction may be present, however the difference in energy between the $n_\sigma(O5)$ energy and the $\sigma^*(O1-H7)$ energy is considerably smaller than the corresponding energy difference between $n_\sigma(O5)$ and $\sigma^*(C-O1)$ of apecMe (0.15688 a.u.). This large difference in donor and acceptor orbital energies decreases the stabilization energy and limits the overall stabilization from this interactions (Figure 8) (*21*). The geometry comparison for apecMe and eqecMe is summarized in Table XIV.

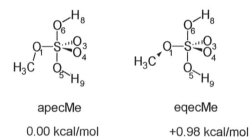

apecMe   eqecMe

0.00 kcal/mol   +0.98 kcal/mol

Figure 8. Structures of the Methyl Substituted Pentaoxasulfuranes and Relative Energy (RHF/3-21+G(*))

Energy, Bond Lengths and Bond Angles. Substitution of the equatorial proton by a methyl group has a drastic effect on the energy of the apecMe and eqecMe species. In fact, the barrier to rotation is now only 0.98 kcal/mol. The origin of this change in energy can be rationalized by steric repulsion in the equatorial plane, loss of hydrogen bond stabilization and increased CT stabilization in apecMe. The transition state structure for the proton substituted series occurs at the equatorial dihedral angle of 90°, however, the eqecMe transition state is at 75°. A small rotation from eqecMe leads to a local energy minimum at 90°, 0.039 kcal/mol lower in energy than the transition state.

The most notable changes in bond length again are the two apical bonds. Table XIV shows clearly, the S-O5 bond length changes by a much smaller degree than in the proton series. Although there is still a bond lengthening effect, the magnitude is much smaller, in fact, the bond shortening of the S-O6 bond is twice as dramatic indicating a significant change in the CT interactions. Also of note is the opposite effect on the C-O1 bond length when compared to the O1-H7 bond length. The change is of similar magnitude but of opposite sign, suggesting the complete loss of stabilization from the hydrogen bonding interaction.

There are major differences in the bond angle changes in the methyl substituted series compared to the proton substituted series. Table XIV outlines the four major differences. First, where the S-O1-H7 bond angle shortened by almost 5° on rotation to the apec species, the S-O1-C bond angle expands greatly, most likely due to steric strain, by 10.6°, to 129.15°. This expansion in bond angle causes a reversal of the bond angles between O1-S and the two apical bonds. The O1-S-O5 bond angle increases by just under half the magnitude of the change in the proton series with the opposite sign. Similarly, the O1-S-O6 bond angle decreases slightly in the apecMe species by 1.5°. The major change in the apical bond angle, a measure of distortion in the TBP structure, is opposite in sign to the proton series and much smaller indicating no significant change in hydridization at sulfur. The affects of these different geometric trends upon the CT network of interactions is most likely responsible for the energetic differences in the proton substituted series (apec - eqec) compared to the methyl series (apecMe - eqecMe). A full NBO analysis must be compared to decipher the changes in interactions.

Table XIV. Bond Length and Bond Angle Changes in the Methyl Substituted Pentaoxasulfurane Compounds: Comparison to Proton Series[a]

Bond Length and Angle	apecMe	eqecMe	Δ(eqecMe-apecMe)	Δ(eqec-apec)
S-O1	1.607	1.613	0.006	0.016
S-O3	1.482	1.470	-0.012	0.001
S-O4	1.482	1.488	0.006	0.008
S-O5	1.773	1.757	-0.016	-0.090
S-O6	1.715	1.746	0.031	0.052
C(H7)-O1	1.456	1.457	0.001	-0.002
O1-S-O5	84.83	83.29	-1.54	3.75
O1-S-O6	82.19	83.70	1.51	-0.44
S-O1-C(H7)	129.15	118.56	-10.59	4.65
O5-S-O6	167.02	166.56	-0.46	3.31

a) RHF/6-31+G*//3-21+G(*) level; bond lengths in Å; bond angles in degrees

NBO Analysis. A simple comparison of the major CT interactions in the proton substituted series to the methyl substituted series is possible and is summarized in Table XV. In the proton series, there is a large amount of stabilization into the S-O5σ* orbital from geminal donor orbitals. Similarly, there was a large destabilization of the apec over eqec species by delocalization into the S-O6σ* orbital from geminal donors. In the methyl series, this interaction is very much smaller but it leads to the same overall destabilization.

The main feature to note in the vicinal interactions is again, the conspicuous absense of the n_σ(O1)→σ*(S-O5) interaction. This again argues against such an interaction being of any importance in TBP sulfurane structure or energy. Notable also, is the loss of the hydrogen-bonding interaction by methyl substitution. The C-O1 anti-bonding orbital does indeed undergo CT with $n_\sigma(O5)$ however, the stabilization energy has undergone almost an eight-fold decrease to only 1.15 kcal/mol. The combination of all of the above changes has lead to a much smaller stabilization of apecMe over eqecMe, relative to the proton series.

Table XV. Important Second Order Perturbational Energies for the Methyl Substituted Sulfurane Series[a]

Acceptor σ^*	Donor Orbital	Energy apecMe	ΔE_{Me} (ΔE_H[b])	Energy EqecMe
Geminal				
S-O1	S-O5 σ	33.55		31.22
	S-O6 σ	30.35		31.21
	Total		-1.47 (-1.09)	
S-O3	S-O5 σ	22.12		22.36
	S-O6 σ	23.12		22.57
			+0.46 (-8.68)	
S-O4	S-O5 σ	22.38		23.27
	S-O6 σ	23.11		23.45
			+1.23 (-1.58)	
S-O5	Total	166.64	-2.87 (-12.46)	163.53
S-O6	Total	154.27	+7.26 (+16.38)	161.53
Vicinal				
S-O1	Total	75.63	+0.06 (+4.00)	75.69
S-O3	Total	33.14	+2.82 (-11.75)	35.96
S-O4	Total	43.68	-1.19 (-1.91)	42.49
S-O5	Total	37.39	+3.12 (-2.84)	40.51
S-O6	Total	37.95	+1.92 (+0.96)	40.66
Remote				
C-O1	O5 n_σ	1.33	-1.33 (-8.90)	0.00

a) RHF/6-31+G*//3-21+G(*) level; All Energies in kcal/mol; b) Values in brackets are for the corresponding interaction in the proton series in kcal/mol

Conclusions

A stereoelectronic effect is present in TBP pentaoxasulfurane intermediates. In some respects, this effect resembles that observed in phosphoranes (3). The effect is correlated with rotation about the equatorial S-OH bond and may be decomposed into three contributions.

1. The lone pair on the apical oxygen interacts with the anti-bonding orbital of the equatorial OH bond leading to a large stabilization of the order of 10 kcal/mol consistent with internal hydrogen bonding.

2. This hydrogen bond pulls the equatorial S-O bond closer to the apical bond causing an increase in the geminal delocalization from the S-O equatorial bonding orbital to the apical anti-bonding orbital ($\sigma \rightarrow \sigma^*$).

3. This in turn causes a large increase in the S-O apical bond length. Thus, internal hydrogen bonding and geminal interactions are the main features controlling both geometry and energy (Figure 9).

Calculations with the methyl substituted series show a number of dramatic changes confirming this analysis. First, the loss of a stabilizing hydrogen bond in apecMe leads to an increase in overall energy such that the equatorial isomer (eqecMe) is now only slightly higher in energy. This loss of stabilization also appears in the individual geminal, vicinal and remote CT interactions where it is clear, most of the stabilizing features of the proton substituted series have been lost. Secondly, the large change in S-O-C bond angle suggests a large steric effect. This is further supported by the location of the transition state at a torsion of 75° instead of 90°. Release of steric strain may partly be responsible for apical S-O bond lengthening (although lengthening is much reduced relative to the proton series).

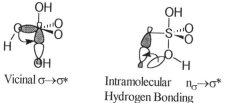

Vicinal $\sigma \rightarrow \sigma^*$ Intramolecular $n_\sigma \rightarrow \sigma^*$
                                               Hydrogen Bonding

Figure 9. Graphical Representation of the Dominant Intramolecular Hydrogen Bonding and Vicinal CT Interactions.

Future Direction. The use of second order perturbational energy is acceptable as a starting point for deciphering these interactions. A complete investigation of both the proton and methyl substituted series by the deletion of important interactions and subsequent geometry optimization must be performed to further quantify these results. The importance of CT interactions makles use of higher basis sets and electron correlation desirable.

Acknowledgments. Drs. P. Petillo and L. Lerner are thanked for the preprint of their work. Ruby Nagelkerke is thanked for helpful discussions. Acknowledgement is made to the donors of the Petroleum Research Fund administered by the ACS, for support of this research.

Literature Cited.
1) Jackson, R.J.; Busch, S.J. and Cardin, A.D. *Physiol. Rev.* **1991** *71* 481
2) Thatcher, G.R.J. and Kluger, R. *Adv. Phys. Org. Chem.* **1989,** *25,* pp99-256
3) Gorenstein, D.G.; Luxton, B.A. and Findlay, J.B. *J. Am. Chem. Soc.* **1979,** *101,* pp.5869-5875

4) Uchimaru, T.; Tanabe, K.; Nishikawa, S. and Taira, K. *J. Am. Chem. Soc.* **1991** *113* pp.4351-4353

5) Pietro, W.J.; Francl, M.M.; Hehre, W.J.; DeFrees, D.J.; Pople, J.A. and Binkley, J.S. *J. Am. Chem. Soc.* **1982**, *104*, p5039

6) Clark, T.; Chandrasekhar, J.; Spitznagel, G.W. and Schleyer, P.v.R. *J. Comput. Chem.* **1983**, *4*, p.294

7) Frisch, M.J.; Pople, J.A. and Binkley, J.S. *J. Chem. Phys.* **1984**, *80*, p.3265

8) Latájka, Z. and Scheiner, S. *Chem. Phys. Lett.* **1984**, *105*, p.435

9) HONDO-8: QCPE #544, Univ. Indiana, Bloomington, Ind.

10) Villar, H.O. and Dupuis, M. *Chem. Phys. Lett.* **1987**, 142, p.59

11) Surfer v.2.0, Golden Software, Golden, Colorado, 1989

12) Spartan v.2.1, Wavefunction, Inc. 18401 Von Karman, #210, Irvine, CA, 92715

13) Foster, J.P. and Weinhold, F. *J. Am. Chem. Soc.* **1980**, *102*, p7211

14) Reed, A.E. and Weinhold, F. *J. Chem Phys.* **1983**, *78*, p.4066

15) Reed, A.E. ; Weinstock, R.B. and Weinhold, F. *J. Chem Phys.* **1985**, *83*, p.735

16) Carpenter, J.E. and Weinhold,F. *J. Mol. Struc. (Theochem)* **1988**, *169*, p.41

17) Gaussian 92, Revision C.: Frisch, M.J.; Trucks, G.W.; Head-Gordon, M.; Gill, P.M.W.; Wong, M.W.; Foresman, J.B.; Johnson, B.G.; Schlegel, H.B.; Robb, M.A.; Replogle, E.S.; Gomperts, R.; Andes, J.L.; Raghavachari, K.; Binkley, J.S.; Gonzalez, C.; Martin, R.L.; Fox, D.J.; DeFrees, D.J.; Baker, J.; Stewart, J.J.P. and Pople, J.A.; Gaussian, Inc. Pittsburgh, PA, 1992

18) Chirlian, L.E. and Francl,M.M. *J. Comput. Chem.* **1987**, *8*, p.894

19) Grein, F. and Deslongchamps, P. *Can. J. Chem.* **1992**, *70*, p.604

20) Reed A.E., Schleyer P v R., *Inorg. Chem.*, **1988**, *27*, 3969.

21) Fleming, I. *Frontier Orbitals and Organic Chemical Reactions*; John Wiley and Sons: Toronto, **1976**, pp12-16

RECEIVED August 13, 1993

Chapter 15

O–C–N Anomeric Effect in Nucleosides
A Major Factor Underlying the Experimentally Observed Eastern Barrier to Pseudorotation

Ravi K. Jalluri, Young H. Yuh, and E. Will Taylor[1]

Computational Center for Molecular Structure and Design, University of Georgia, Athens, GA 30602–2352

The anomeric effect in nucleosides has been little studied, probably because its contribution cannot be measured from the observable equilibrium between the preferred N and S type conformations. Theoretically, the anomeric effect is expected to be most and least favorable in the W and E type furanose conformations, respectively, which are actually barriers to interconversion between N and S. Suspecting that an O-C-N anomeric effect may nonetheless contribute to the conformational energetics of nucleosides, we have calculated the pseudorotational energy profile for various nucleoside analogs using the Tripos and Kollman force fields (Sybyl 5.5 implementation). The default parameter sets fail to reproduce the eastern barrier, the existence of which has been demonstrated by NMR studies and quantum mechanics calculations, as well as having been inferred from the distribution of crystal conformers of nucleosides. Addition of the appropriate parameters for the anomeric effect (determined by MP2/6-31G* *ab initio* calculations) is necessary to reproduce the known pseudorotational barrier to N-S interconversion using molecular mechanics. The addition of these or equivalent anomeric parameters to the macromolecular force fields in common use is probably essential for realistic simulations of nucleic acids, since without them interconversion between N and S furanose conformations is virtually unimpeded, leading to the disruption of unrestrained helical structures during molecular dynamics simulations.

The anomeric effect originates in the tendency of electron lone pairs to orient *anti* to an electronegative heteroatom (see refs. *1* and *2* and other papers in this volume for reviews). The O-C-O anomeric effect has been very widely studied in carbohydrates and related compounds, and is known to play a significant role in the conformational preferences of glycosides. The analogous O-C-N anomeric effect in nucleosides involves the interaction between the lone pairs of the endocyclic furanose oxygen atom (O4') and an aromatic nitrogen atom of the purine or pyrimidine base that

[1]Corresponding author

0097–6156/93/0539–0277$06.00/0

is attached to the anomeric carbon (C1') of the furanose ring (Figure 1), which is ribose in RNA, or 2'-deoxyribose in DNA.

Since the stereochemical properties of the furanose ring are critically important in determining the conformation and behavior of nucleosides, nucleotides, and nucleic acids, it may seem somewhat surprising that the role of the anomeric effect in these compounds has never been quantitatively investigated. In fact, out of a vast literature on nucleoside and nucleotide conformations, references to any actual or potential role for the anomeric effect are quite rare, with a few exceptions (2-4). The reasons for this neglect may include the comparative complexity of the conformational properties of 5-membered rings as compared to 6-membered rings, as well as the lack of any simple experimental method for quantifying the contribution of the anomeric effect to the observable distribution of conformers. Most significantly, however, the anomeric effect has probably been often ignored for nucleosides because it is most and least favourable in the very regions of pseudorotational space which are known to be sterically and energetically the least accessible ("west" and "east" - see Figure 2), which explains the difficulty in measuring it experimentally. Perhaps because the contribution of the anomeric effect to the equilibrium between the two preferred nucleoside conformations ("north" and "south") is ambivalent, its neglect is somewhat understandable.

Given this situation, it is clear that an adequate dissection of the contribution of the anomeric effect to the conformational energetics of nucleosides can probably only be accomplished by means of computational chemistry methods, which is the approach taken in this paper. Using both molecular and quantum mechanics calculations, we will demonstrate that the anomeric effect is a very significant contributor to the pseudorotational energy profile of nucleosides, and is apparently a major factor underlying the experimentally observed eastern barrier. That the anomeric effect should have some role in nucleosides is not surprising, since, as Kirby has pointed out (2), it is clearly the basis of certain experimental results that have demonstrated a dependence of nucleoside C1'-O4' and N-glycosyl bond lengths upon both furanose conformation (5) and electronegativity of base substituents (6). However, before we can investigate this role in more detail, it is first necessary to briefly review the various conformational descriptors for nucleosides, and the pseudorotation cycle itself.

Parameters Used in Describing the Conformations of Nucleosides

Assuming that the amplitude (τ_m) of furanose ring puckering is constant, nucleoside conformations can be fully described using only three parameters:

i) the glycosidic angle (χ), which is *syn* when the pyrimidine C2 carbonyl or purine N3 is over the sugar ring, *anti* when the carbonyl or N3 is opposite the sugar ring (Figure 1);

ii) the orientation of O5' relative to C3' (γ), either +*synclinal* (+*sc*), -*synclinal* (-*sc*) or *antiperiplanar* (*ap*);

iii) the pseudorotational phase angle, P (Figure 2). This parameter (7) describes the specific puckering state of the furanose ring of the nucleoside, which in turn depends on the values of all five of the endocyclic torsional angles (τ_0 through τ_4). Thus, P is of extreme practical importance, as it reduces a set of interrelated variables

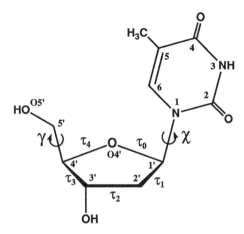

Figure 1. General scheme of numbering of atoms and definition of torsional angles in nucleosides, shown for thymidine; χ is shown in the preferred *anti* conformation. There are three preferred staggered rotamers of γ, defined relative to C3': ±*sc* and *ap*.

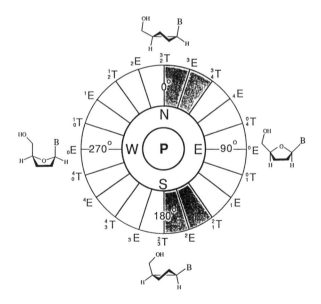

Figure 2. The pseudorotation cycle of the furanose ring in nucleosides. In this figure, E and T refer to "envelope" and "twist", respectively; superscripts refer to a ring atom that is endo, subscripts to a ring atom that is exo. Thus ^2E means "C2'-endo", a typical S type conformation. The shaded regions approximately represent the preferred N and S conformational ranges for natural nucleoside analogs.

(τ_i) to a single parameter. P can be calculated from the intracyclic torsional angles τ_i by the following formula:

$$\tan P = (\tau_1 + \tau_4 - \tau_0 - \tau_3)/3.078 \cdot \tau_2$$

Certainly, the first two parameters (χ and γ) are very important for understanding the conformational behavior of nucleosides; however, in this study we will concentrate on the energy changes associated with variations in P, because it is changes in furanose ring puckering that alter the torsional relationship between the atoms involved in the O-C-N anomeric effect.

Although it ranges from 0 to 360, P does not correspond to any single real torsional angle; it is merely a convenient mathematical convention used to describe the cyclic pathway for interconversion between the 20 possible envelope and twist conformations of the 5-membered ring (Figure 2). The highly strained planar conformation is believed to be avoided entirely, under normal conditions. The furanose ring interconverts between different conformations by passing sequentially through the various intervening envelope and twist conformations of the pseudorotation cycle.

In nucleosides, P is largely restricted to several small preferred regions (7), corresponding to conformations of the so-called north or N type (P ≈ 0 to 36) and south or S type (P ≈ 144 to 180). Interconversion between these two can occur via either the W (P=270, O4'-exo) or E (P=90, O4'-endo) type furanose conformations, which are actually barriers to interconversion between the more favored N and S type conformations. These energy minima at N and S, and barriers at W and E, are precisely what one would predict for nucleosides using classical stereochemistry (4,8), being largely due to the specific location of the heteroatom in the tetrahydrofuran ring. The barriers at W and E come in part from the eclipsing of substituents on C2' and C3'; in β-nucleosides the western barrier is higher, due to the diaxial interaction of the bulky base and 5'-carbinol groups in the O4'-exo conformation. It should also be noted that the N and S furanose ring puckering modes are characteristic of the A and B types of double helix, respectively.

The Anomeric Effect and Nucleoside Conformation

A stereochemical analysis (Figure 3) of the 1,4 interactions between the lone pairs (LP) of the furanose oxygen (O4') and the aromatic nitrogen of the purine (N9) or pyrimidine (N1) base in different conformations suggests that the oxygen lone pair interactions should be optimal (LP *anti* to N, C4' *gauche* to N) in O4'-exo (W type) conformations, and least optimal (C4' *anti* to N, LP *gauche* to N) in O4'-endo (E type) conformations. More precisely, these conformations are O4'-exo/C1'-endo (P = 288, W type) and O4'-endo/C1'-exo (P = 108, E type).

Thus, the O-C-N anomeric effect, if present, should act to reduce the western barrier to pseudorotation, and possibly increase the eastern barrier. The question then becomes, what is the magnitude of the O-C-N anomeric effect, if it exists, and, particularly, how large is the effect when this type of aromatic nitrogen is involved? It is also important to consider any available evidence regarding the height of the pseudorotational barrier, and ask if it is accounted for by other intramolecular forces, such as those used in molecular mechanics potential energy functions.   Any discrepancies may relate to a contribution by the anomeric effect.   In the absence of any useful experimental data, high level quantum mechanics calculations will have to be the final arbiter as to the magnitude of the effect, and will be used for the parameterization of molecular mechanics force fields.

Preliminary Evidence for an O-C-N Anomeric Effect

The importance of computational chemistry methods to the resolution of this question cannot be overemphasized. In addition to the stereochemical considerations described above and illustrated in Figure 3, which led us to speculate that the anomeric effect might play an important role in nucleosides (*4*), it was the ability to compute the full pseudorotational energy profile using molecular mechanics (see section on *Computational Methods*) that made us realize that the default parameter sets of several common force fields (Tripos and Kollman), which are not parameterized for an O-C-N anomeric effect, fail to produce a pseudorotational profile consistent with what has been suggested by the distribution of X-ray crystal conformers, and by previous quantum mechanics calculations (reviewed in *8*).   The key observation is that, *if parameters for the anomeric effect are not included, there is essentially no eastern barrier*, or only a very small one (Figures 4 and 5).   This is consistent with the theoretical analysis given in the previous section, which suggests that the anomeric effect could act to increase the eastern barrier. NMR experiments have shown that, in solution, the barrier to interconversion between N and S conformations is quite high, on the order of about 5 kcal/mole (*9*). Thus, since the western barrier must be higher than that in the east (for reasons described above), the eastern barrier must be substantial – certainly higher than that predicted by the Tripos or Kollman force fields with default parameter sets.

Before undertaking a rigorous parameterization of the O-C-N anomeric effect for the Kollman and MM3 force fields, using *ab initio* calculations on model compounds, we first performed semiempirical quantum mechanics calculations on the *gauche* vs. *anti* conformers of 1-(methoxymethyl)uracil, using the AM1 method, in order to get some idea as to whether the O-C-N anomeric effect in such a system was sufficiently large to account for a substantial portion of the eastern pseudorotational barrier in nucleosides.   This method predicted a ΔE of about 4 kcal/mole, in favor of the *gauche* conformer, which has an oxygen lone pair *anti* to N1.   As expected, this was consistent with a very significant O-C-N anomeric effect.   Several recently published *ab initio* studies (*10-13*) of the anomeric effect in O-C-N containing compounds, all of aliphatic type amines (aminomethanol derivatives), also confirm the existence of a significant anomeric effect in such compounds. These theoretical results encouraged us to proceed with a detailed study of the contribution of the anomeric effect to the pseudorotational potential of nucleosides.

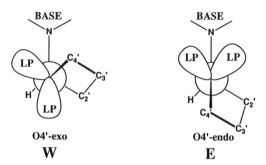

Figure 3.   The orientation of the electron lone pairs (LP) of the endocyclic oxygen (O4') in the W and E type furanose conformations.   The anomeric effect favors W over E, because it places a lone pair *anti* to nitrogen.

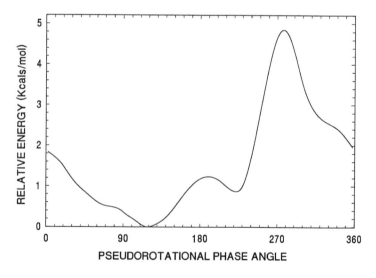

Figure 4.   Pseudorotational energy profile for thymidine, calculated with the default parameters of the Tripos force field.   The minimum is located near the east, where there should be a barrier.

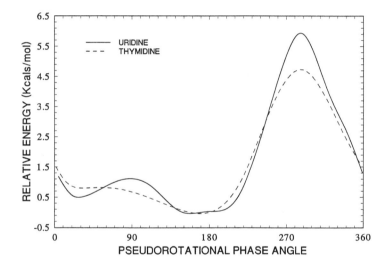

Figure 5. Pseudorotational energy profile for thymidine and uridine, calculated with the default parameters of the Kollman (AMBER) force field (Sybyl 5.5 implementation).

Computational Methods

Programs and force fields used. Molecular mechanics calculations were performed using several force fields, including the Tripos force field (*14*) in Sybyl 5.5 (Tripos Associates, St. Louis), and the Sybyl implementation of the Kollman all atom force field (*15*), which, in version 5.5, is an accurate implementation of the Kollman potential energy function. The MM3 program (Vax version) was kindly provided by Dr. N. L. Allinger. The Gaussian 90 program (Gaussian, Inc. 4415 Fifth Avenue, Pittsburgh, PA 15213) was used for all the *ab initio* calculations. Sybyl computations were run on an SGI IRIS 4D/240 workstation; Gaussian 90 calculations were done on an IBM RS/6320.

General considerations in molecular mechanics energy minimizations. For the conformational analyses described in the following sections, partial charges for the molecular mechanics Coulombic potential were obtained by AM1 calculations on structures that were first minimized without electrostatics. Molecular mechanics calculations used for parameterization were conducted with a dielectric constant of 1, and constant dielectric model, because they are being compared to *ab initio* results. Molecular mechanics calculations used for determination of pseudorotational energy profiles were conducted with a dielectric constant of $4R$, where R is the distance in Å between the charge centers for the Coulombic potential. This distant-dependent dielectric model has been shown to be a good approximation of hydrated crystal or aqueous environments (*16*). All minimizations were terminated when the energy change per iteration fell below 0.0001 kcal/mole.

A semi-automated strategy for the calculation of the complete pseudorotational energy profile. These energy profiles are obtained by using the GRID SEARCH option in Sybyl, and driving one endocyclic dihedral angle of the furanose ring by an increment of 3°, from +39° to -39° (the approximate value of τ_m). By using grid search, at each increment of the dihedral angle the molecule is completely optimized, except for the dihedral angle being driven. Each individual grid search will give the energy profile for only a part of the pseudorotational pathway. From the calculated and tabulated data of the first grid search, a conformer is selected to continue the study by driving another endocyclic dihedral angle, such that it will partly overlap or continue the energy profile from where the previous grid search ended. Similar grid searches are performed (usually a total of four or five is required) until the entire pseudorotation pathway has been traversed. Since it has been shown that in pyrimidine nucleosides the *anti* conformation of χ is preferred over *syn* by about 1 kcal/mol (*17*), all searches are conducted with the base *anti*, but allowed to freely move within the *anti* minimum energy well during the searches. This consistently produces pyrimidine base conformations (χ values) that are in agreement with previously demonstrated preferred values at different values of P. Since, due to the *gauche* effect, γ is known to strongly prefer the +*sc* and *ap* orientations over -*sc* (*4,8,17,18*), for each pseudorotational profile, two separate sets of grid searches are performed, one with γ in +*sc*, and one with γ in *ap*, and the two databases of conformations merged before the data analysis. This is done in order to ensure that the minima are accurately determined. Data is analysed using the TABLE functions of Sybyl, which permits the automated calculation of various parameters such as the τ_i and P, and the graphical analysis of the relationships between these variables and the calculated energy values.

Molecular mechanics parameterization using ab initio results on model compounds.   Torsional parameters (V_1, V_2 and V_3) for the C-O-C-N dihedral angle (anomeric effect) were obtained by harmonic analysis (Fourier transformation) of the difference curve from the conformational analyses, which is the difference between the Gaussian results and the molecular mechanics results obtained with the torsional parameters for C-O-C-N set to zero, and all other parameters at default values.   The V_i parameters are calculated by a least squares fit of the difference curve to the torsional potential function.   Although force fields like Allinger's MM2 and MM3, and Kollman's AMBER force field, can utilize all three of the V_1, V_2 and V_3 parameters, the Tripos force field allows only one parameter for each dihedral atom combination, while the Sybyl implementation of the Kollman force field allows two (probably because AMBER never uses more than two, although it permits three).   Thus, we continued our study using the Kollman force field, and determined parameters suitable for both the Sybyl and AMBER implementations.   For the 2-parameter solutions, the torsional parameters are repeatedly recalculated with each of the V_i parameters in turn constrained to zero; the solution giving the best least squares fit to the difference curve is the optimal parameter set.   Two different sets of torsional parameters were calculated for both the AMBER and Sybyl implementations of the Kollman force field, one for acyclic and the other for the cyclic compounds, because in cyclic compounds (e.g. conventional nucleosides) the C-O-C-N torsion is constrained to a range somewhere between 60° and 180°; this additional constraint permits a more accurate fit to the *ab initio* data in the range of interest.

Results and Discussion

Nucleoside pseudorotational potentials by molecular mechanics calculations without including parameters for the anomeric effect.   The pseudorotational energy profiles obtained in our initial runs with the Tripos force field consistently show a global minimum shifted from the usual N and S minima established by experiment, as well as some previous theoretical studies.   The profile for thymidine (Figure 4) calculated with this force field has a global minimum at C1'-exo (P $\approx$ 125), whereas previous experimental and theoretical studies have suggested that the minimum is probably an S type conformation (C2'-endo/C3'-exo, P $\approx$ 180).   Thus, the unmodified Tripos force field appears to have significant deficiencies when applied to the study of nucleosides.   Part of this problem appears to originate in inappropriate V_3 torsional parameters for ether C-O bonds (e.g. in the furanose ring), which appear to be too high by a factor of about 2.   This problem is immediately noticeable in the results given in reference *14*.   Data on the performance of the Tripos force field in calculating rotational barriers in alcohols and ethers (reference *14*, Table XI) suggest that a simple reduction of the * C.3 O.3 * torsional parameter (k = 1.2 kcal) by about half would greatly improve the performance on such compounds.   However, correction of this V_3

parameter alone is insufficient to account for the "missing" eastern barrier to pseudorotation observed with this force field (Figure 4).

Pseudorotational energy profiles were then studied using the unmodified Kollman all-atom force field (Sybyl 5.5 implementation), which showed a somewhat better profile for thymidine (Figure 5A) as compared to that obtained with the Tripos force field, giving an S-type conformation (C2'-endo/C3'-exo, P=180) as the global minimum, but nonetheless having a rather flat energy minimum extending all the way from about P=20 to about P=200. Thus, as observed with the Tripos force field, the well established eastern barrier of the pseudorotation pathway is not present. The calculation for uridine (Figure 5B) shows a similar profile with a negligible eastern barrier of about 0.5 kcal/mol. Thus, even the Kollman force field has significant deficiencies in describing nucleoside conformation. These deficiencies must involve either incorrect or missing parameters, and/or inherent limitations of the potential energy function. Since the default parameter sets of both force fields fail to account in any way for the anomeric effect, this suggests that the anomeric effect may contribute to the eastern barrier.

Parameterization of the Kollman force field for the O-C-N anomeric effect. A conformational analysis of the C-O-C-N torsional potential for the model compound methoxymethylamine using the Kollman force field with default parameters suggests that the *anti* conformation is prefered over *gauche* by 0.76 kcal/mol (Table I, Figure 1). This result is counter to that expected by the anomeric effect, suggesting that appropriate torsional parameters would have to be added to the force field in order to accurately model this effect. A similar calculation for methoxymethylamine using semiempirical quantum mechanics (AM1) gives the expected opposite result that *gauche* is favored over *anti* by about 1.2 kcal/mol (Table I). This preference for the *gauche* conformation by methoxymethylamine is consistent with an O-C-N anomeric effect, since it places an oxygen lone pair *anti* to the amino nitrogen.

For the molecular mechanics parameterization, a conformational analysis of methoxymethyl-amine by the Gaussian *ab initio* method was performed, using the 6-31G* polarization basis set, with MP2 correction for electron correlation effects (Table I; Figure 6). The Gaussian result suggests the preference for *gauche* is even greater than that predicted by the AM1 method, with a ΔE (*anti - gauche*) of about 3.0 kcal/mol. This result is consistent with recently published *ab initio* results for related compounds, that used similar or lower basis sets (*10-13*). However, there is a legitimate concern that this data obtained using an sp3 nitrogen may not give an adequate approximation of the systems of interest (nucleosides), which have an sp2 hybridized aromatic nitrogen attached to the anomeric carbon.

Regarding this point, before undertaking any of the *ab initio* studies, we had performed a conformational analysis by the semiempirical AM1 method of the C-O-C-N torsional potential with a more complex model compound having an aromatic nitrogen, 1-(methoxymethyl)uracil, which suggested that *gauche* is prefered over *anti* by about 4 kcal/mol, a significantly larger effect than what AM1 predicted for the analogous aliphatic amine (ΔE = 1.2 kcal/mol), but close to the *ab initio* value of 3.0 for the aliphatic system. Thus we were confident that the O-C-N anomeric effect is at least semi-quantitatively similar in both aliphatic and aromatic systems. Due to the

Table I. Relative energies (Kcal/mol) from AM1, Gaussian (MP2/6-31G*), and the Kollman force field (with default parameters, and anomeric parameters for acyclic and cyclic compounds), for the model compound methoxymethylamine used in the molecular mechanics parameterization. Missing values in the last column were not fitted, since these torsional angles are not possible in cyclic compounds

TORSION C-O-C-N	MOPAC (AM1)	GAUSSIAN (6-31G*)	KOLLMAN (UNMODIFIED)	KOLLMAN (ACYCLIC)	KOLLMAN (CYCLIC)
0	3.04	5.98	3.75	3.99	–
30	1.86	3.12	2.81	2.05	–
60	0.11	0	0.76	0	0
90	0	1.64	0.84	1.69	1.99
120	1.16	4.57	1.52	4.38	4.99
150	1.52	4.28	0.76	3.53	4.55
180	1.35	3.01	0	2.14	3.37
ΔE (anti-gauche)	1.24	3.01	-0.76	2.14	3.37

extensive computer time required for optimization of even a single conformation of 1-(methoxymethyl)uracil by the *ab initio* method, we had to limit our study to comparing the energy differences between the *anti* and *gauche* conformers. Even the *ab initio* study of only these two conformers took a total of over seven weeks on a dedicated IBM RS/6320 workstation (*19*). First, geometry was optimized up to the HF/4-31G* basis set, followed by a single point calculation with the HF/6-31G* basis set; this gave the result that *gauche* is preferred over *anti* by 2.7 kcal/mol, which is quite close to the 3.0 kcal/mol result obtained with an aliphatic amine (Table I, Figure 6). Thus, the use of the Gaussian data obtained from the more feasible calculations on methoxymethylamine to parameterize the molecular mechanics methods seems justified. However, the numbers we have used may still underestimate the magnitude of the effect in the aromatic system, since for both types of nitrogen the ΔE seemed to be increasing with higher basis sets.

Several C-O-C-N anomeric effect torsional parameter sets for the Kollman force field were calculated, optimized for slightly different situations. The complete 3-term (V_1, V_2 and V_3) solutions (Table II) are suitable for use in the AMBER program, or other implementations of the Kollman force field that permit the use of more than two torsional parameters per dihedral atom combination. Since the Sybyl 5.5 implementation permits only two torsional parameters per dihedral combination, we generated additional parameter sets with that constraint. In both cases, we provide two parameter sets, optimized for acyclic and cyclic systems, the latter being most appropriate for conventional nucleosides.

Using the Kollman force field with the anomeric parameters optimized for acyclic compounds, the torsional potential for methoxymethylamine has a ΔE (*anti - gauche*) of 2.1 kcal/mol (Table I), as compared to the *ab initio* result of 3.0; this is the best that can be expected given the limitation of using only 2 torsional parameters, optimized to fit the full 360 rotation possible in acyclic compounds. When the torsional parameters are optimized to fit the *ab initio* data only within the range of 60° to 180° that is possible for cyclic compounds, the molecular mechanics ΔE (*anti - gauche*) is about 3.4 kcal/mol (Table I), which is closer to the *ab initio* ΔE value of 3.0, and the fit between the molecular mechanics and *ab initio* curves in that range is quite good (Figure 7). This consistently good fit over the entire range possible in a cyclic nucleoside suggests that these parameters will give a good representation of the contribution of the O-C-N anomeric effect in calculations of the complete pseudorotational energy profile for nucleoside analogs.

Nucleoside pseudorotational potentials by molecular mechanics calculations including parameters for the anomeric effect. Using the Kollman force field with the C-O-C-N anomeric torsional parameters optimized for cyclic compounds, the pseudorotational energy profile of uridine (Figure 8) is now much more in line with the expected energy profile, with distinct energy minima in the N and S regions of the pseudorotational pathway that are consistent with the regions established by X-ray crystallographic studies. Most significantly, however, an eastern pseudorotational barrier of about 3.0 kcal/mol is now present; this barrier was barely detectable with the default Kollman force field parameters. Note that the western energy barrier is somewhat lower than that obtained without the anomeric parameters; this is because

METHOXYMETHYLAMINE

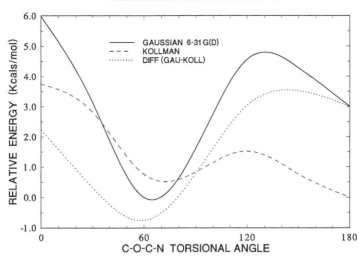

Figure 6. Comparision of torsional potentials for methoxymethylamine calculated with the Kollman force field (using default parameters, i.e. neglecting the O-C-N anomeric effect) and the Gaussian *ab initio* method (MP2/6-31G*), and their difference curve.

Table II. Torsional parameters for the anomeric effect (C-O-C-N dihedral angle). For this table only, γ refers to the phase parameter, which is either 0 or 180

KOLLMAN FORCE FIELD CT - OS - CT - N* Torsional Parameters	V_1/γ	V_2/γ	V_3/γ
SYBYL IMPLEMENTATION			
a) For acyclic systems	3.49/180	------	1.23/0
b) For cyclic systems	5.14/180	------	1.01/0
OTHER AMBER IMPLEMENTATIONS			
a) For acyclic systems	3.70/180	0.50/0	1.02/0
b) For cyclic systems	5.10/180	0.05/0	0.98/0

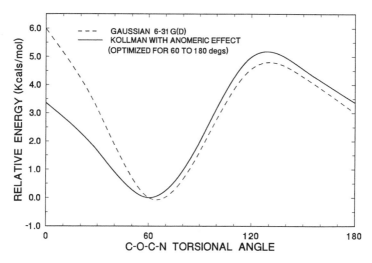

Figure 7.   Comparision of the torsional potential for methoxymethylamine calculated with the Kollman force field, using O-C-N anomeric parameters for cyclic compounds (Table II), and the Gaussian *ab initio* MP2/6-31G* results used for the parameterization (Figure 6). The parameters were optimized for the best fit in the range from 60-180°, as limits for this angle in cyclic nucleosides are within that range.

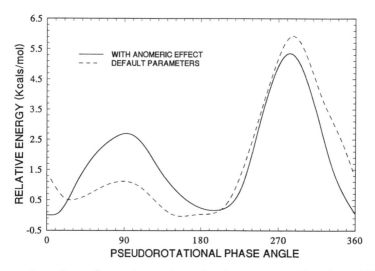

Figure 8.   Comparison of pseudorotational energy profiles for uridine, calculated using the Kollman force field with default parameters, and with anomeric parameters optimized for cyclic compounds.

the anomeric effect is most favorable in W type conformations, and thus tends to partially offset the unfavorable steric interactions that are responsible for much of the western barrier.

A similar conformational analysis for thymidine, using the "cyclic" anomeric parameters, resulted in the pseudorotational energy profile shown in Figure 9. The eastern barrier in the case of thymidine is about 1 kcal/mol lower than that obtained for uridine; this difference between the thymidine and uridine curves in the eastern region arises primarily from the presence of the 2'-OH in uridine, which has a somewhat larger steric interaction with the 3'-OH than that produced by the 2'-H of thymidine. These 2' and 3' substituents are eclipsed in both W and E type conformations.

There is at least tentative experimental evidence that the eastern pseudorotational energy barrier may be even higher than that suggested by the results shown in Figures 8 and 9. From solution C13-NMR relaxation studies, Roder et al. (9) determined that for ribonucleosides the energy of activation for interconversion between the N and S type conformations is on the order of 5 kcal/mol, possibly even higher in pyrimidine nucleosides due to the possibility of H-bonding between the 5'-OH and the base. Certainly, it is difficult to compare a molecular mechanics calculation on an isolated molecule with experimental results obtained in an aqueous environment, with the added factors of hydration, H-bonding, dielectric damping, and entropy contributing to the latter measurement. Our results suggest that the anomeric effect is responsible for a substantial portion of the eastern barrier. However, the fact that even with this included, that barrier is only about 3 kcal/mol in uridine suggests that, in addition to the anomeric effect some other factors or parameters, which are yet to be determined, may contribute to eastern barrier of the pseudorotational pathway. One such factor may be that in the Kollman force field, the V_3 torsional parameters for ether C-O bonds are possibly too high; reducing these by about one third (from .77 to .5, more in line with MM2) gives a better shape to the N and S minima, and also slightly increases the height of the eastern barrier (results not shown). Although the default parameters were apparently chosen to fit experimental ΔE values between *gauche* and *anti* conformers of simple ethers (15), it is not clear to what extent torsional barriers were considered in the original parameterization.

Conclusions

The results presented here demonstrate that inclusion of parameters for the O-C-N anomeric effect in molecular mechanics calculations produces results more in line with experimental data, suggesting that the anomeric effect plays a significant role in shaping the pseudorotational potential of nucleosides. The effect appears to be a major factor underlying the eastern barrier. Because it is most favorable in the W-type O4'-endo conformation, it tends to significantly lower the western barrier, making it lower than what has been previously suggested by molecular mechanics studies that failed to take the anomeric effect into account (e.g. 17). The anomeric effect thus acts to equalize the barriers at the W and E, and may make interconversion by the western pathway a somewhat more common event than has been previously supposed. The prevailing belief has been that interconversion between the N and S type conformations is almost exclusively via the eastern route, because the western barrier is so much higher. Our results for uridine (Figure 8) suggest that the difference between the barriers at W and E may be less than 3 kcal/mol.

THYMIDINE (KOLLMAN FORCE FIELD)

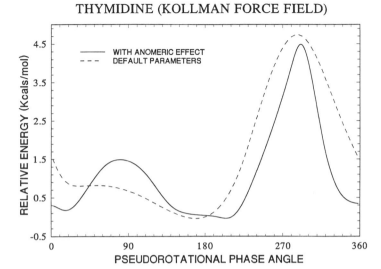

Figure 9.   Comparison of pseudorotational energy profiles for thymidine, calculated using the Kollman force field with default parameters, and with anomeric parameters optimized for cyclic compounds.

Several points may be made about the limitations and remedies for the molecular mechanics methods we have used here.   The Tripos force field fails to produce the known eastern barrier for several reasons.   First, the V_3 torsional parameters for the ether C-O bond appear to be too high by a factor of two; in addition, the O-C-N anomeric effect parameters must also be included to reproduce the eastern barrier.   The Kollman force field suffers from essentially similar though somewhat less severe deficiencies.   Modification of the default parameters, or addition of appropriate energy correction terms in the case of Tripos, permits more realistic modeling of nucleosides with these force fields.   Such parameter sets, optimized for either cyclic or acyclic nuleoside analogs, have been developed for the Kollman force field in this study.   However, inherent limitations in the force field potential energy functions (particularly for the Tripos force field) suggest that these corrective measures will be of limited usefulness. Rigorous parametrization of the more acurate MM3 force field should largely overcome these limitations.   We have an interim MM3 parameter set for the O-C-N anomeric effect, and are working on the additional parameters that will be necessary to study nucleosides using MM3.

Finally, it is important to emphasize the significance of these results for any studies involving molecular mechanics and dynamics of nucleic acids. It is commonly found that, with the macromolecular force fields in common use, there is a tendency for nucleic acid structures to become disrupted during extended molecular dynamics simulations.   Our results clearly show that, in the absence of these (or equivalent) anomeric parameters, interconversion between N and S furanose conformations would be virtually unimpeded, which, for example, could be a factor leading to the disruption of unrestrained helical structures during dynamics simulations, due to localized transitions between A and B type backbone conformations.   We are currently investigating the use of these parameters in molecular dynamics to see if this is indeed a factor underlying this problem.

Acknowledgments

This work was supported by United States Public Health Service Grant AI 30392, from the National Institute of Allergy and Infectious Diseases.   We would also like to thank Dr. David Stewart of University Computing and Networking Services for his generous technical support.

Literature Cited

1.   *Anomeric Effect: Origin and Consequences*, Szarek, W. A.; Horton, D., Eds. ACS Symposium Series 87, American Chemical Society, Washington, D.C., 1979.

2.   Kirby, A. J. *The Anomeric Effect and Related Stereoelectronic Effects at Oxygen*, Springer-Verlag, Berlin, 1983.

3.   Koole, L. H.; Buck, H. M.; Nyilas, A.; Chattopadhyaya, J. *Can. J. Chem.* **1987**, *65*, 2089-2094.

4.   Taylor, E. W.; Van Roey, P.; Schinazi, R. F.; Chu, C. K. *Antiviral Chem. Chemother.* **1990**, *1*, 163-173.

5.   Lo, A.; Shefter, E.; Cochran, T. G. *J. Pharm. Sci.* **1975**, *64*, 1707-1710.

6.   Egert, E.; Linder, H. J.; Hillen, W.; Bohn, M. C.   *J. Am. Chem. Soc.* **1980**, *102*, 3707-3713.

7.   Altona C.; Sundaralingam, M. *J. Am. Chem. Soc.* **1972**, *94*, 8205-8211.

8.   Saenger, W. *Principles of Nucleic Acid Structure*, Springer-Verlag, New York, N.Y., 1984.

9.   Roder, O.; Ludemann, H.; Von Golammer, E. *Eur. J. Biochem.* **1975**, *53*, 517-524.

10.   Gren, F.; Deslongchamps, P. *Can. J. Chem.* **1992**, *70*, 604-611.

11.   Fernandez, B.; Rios, M. A.; Carballeira, L.   *J. Comp. Chem.* **1991**, *12*, 78-90.

12.   Carballeira, L; Fernandez, B.; Rios, M. A.   *J. Mol. Struc.* **1990**, *209*, 201-209.

13.   Krol, M. C.; Huige, C. J. M.; Altona, C.   *J. Comp. Chem.* **1990**, *11*, 765-790.

14.   Clark, M.; Cramer III, R. D.; Van Opdenbosch, N.   *J. Comp. Chem.* **1989**, *10*, 982-1012.

15.   Weiner, S. J.; Kollman, P. A.; Nguyen, D. T.; Case, D. A.   *J. Comp. Chem.* **1986**, *7*, 230-252.

16.   Pickersgill, R.W. *Protein Eng.* **1988**, *2*, 247-248.

17.   Pearlman, D. A.; Kim, S-H.   *J. Biomolec. Struc. Dynam.* **1985**, *3*, 99-125.

18.   Olson, W. K. *J. Am. Chem. Soc.* **1982**, *104*, 278-286.

19.   Jalluri, R.K.; Yuh, Y.H.; Taylor, E.W., manuscript in preparation.

RECEIVED March 30, 1993

INDEXES

Author Index

Anderson, Gary, 227
Andrews, C. Webster, 114
Bowen, J. Phillip, 114
Cameron, Dale R., 256
DeShong, Philip, 227
Deslongchamps, P., 26
Edward, John T., 1
Fraser-Reid, Bert, 114
Grein, F., 205
Jalluri, Ravi K., 277
Kirby, Anthony J., 55
Laidig, K. E., 176
Le, Thuy X., 227
Lerner, Laura E., 156
Lessen, Thomas A., 227
Leung, Ronald Y. N., 126
Lim, Carmay, 240

Ma, J., 176
Perrin, Charles L., 70
Petillo, Peter A., 156
Pinto, B. Mario, 126
Sidler, D. Rick, 227
Sinnott, Michael L., 97
Slough, Greg A., 227
Taylor, E. Will, 277
Thatcher, Gregory R. J., 6,256
Tole, Philip, 240
Vöhler, Markus, 227
von Philipsborn, Wolfgang, 227
Werstiuk, N. H., 176
Williams, Nicholas H., 55
Yuh, Young H., 277
Zerbe, Oliver, 227

Affiliation Index

Burroughs Wellcome, 114
Duke University, 114
McGill University, 1
McMaster University, 176
Pennsylvania State University, 227
Queen's University, 6,256
Simon Fraser University, 126
Université de Sherbrooke, 26
University Chemical Laboratory, 55

University of California—San Diego, 70
University of Georgia, 114,277
University of Illinois at Chicago, 97
University of Maryland, 227
University of New Brunswick, 205
University of Toronto, 240
University of Wisconsin—Madison, 156
University of Zurich, 227

Subject Index

A

Acetal(s)
anomeric and reverse anomeric effect,
205–225
biomolecularity of reactions, 101–105
Acetal hydrolysis mechanism
antiperiplanar lone pair hypothesis, 16
antiperiplanar vs. synperiplanar
hypothesis, 41

Acetal hydrolysis mechanism—*Continued*
1,7-dioxaspiro[5.5]undecane formation,
31–33
importance, 30
kinetic cyclization of hydroxyenol ether,
35–38
mild acid cyclization of bicyclic hydroxypropyl acetal, 38–41
principle of least motion, 41–43
solvent effect, 38*t*

Acetal hydrolysis mechanism—*Continued*
transition states, 34–41
Acetal oxygen basicity, antiperiplanar
lone pair hypothesis, 16
N-Acetylneuraminic acid glycosides,
transition-state geometry of nonenzymic
reactions, 108–109
Alkylmanganese pentacarbonyl complexes,
synthesis of carbonyl derivatives, 227–228
2-Alkylthio derivatives of 1,1-dimethoxy-
ethane, conformational analysis, 132–133
Amidine hydrolysis, antiperiplanar lone pair
hypothesis, 16–17
Anomeric and *gauche* effect interplay in
substituted 1,4-diheterocyclohexanes
additivity of orbital interactions,
149,151–152
equilibrium data, 144*t*,148
orbital interaction component,
145*t*,148–150*t*
Anomeric destabilization, description, 8
Anomeric effect
bonding n → σ* interactions, 55–57
carbohydrates, 277
description and definition, 6–7,70–71
electrostatic effects, 55–57
geometries and reactivities, changes,
176–177
nucleosides, 277–278
OAc groups, 59–61
origin, 79–84,156–169
postulation, 1–4
quantitative modeling, 169–173
structure and conformation, 7–14
theoretical studies, 205–206
transition-state structure and reactivity,
14–19
Anomeric effect in acetals
energy analysis
neutral systems, 206–213
protonated systems, 214–219
energy parameters, 223–225
nature, 222–223
π-bonding model
NH$_2$ systems, 221*t*,222
OH systems, 219–221
studies, 205–206
Anomeric effect in dimethoxymethane,
application of quantum theory of atoms
in molecules, 176–204

Anomeric groups, OAr effects, 57–59
Anomeric stabilization, definition, 8
Antiperiplanar, description, 278
Antiperiplanar lone pair hypothesis
acetal hydrolysis mechanism, 16
acetal oxygen basicity, 16
amidine hydrolysis, 16–17
concept, 14–15
orthoester hydrolysis, 17–18
principle of least nuclear motion in
acetals, comparison, 15
questions, 19
synperiplanar lone pair hypothesis,
comparison, 16,18
stereoelectronic control in phosphoryl
transfer, 19–22
See also Stereoelectronic control
Apical ligands, definition, 258
Atom electron populations, dimethoxymethane conformers, 180,181*t*
Attractive *gauche* effect, definition, 127

B

Barton, anomeric effect postulation, 1,3–4
Bicyclic hydroxypropyl acetal, acid
cyclization, 38–41
Biomolecularity of reactions of acetals and
glycosides, evidence, 101–105
Bond critical point properties, dimethoxymethane conformers, 181–184
Bonding n → σ* interactions
anomeric effect, 57–61
description, 55–57
gauche effect, 63–68
reverse anomeric effect, 61–63

C

Computational methods for O–C–N anomeric
effect studies in nucleosides, 284–285
Conformation of anomeric and
stereoelectronic effects
computational studies, 12–13
definition and clarification, 7–9
experimental evidence, 9–10
lone pairs of electrons, 11–12
second and third row effects, 13–14
theoretical rationale, 10–11

Conformational behavior, studies, 126–127
Conformationally restrained pyranosides, hydrolysis, 116,117*f*
Corey, role in anomeric effect postulation, 2
Cyclic amidine hydrolysis, test of stereoelectronic control, 85–90
Cyclic orthoester hydrolysis, 47–53

D

1,2-Dihetero-substituted ethanes, *gauche* conformations, 127–128
Dimethoxymethane, protonated, *See* Protonated dimethoxymethane
Dimethoxymethane conformers
atom electron populations, 180,181*t*
atom energies, 180–182*t*
bond critical point properties, 181–184
component energies, 179,180*t*
geometric bond lengths, 179
1,8-Dioxadecalin-derived systems, axial–equatorial rate ratios, 100–101
1,7-Dioxaspiro[5.5]undecane formation, acetal hydrolysis mechanism, 31–33
1,2-Disubstituted cyclohexanes, *gauche* effects, 127–129
Dominant interactions, pentaoxysulfuranes, 271*t*,272

E

2e⁻ stabilization and 4e⁻destabilization model of anomeric effect
development, 162
origin of anomeric effect, 156,157*f*
splitting, 162–163
Edward, John T., history of anomeric effect postulation, 1–4
Edward–Lemieux effect, 7,22
Electron density, definition 162
Electrostatic component, anomeric and *gauche* effects, 126–152
Electrostatic effects
anomeric effect, 57–61
description, 55–57
gauche effect, 63–68
reverse anomeric effect, 61–63
Endo and *exo* anomeric interactions for 2-substituted diheterocyclohexanes
component analysis, 142,145*t*

Endo and *exo* anomeric interactions for 2-substituted diheterocyclohexanes— *Continued*
experimental conformational free energies, 142,144*t*
O–N interactions, 142,146–147
O–O interactions, 142,145,147
reverse anomeric effect, 146
S–N interactions, 148
S–O interactions, 146,148
Endo anomeric effect, 8,132,133*f*
Energy analysis for neutral acetal systems
energy decomposition model, 206–208*t*
NH₂ rotation model, 211–213*t*
OH rotation model, 208–211
structures, 206–207
Energy analysis for protonated acetal systems
energy decomposition model, 214*t*,215
NH₂ rotation model, 217–219
OH rotation model, 215–217
Energy decomposition model analysis
neutral systems of acetals, 206–208*t*
protonated systems of acetals, 214*t*,215
Energy of system, definition, 162
Enzymic reactions, determination of transition-state geometry, 109–111
Epimeric pairs of tetrahydropyranyl derivatives
antistereoelectronic behavior, 99
1,8-dioxadecalin-derived systems, generation or decomposition, 100–101
reaction profile, 98,99*f*
Eqec, definition, 258
Equatorial plane, definition, 258
Equatorial rotation of pentaoxysulfuranes using partially optimized torsional scan
bond angle and length, 263,264*t*
charges, 264,265*t*
comparison with HOSO₂⁺, 262*f*,263
energy, 261,262*f*
Escherichia coli β-galactosidase-catalyzed reactions, transition-state geometry, 110
Exo anomeric effect, 8,132,133*f*

F

Fieser, role in anomeric effect postulation, 3
Fock matrix, regions in natural bond orbital basis, 161*f*

G

β-Galactosidase-catalyzed reactions,
    transition-state geometry, 110
Gauche and anomeric effect interplay in
    substituted 1,4-diheterocyclohexanes
    additivity of orbital interactions,
        149,151–152
    equilibrium data, 144t,148
    orbital interaction component,145t,148–150t
Gauche effect
    definition, 8,127
    energy-level diagram
        through-space and through-bond
            orbital interactions, 129,131f
        two-orbital, four-electron destabilizing
            orbital interaction, 129,130f
    examples, 127–130
    explanation, 63–68
Gauche effects in 5-substituted 1,3-dihetero-
    cyclohexanes
    component analysis, 136,138t
    Endo and exo anomeric interactions, 136
    equilibrium data, 136,137t
    orbital interaction component analysis,
        136,139–144
Geminal interactions, 267–269t
Generalized anomeric effect, definition, 7–8
Geometry-dependent kinetic isotope
    effects, determination of transition-state
    geometry, 105–107
Glucopyranosyl derivatives
    bimolecularity of reactions, 102–105
    kinetic isotope effects in reactions, 103t
    transition state, 104–105
    transition-state geometry of nonenzymic
        reactions, 107–108
Glucopyranosylamines
    α and β anomers, equilibrium, 76,77t
    anomeric equilibrium, 72–73
    ^{13}C-NMR parameters, 75
    free energy change for β to α anomer
        conversion, 76,77t
    ^{1}H-NMR parameters, 73t–76
    reverse anomeric effect, 78t,79
Glycoside(s), biomolecularity of reactions,
    101–105
Glycoside cleavage
    ab initio studies with protonated dimeth-
        oxymethane, 116,118–122

Glycoside cleavage—Continued
    advancement to transition state for
        cleavage of dimethoxymethane, 114
    dimethoxymethane study, relevance to
        pyranosides, 123
    hydrolysis of conformationally restrained
        pyranosides, 116,117f
    sp^2 vs. sp^3 oxygen hybridization effect,
        124f,125
    stereoelectronic control, 114–116
    stereoelectronic requirements for cleavage,
        123–124
Glycoside hydrolysis
    reaction coordinates and conformational
        changes, 43–44
    substituent vs. rate, 45
    transition state, 44–47
Glycoside(s) of N-acetylneuraminic acid,
    transition-state geometry of nonenzymic
    reactions, 108–109
Glycosidic angle, description, 278
Glycosyl transferring enzyme catalyzed
    reactions, determination of transition-
    state geometry, 109–111
Glycosylmanganese complex insertion
    processes
    achimeric stabilization of transition state
        by C-2 alkoxy substituent, 232f
    carbon–metal bond orientation vs. rate of
        migratory insertion, 235–238
    migratory insertion, relative rates, 231t
    orientation of lone pairs of ring oxygen
        vs. rate of migratory insertion,
        232–233,235
    solution conformations of anomers, 233–235
    structures, 230

H

Hassel, anomeric effect postulation, 2–3
Haworth, W. N., role in anomeric effect
    postulation, 2
Hydrolysis
    conformationally restrained pyranosides,
        116,117f
    cyclic orthoesters, 47–53
    glycosides, 30–47
Hydrolysis of tetrahydropyranyl derivatives,
    axial–equatorial rate ratios, 97–101

K

Kinetic anomeric effect
concept, 14–15
lack of evidence in acetal derivative
reactions, 97–111
Kinetic isotope effects, geometry dependent,
determination of transition-state
geometry, 105–107
Kollman force field parameterization,
O–C–N anomeric effect in nucleosides,
286–290
Kreevoy, anomeric effect postulation, 3
Kubo effect, definition, 22

L

Lemieux, Raymond, role in anomeric effect
postulation, 3
Lone pairs of electrons, role in anomeric
and stereoelectronic effects, 11–12
Long-range transition states, stereoelectronic
control of trigonal–bipyramidal
phosphoesters, 248–250

M

Methoxy exchange, stereoelectronic control,
90–93f
2-Methoxy-cis-4,6-dimethyl-1,3-dioxanes,
stereoelectronic control, 90–93f
2-Methoxy-1,3-
dimethylhexahydropyrimidine, 80–81
Methyl ethylene phosphate, hydrolysis, 20
Methyl-substituted sulfuranes, 272,273t
Methyl vinyl ether conformers
atom electron populations, 201t,202
bond critical point properties, 202t
component energies, 198,201t
geometric bond lengths, 198t
nonbonded charge concentrations,
184–198,202–203
Migratory insertion process, glycosylmanganese complexes, 230–238

N

n → σ* interactions
bonding, See Bonding n →σ* interactions

n → σ* interactions—Continued
glycoside cleavage
ab initio studies with protonated dimethoxymethane, 116,118–122
advancement to transition state for
cleavage of dimethoxymethane, 114
dimethoxymethane study, relevance to
pyranosides, 123
hydrolysis of conformationally restrained
pyranosides, 116,117f
sp² vs. sp³ oxygen hybridization effect,
124f,125
stereoelectronic control, 114–116
stereoelectronic requirements for cleavage,
123–124
Natural atomic orbitals, concept, 161
Natural bond orbital analysis
anomeric effect
concept, 159–161
deletion energies, 163–165
electronic energies, 163–165
methodology, 173
NOSTAR geometries, 165–169
oxygen-centered pure-p lone pair
interacting with antiperiplanar
σ*_{CO}, 158,159f
methyl-substituted sulfuranes, 273,274t
pentaoxysulfuranes
antibonding orbital populations,
265,266t
bonding orbital populations, 265,266t
description, 264
oxygen lone pair populations, 266,267t
Natural orbitals, concept, 161
Neuraminidase-catalyzed reactions,
transition-state geometry, 110–111
Neutral systems of acetals, energy analysis,
206–213
NH₂ rotation model, energy analysis
neutral systems of acetals, 211–213t
protonated systems of acetals, 217–219
Nonenzymic reactions
glucopyranosyl derivatives,
determination of transition-state
geometry, 107–108
glycosides of N-acetylneuraminic acid,
transition-state geometry, 108–109
NOSTAR calculation, 165–169
Nucleosides
atom numbering, 277–279f

Nucleosides—*Continued*
electron lone pairs of endocyclic oxygen, orientation, 280–282*f*
O–C–N anomeric effect, preliminary evidence, 281–283*f*
parameters describing conformations, 278,280
pseudorotation cycle of furanose ring, 278,279*f*
torsional angles, 277–279*f*

O

O–C–N anomeric effect in nucleosides
atom numbering, 277–279*f*
computational methods, 284–285
function, 277–278
Kollman force field parameterization, 286–290
parameters describing conformations, 278,280
primary evidence, 281–283*f*
pseudorotation cycle of furanose ring, 278,279*f*
pseudorotational potentials, 285–286,288–292
relationship to conformation, 280–282*f*
torsional angles, 277–279*f*
O–N interactions, *Endo* and *exo* anomeric interactions for 2-substituted dihetero-cyclohexanes, 142,146–147
O–O interactions, *Endo* and *exo* anomeric interactions for 2-substituted dihetero-cyclohexanes, 142,145,147
OAc groups, anomeric effects, 59–61
OAr groups, anomeric effects, 57–59
OH rotation model, energy analysis
neutral systems of acetals, 208–211
protonated systems of acetals, 215–217
Oligosaccharides, roles in biochemical processes, 114
Orbital interaction component of 5-substituted 1,3-diheterocyclohexanes
through-bond effects, 136,139,140*f*
through-bond $\sigma \rightarrow \sigma^*$ interactions, 139,141–144
through-space effects, 136,139,140*f*
Origin of anomeric effect
comparison of interactions, 80–81

Origin of anomeric effect—*Continued*
conformations studied, 157,158*f*
destabilization model, 156,157*f*
experimental description, 157,158*f*
geometric changes, 84
influencing factors, relative importance, 81,83*t*,84
$n(O) \rightarrow \sigma^*_{CO}$ interactions, 158–169
proposed explanations, 79–80
stabilization model, 156,157*f*
temperature effect on NMR spectrum, 81,82*f*
Orthoester hydrolysis
antiperiplanar lone-pair hypothesis, 17–18
cyclic, 47–53
Ottar, role in anomeric effect postulation, 2

P

Pentaoxaphosphoranes, $n \rightarrow \sigma^*$ delocalization, 257*f*
Pentaoxysulfuranes, stereoelectronic effects, 256–275
Phosphoryl transfer, stereoelectronic control, 19–22
Principle of least motion, arguments against validity for acetal hydrolysis mechanism, 41–43
Principle of least nuclear motion, 15
Protonated dimethoxymethane
ab initio studies of glycoside cleavage, 116,118–122
advancement to transition state for cleavage, 119,123
Protonated systems of acetals, energy analysis, 214–219
Pseudorotational phase angle, description, 278,280
Pseudorotational potentials, O–C–N anomeric effect in nucleosides, 285–286,288,290–292
Pyranosides, conformationally restrained, hydrolysis, 116,117*f*

Q

Quantitative modeling of anomeric effect
correlation of four lone pair electrons, 169–172*f*

Quantitative modeling of anomeric effect—
  Continued
  importance, 170,173
  methodology, 173
Quantum theory of atoms in molecules,
  anomeric effect in dimethoxymethane
  calculation procedure, 203–204
  conformers, 177–178
  dimethoxymethane conformers, 179
  experimental description, 177,179
  methyl vinyl ether conformers, 180–203

R

Reactivity of trigonal–bipyramidal
  phosphoesters, role of stereoelectronic
  effects in control, 240–254
Reeves, anomeric effect postulation, 2–4
Remote interactions, pentaoxysulfuranes,
  271t,272
Repulsive *gauche* effect, definition, 127
Reverse anomeric effect
  bulky cationic substituent effect on
    conformational behavior, 72–79
  definition, 8
  description, 61
  Endo and *exo* anomeric interactions for
    2-substituted diheterocyclohexanes, 146
  examples, 71
  experimental evidence, 61–63
  explanation, 70
Reverse anomeric studies in acetals
  energy analysis
    neutral systems, 206–213
    protonated systems, 214–219
  energy parameters, 223–225
  π-bonding model
    NH$_2$ systems, 221t,222
    OH systems, 219–221
Ribonuclease, stereoelectronic control in
  phosphoryl transfer, 20–22
Rigid rotation of pentaoxysulfuranes
  bond orders, 258–261
  energy, 258,259f

S

Second-order perturbational analysis of
  pentaoxysulfuranes
  dominant interactions, 271t,272

Second-order perturbational analysis of
  pentaoxysulfuranes—*Continued*
  geminal interactions, 267–269t
  vicinal interactions, 267–271t
Sequential insertion processes
  condensation reaction, 229
  C-glycosyl derivatives, 227–228
  rate of manganocycle production vs. CO
    insertion into anomeric carbon–metal
    bond, 229–230
Shafizadeh, anomeric effect postulation, 3
S–N interactions, *Endo* and *exo* anomeric
  interactions for 2-substituted diheterocyclohexanes, 148
S–O interactions, *Endo* and *exo* anomeric
  interactions for 2-substituted diheterocyclohexanes, 146,148
Spiro acetals
  formation mechanism, 31–33
  transition states, 34–38
Stereoelectronic control
  description, 84
  evidence, 85
  methoxy exchange, 90–93f
  predicted vs. experimental results for
    acetal derivatives, 97–111
  test using cyclic amidine hydrolysis, 85–90
  theory, 97
  trigonal–bipyramidal phosphoesters
    bond angles and lengths, 242,244t,245t
    CHELP charges, 242,247t
    dihedral angles, 242,246t
    energies, 242,243t
    experimental procedure, 242
    long-range transition states, 248–250
    Mulliken atomic charges, 242,247t
    nomenclature, 242
    phosphorane complexes from methyl
      ethylene phosphate alkaline hydrolysis,
      250,251f
    phosphorane complexes with basal NH
      group, 252–254
    previous studies, 240–242
    thermodynamic parameters, 242,243t
    X–C–Y systems, identification, 240
Stereoelectronic effects
  applications, 6
  control in phosphoryl transfer, 19–22
  definitions, 7
  structure and conformation, 7–14

Stereoelectronic effects—*Continued*
transition-state structure and reactivity,
14–19
Stereoelectronic effects in hydrolysis
acetal hydrolysis mechanism, 30–47
cyclic orthoester hydrolysis, 47–53
experimental evidence, 26–28,30
rate of hydrolysis, explanation using early
transition state, 28–29
proposed hydrolysis pathways, 30
Stereoelectronic effects in pentaoxysulfuranes
dominant intramolecular hydrogen bonding
and vicinal charge transfer
interactions, 275*f*
equatorial rotation using partially
optimized torsional scan, 261–265
equatorial substitution effects, 272*f*–274*t*
experimental procedure, 257,258*f*
future work, 275
natural bond orbital analysis, 264–267*t*
pentaoxasulfurane trigonal bipyramidal
structures, 258*f*
reaction scheme, 256,257*f*
rigid rotation, 258–261
second-order perturbational analysis,
267–272
Stereoelectronic stabilizing interactions, 6
Steric component, anomeric and *gauche*
effects, 126–152
Structure of anomeric and stereoelectronic
effects
computational studies, 12–13
definition and clarification, 7–9
experimental evidence, 9–10
lone pairs of electrons, 11–12
second and third row effects, 13–14
theoretical rationale, 10–11
Structure of trigonal–bipyramidal
phosphoesters, role of stereoelectronic
effects in control, 240–254
Substituted diheterocyclohexanes
conformational effects, 127–128
endo and *exo* anomeric interactions,
142,144–148
gauche effects, 129–130
interplay of anomeric and *gauche* effects,
144,148–152
5-Substituted 1,3-dioxanes, attractive *gauche*
effect, 127–128
Sulfate esters, reaction scheme, 256,257*f*

Sulfuryl group transfer, 256–275
Syn and ±synclinal, description, 278
Synperiplanar lone-pair hypothesis, 16,18

T

Taft, role in anomeric effect postulation, 3
Tetrahydropyranyl derivatives, axial–
equatorial rate ratios in hydrolysis, 97–101
Thompson, anomeric effect postulation, 3
Through-bond effects, orbital interaction
component of 5-substituted 1,3-dihetero-
cyclohexanes, 136,139,140*f*
Through-bond $\sigma \rightarrow \sigma^*$ interactions, orbital
interaction component of 5-substituted
1,3-diheterocyclohexanes, 139,141–144
Through-space effects, orbital interaction
component of 5-substituted 1,3-dihetero-
cyclohexanes, 136,139,140*f*
Transition-state geometry
determination from geometry-dependent
kinetic isotope effects, 105–107
enzymic reactions, 109–111
nonenzymic reactions
glucopyranosyl derivatives, 107–108
glycosides of *N*-acetylneuraminic acid,
108–109
Transition-state structure and reactivity,
anomeric and stereoelectronic effects
acetal hydrolysis, 16
acetal oxygen basicity, 16
amidine hydrolysis, 17–18
antiperiplanar lone-pair hypothesis, 14–19
orthoester hydrolysis, 17–18
principle of least nuclear motion in
acetals, 15
synperiplanar lone-pair hypothesis, 16
Trigonal–bipyramidal phosphoesters, role
of stereoelectronic effects in control
of structure and reactivity, 240–254
3,7,9-Triheterobicyclo[3.3.1]nonanes, lone-
pair interactions, 129–130

V

Vibrio cholerae neuraminidase catalyzed
reactions, transition-state geometry,
110–111
Vicinal interactions, 267–271*t*

X

X–C–Z anomeric effect
  endo and *exo* anomeric effects, 132,133*f*
  endo and *exo* interactions, 142,144–148
  interplay with *gauche* effects, 144,148–152
  methodology, 132,134–135
3-X-substituted 1,5-benzodioxepins, confor-
    mational study, 129,131

Y

Y–C–C–Z *gauche* effect
  attractive and repulsive effects, 127–133
  interplay with anomeric effects,
    144,148–152
  methodology, 132,134–135
  orbital interaction component,
    136,139–144

Production: Meg Marshall
Indexing: Deborah H. Steiner
Acquisition: Anne Wilson

Printed and bound by Maple Press, York, PA

Highlights from ACS Books

Good Laboratory Practice Standards: Applications for Field and Laboratory Studies
Edited by Willa Y. Garner, Maureen S. Barge, and James P. Ussary
ACS Professional Reference Book; 572 pp; clothbound ISBN 0–8412–2192–8

Silent Spring Revisited
Edited by Gino J. Marco, Robert M. Hollingworth, and William Durham
214 pp; clothbound ISBN 0–8412–0980–4; paperback ISBN 0–8412–0981–2

The Microkinetics of Heterogeneous Catalysis
By James A. Dumesic, Dale F. Rudd, Luis M. Aparicio, James E. Rekoske,
and Andrés A. Treviño
ACS Professional Reference Book; 316 pp; clothbound ISBN 0–8412–2214–2

Helping Your Child Learn Science
By Nancy Paulu with Margery Martin; Illustrated by Margaret Scott
58 pp; paperback ISBN 0–8412–2626–1

Handbook of Chemical Property Estimation Methods
By Warren J. Lyman, William F. Reehl, and David H. Rosenblatt
960 pp; clothbound ISBN 0–8412–1761–0

Understanding Chemical Patents: A Guide for the Inventor
By John T. Maynard and Howard M. Peters
184 pp; clothbound ISBN 0–8412–1997–4; paperback ISBN 0–8412–1998–2

Spectroscopy of Polymers
By Jack L. Koenig
ACS Professional Reference Book; 328 pp;
clothbound ISBN 0–8412–1904–4; paperback ISBN 0–8412–1924–9

Harnessing Biotechnology for the 21st Century
Edited by Michael R. Ladisch and Arindam Bose
Conference Proceedings Series; 612 pp;
clothbound ISBN 0–8412–2477–3

From Caveman to Chemist: Circumstances and Achievements
By Hugh W. Salzberg
300 pp; clothbound ISBN 0–8412–1786–6; paperback ISBN 0–8412–1787–4

The Green Flame: Surviving Government Secrecy
By Andrew Dequasie
300 pp; clothbound ISBN 0–8412–1857–9

For further information and a free catalog of ACS books, contact:
American Chemical Society
Distribution Office, Department 225
1155 16th Street, NW, Washington, DC 20036
Telephone 800–227–5558